I0072294

TC 4-15.51 (FM 55-501)

Marine Crewman's Handbook

MAY 2013

DISTRIBUTION RESTRICTION: Approved for public release; distribution is unlimited.

Headquarters, Department of the Army

Training Circular
No. 4-15.51 (FM 55-501)

Headquarters
Department of the Army
Washington, D.C., 3 May 2013

Marine Crewman's Handbook

Contents

Page

PREFACE............ xvii
Chapter 1   INTRODUCTION TO ARMY WATERCRAFT.. 1-1
            Introduction.. 1-1
            Watercraft Operations 1-1
            Mission.. 1-1
            Harbor Operations 1-1
            Intra-Island and Coastal Operations..................................... 1-1
            Logistics-over-the-Shore Operations..................................... 1-1
            World Wide Missions 1-2
Chapter 2   SHIPBOARD LIFE 2-1
            Introduction.. 2-1
            Marine Qualification... 2-1
            Army Regulation 56-9 2-1
            Marine Qualification Board (MQB) 2-1
            Marine Service Record (DA Form 3068-1)................................ 2-1
            Skill Levels of Watercraft Operators....................................... 2-1
            Shipboard Customs and Courtesies.. 2-3
            Flying the National Ensign.. 2-3
            Half-Masting the Ensign 2-4
            Church Pennant.. 2-4
            Union Jack... 2-4
            Transportation Corps Flag.. 2-5
            Dressing and Full Dressing Ship 2-5
            Salutes... 2-6
            Deck Watches 2-6
            Watch List and Quarters List 2-6
            Watch Duties: Class A Vessels 2-6
            Standing Night Watch.. 2-8

DISTRIBUTION RESTRICTION: Approved for public release: distribution is unlimited.
*This publication supercedes FM 55-501, 1 December 1999.

Rules for Developing and Maintaining Dark Adaptation 2-9
Anchor Watch and Fire Watch 2-9
Gangway Watch in Port 2-9
Watch Duties 2-9
Logbooks 2-10
Shipboard Sanitation 2-12
Personal Hygiene 2-12
Taking Potable Water Aboard Ship 2-12
Pollution Control 2-13
Oil Transfer Procedures 2-14

Chapter 3   VESSEL TERMS AND DEFINITIONS 3-1
Introduction 3-1
Nautical Terminology 3-1
Structural Parts of the Hull 3-1
External Parts of the Hull 3-3
Names of Decks 3-3
Shipboard Directions and Locations 3-4
Shipboard Measurements 3-5
Procedure for Reading Draft Marks 3-7
Weight Tonnage Terms 3-8
Categories of Ship's Deck Gear 3-8

Chapter 4   BOAT HANDLING 4-1
Introduction 4-1
Forces Affecting Boat Handling 4-1
Design of the Vessel 4-1
Power 4-1
Propeller Action 4-1
Other Factors Affecting Control 4-3
Standard Steering Commands 4-3
Handling Characteristics of Single- and Twin-Screw Vessels 4-4
Handling Single-Screw Vessels 4-4
Handling Twin-Screw Vessels 4-8
Turning in a Limited Space 4-8
Docking and Undocking 4-8
Mooring Lines 4-9
Using the Lines 4-9
Dipping the Eye 4-9
Stopping Off a Mooring Line 4-9
Docking (Single-Screw Vessel) 4-10
Making Landings 4-10
Making Use of the Current 4-11
Tying Up to a Pier (Heavy Weather Procedures) 4-11
Getting Underway from a Pier 4-11
Handling Grounded Harbor Craft 4-12
Sallying 4-12
Flooding the Vessel 4-12
Removing Ballast 4-12

Kedging... 4-12
Shifting Cargo 4-13
Heavy Weather Measures 4-16
Standard Precautions 4-16
Use of Sea Anchor 4-17

Chapter 5   **CHARTS AND PUBLICATIONS**.. 5-1
Introduction 5-1
Care of Charts 5-1
Chart Portfolios 5-1
Correcting a Chart 5-2
Chart/Publication Correction Record Card 5-3
Chart Correction Techniques... 5-5
Requisitioning Procedures for Charts.. 5-5
Publications 5-7

Chapter 6   **SHIPBOARD COMMUNICATIONS** 6-1
Introduction 6-1
Shipboard Tactical and Marine Radios 6-1
Marine Radios 6-3
Global Marine Distress and Safety System.. 6-4
How to Talk on a Radio 6-4
How to Call and Reply 6-5
Useful Operating Frequencies.. 6-5
Phonetic Alphabet 6-6
Pronunciation of Numbers 6-6
Army Key Management System (AKMS) 6-6
Signaling by International Code Flags.. 6-8
Flashing Light Signals 6-9
Procedure Signs (Prosigns).. 6-11
Sending by Flashing Light 6-11
Hints for Better Signaling.. 6-11
Distress, Urgent, and Special Signals 6-12
Frequencies.. 6-12
Distress Procedures 6-12
Visual International Distress Signals 6-13
Priority of Messages 6-14
Urgent Messages 6-14
Cancellation of Urgent Traffic 6-15
Safety Messages 6-15
Safety Call and Message... 6-15
Weather Broadcasts 6-15
Publications 6-15
Emergency Radiotelephone Procedures.. 6-16

Chapter 7   **MARINE EMERGENCIES**... 7-1
Introduction 7-1
Station Bill.. 7-1
Emergency Duties 7-2
Abandon Ship Procedures... 7-2

Drownproofing..7-4
Travel Stroke...7-6
Swimming Through a Thick Oil Fire...7-6
Swimming Through a Thin Oil Fire 7-8
Cold Water Victims..7-9
Cold Water Survival 7-11
Man Overboard Procedures..7-13
Life Rafts...7-16
Life Raft Survival Equipment..7-22
Life Raft Patching...7-23
Signaling...7-25
Righting an Overturned Life Raft 7-26
Survival Aboard a Life Raft 7-29
Search Air Rescue 7-33
Types of Distress Signals 7-33
How Aircraft Direct Ships to Distress Scene..7-34
Surface Ships Action..7-34
Assisting an Aircraft that May Ditch 7-35
Preparation for Medical Evacuation of Personnel from Your Ship 7-35
Hoist Operations 7-36
Helicopter Hoist Procedures 7-36
Shipboard CBRNE Defense..7-39
Types of Nuclear Bursts...7-39
Underwater Shock..7-40
Thermal Radiation..7-40
Nuclear Radiation...7-40
Initial Nuclear Radiation 7-41
Nuclear Radiation Injury...7-41
Fallout 7-41
Protective Shielding 7-42
Preventive Measures (Before Attack) 7-42
Preventive Measures 7-43
Radiological Decontamination 7-44

Chapter 8   MARLINESPIKE SEAMANSHIP..8-1
Introduction 8-1
Fiber Line 8-1
Cordage 8-3
Inspection...8-3
Uncoiling New Line 8-4
Stowing Line...8-5
Whipping a Line 8-6
Knots, Bends, and Hitches...8-8
Splicing Three-Strand Fiber Line 8-17
Back Splice with a Crown Knot 8-17
Splice 8-19
Wire Rope 8-22
Measurement 8-23

Safe Work Load and Breaking Strength 8-24
Inspection 8-24
Unreeling 8-25
Seizing 8-25
Cutting 8-26
Coiling 8-27
Size of Sheaves and Drums 8-27
Lubrication 8-27
Reversing Ends 8-28
Storage 8-28
Putting an Eye in Wire Rope 8-28
Tools Used for Splicing 8-28
Temporary Eye Using Wire Rope Clip 8-29
The Hasty Eye ("Molly Hogan") Splice 8-31
The Liverpool Splice 8-32
Splicing 2-in-1 Double-Braided Nylon Line (Samson 2-in-1 Braid-Splicing
Principles) 8-34
Special Terms 8-34
Special Tools and Techniques 8-36
Standard Eye Splice 8-38
End-for-End Splice 8-44

Chapter 9 **DECK MAINTENANCE** 9-1
Introduction 9-1
Preventive Maintenance 9-1
Hand Tools and Their Uses 9-1
Sharpening Scrapers and Chipping Hammers 9-2
Power Tools 9-3
Safety Precautions 9-5
Paint 9-5
Paint Removers 9-8
Preparation of Aluminum Surfaces 9-9
Preparation of Galvanized Steel Surfaces 9-9
Electrical Equipment 9-10
Paint Preparation 9-10
Painting by Brush 9-11
Brushes 9-12
Paint Application 9-14
Paint Rollers 9-17
Spray Guns and Their Uses 9-18
Operation of the Spray Gun 9-23
Care of the Spray Gun 9-26
Respirators 9-29
Safety Precautions 9-30
Estimating Paint Jobs 9-31
Maintenance of Ship's Rigging and Deck Machinery 9-31

Chapter 10 **BEACHING AND RETRACTING OPERATIONS** 10-1
Introduction 10-1

Rules for Landing Operations 10-1
Surf Action.. 10-1
Preparing to Hit the Beach.. 10-2
Beaching Hazards.. 10-3
Beaching Procedures... 10-5
Retracting Procedures 10-5
Salvage Procedures... 10-7

Chapter 11 LANDING CRAFT OPERATIONS... 11-1
Introduction 11-1
Loading for Movement Overseas... 11-1
Boat Groups 11-1
Calling Boats Alongside 11-1
Landing Craft Waves 11-2
Types of Formations 11-3
Landing Craft Visual Signals 11-3
Maneuvering Signals 11-6
Hydrographic and Beach Markers 11-6
General Unloading Phase.. 11-7
Cargo Documentation 11-8
Cargo Loading Operations 11-8
Tips on Securing Cargo Aboard Landing Craft.. 11-8

Chapter 12 SEARCH AND RESCUE 12-1
Introduction 12-1
Personal Survival 12-1
Cold Water Survival and Hypothermia.. 12-1
Sector Search Pattern for One Ship 12-2

Chapter 13 TOWING 13-1
Introduction 13-1
Types of Tows.. 13-1
Description of Towing Equipment 13-2
Care of Nylon Towline.. 13-2
New Hawsers 13-2
Coiling and Winding 13-2
Stowage 13-2
Use of Nylon With Other Rope.. 13-3
Characteristics of Nylon Line Under Load 13-3
Alongside Towing.. 13-3
Special Precautions 13-3
Names of Towlines.. 13-3
Stern Towlines 13-4
Care of Wire Rope Towline.. 13-4
Misuse of Wire Rope Towlines 13-5
Capstan.. 13-5
Chafing Gear 13-5
Flounder Plate (Fish Plate) 13-6
Bridles 13-6
Retrieving Line 13-6

Pendants..13-6
Plate Shackle..13-6
Handling Towlines 13-6
Single-Line Lead..13-7
Doubling Up a Towline 13-7
Rigging a Stern Towline 13-8
Towing Alongside (Hip Tow)..13-9
Towlines...13-9
Determining Which Side To Make Up To 13-10
Securing the Towlines 13-10
Shifting the Tow to the Other Side 13-10
Towing Two Barges Alongside...13-11
Push Towing 13-11
Rigging a Push Tow 13-11
Towing Astern (Inland Waters) 13-12
Shifting the Tow From Pushing Ahead or Along Side Towing to Astern........13-12
Towing Lines 13-13
Towing Astern (Open Sea) 13-13
Inspection of Towing Equipment 13-13
Scope of Hawser 13-13
Hawser Watch 13-13
Preventive Maintenance 13-14
Towing in Tandem 13-14

Chapter 14   SEA-BASED CREW GUNNERY TRAINING 14-1
Introduction..14-1
Fire Commands 14-1
Elements of a Fire Command..14-1
Fire Command Terms...14-3
Types of Fire Commands 14-4
Training Aids, Devices, Simulators and Simulations (TADSS) 14-5
Sea-Based Individual Gunnery Qualification...14-6
Stoppage. 14-10
Sea-Based Collective Gunnery Training 14-12
Concept 14-12
Principles of Fire Control 14-12
Fire Control Measures 14-13
Collective Gunnery 14-14
Evaluation..14-15

Chapter 15   RIGGING (SEAMANSHIP)...15-1
Introduction..15-1
Description of Blocks...15-1
Common Cargo Block 15-2
Naming a Block 15-3
Types of Rigs...15-4
Combinations of Blocks and Tackles 15-5
Reeving Blocks and Tackles 15-6
Determining the Mechanical Advantages of Tackles 15-7

Computing Friction 15-8
Computing Breaking Strength and Safe Working Load 15-8
Useful Formulas for Lines 15-8
SWL and BS for Wire Rope 15-9
Computing the Breaking Strength of a Block and Tackle 15-10
Determining Breaking Stress 15-10
Computing Safe Working Load for Hooks, Shackles, and Turnbuckles 15-11
Maintenance and Overhaul of Blocks 15-13
Standing Rigging 15-14
Inspections of Rigging 15-15
Grounding Masts 15-15

Chapter 16   GROUND TACKLE 16-1
Introduction 16-1
Anchors 16-1
Anchor Chain 16-3
Marking the Anchor Chain 16-5
The Anchor Windlass 16-6
Terminology 16-8
Letting Go the Anchor—General Procedures 16-9
Operating the Capstan Anchor Windlass 16-9
Operating the Horizontal Anchor Windlass 16-12
Sequence of Weighing Anchor 16-21
Maintenance 16-23

Chapter 17   DAMAGE CONTROL 17-1
Introduction 17-1
Damage Control Program 17-1
Damage Control Team 17-1
Damage Control Kits 17-1
Purpose of the Damage Control Team 17-1
Shoring 17-2
Clearing the Decks 17-3
Speed 17-3
Preparation 17-3
Shoring Principles 17-3
Bracing 17-5
Use of Carpenter's Square in Shoring 17-8
Plugging 17-11
Wooden Plugs 17-11
Lead Plugs 17-11
Cracks 17-11
Patching 17-12
Pillows, Mattresses, and Blankets 17-12
Hinged Plate Patch 17-13
Bucket Patch 17-13
Use of a Hook Bolt for Securing Patch 17-13
Pipe Repair 17-15
Emergency Damage Control Metallic Pipe Repair Kit 17-16

Description of Materials 17-16

Materials Required for Pipe Patch 17-18

Advantages of the Plastic Patch 17-18

Application of Plastic Patch 17-19

Simple Pipe Patch 17-20

GLOSSARY Glossary-1

    Section I – Acronyms and Abbreviations Glossary-1

    Section II – Terms Glossary-7

REFERENCES References-1

Figures

Figure 2-1. National Ensign 2-3

Figure 2-2. Union Jack 2-4

Figure 2-3. TC Flag 2-5

Figure 2-4. Point System 2-7

Figure 2-5. Tow Riding In/Out of Step 2-8

Figure 3-1. Construction of a Hull 3-2

Figure 3-2. Bulkheads and Decks 3-2

Figure 3-3. External Parts of the Hull 3-3

Figure 3-4. Weather Decks 3-4

Figure 3-5. Locations and Directions Aboard Ship 3-5

Figure 3-6. A Ship's Dimensions 3-6

Figure 3-7. Draft Marks on Bow and Stern of Vessel 3-7

Figure 3-8. Draft Numbers Showing Foot Draft Mark 3-7

Figure 3-9. Various Draft Readings 3-8

Figure 3-10. Standing Rigging Gear 3-9

Figure 3-11. Deck Fittings 3-10

Figure 3-12. Electric Winch 3-10

Figure 3-13. Windlass 3-11

Figure 3-14. Capstan 3-11

Figure 4-1. Propeller Action and Resulting Screw Action 4-2

Figure 4-2. Propeller Pitch 4-2

Figure 4-3. Vessel with Sternway, Propeller Backing, Rudder Amidships 4-4

Figure 4-4. Vessel with Sternway, Propeller Backing, Rudder to Starboard 4-5

Figure 4-5. Vessel with Sternway, Propeller Backing, Rudder to Port Vessel with
Headway Propeller Backing 4-5

Figure 4-6. Vessel with Headway, Propeller Backing, Rudder Amidships 4-5

Figure 4-7. Vessel with Headway, Propeller Backing, Rudder to Starboard 4-6

Figure 4-8. Vessel with Headway, Propeller Backing, Rudder Left Vessel with
Sternway, Propeller Going Ahead 4-6

Figure 4-9. Vessel with Sternway, Propeller Going Ahead, Rudder Amidships 4-7

Figure 4-10. Vessel with Sternway, Propeller Going Ahead, Rudder to Right 4-7
Figure 4-11. Vessel with Sternway, Propeller Going Ahead, Rudder to Left Handling
Twin-Screw Vessels 4-8
Figure 4-12. Names and Locations of Mooring Lines 4-9
Figure 4-13. Stopping Off a Mooring Line.. 4-10
Figure 4-14. Making Use of the Current... 4-11
Figure 4-15. Maneuverability of Towing Vessel (Not Anchored) 4-14
Figure 4-16. Maneuverability of Towing Vessel (with Secured Line)................... 4-14
Figure 4-17. Receiving Line Forward... 4-15
Figure 4-18. Towing Vessel Approaching Bow On in Packed Sand...................... 4-15
Figure 4-19. Sea Anchor 4-17
Figure 5-1. World Sub Regions... 5-2
Figure 5-2. Lower Right-Hand Corner of Chart... 5-2
Figure 5-3. Explanation of Format 5-3
Figure 5-4. Chart/Publication Correction Record (DMAHC-86609)...................... 5-4
Figure 6-1. Urgent Code Flags 6-8
Figure 6-2. Order of Reading Flag Hoists on Yardarm 6-9
Figure 6-3. International Morse Code 6-10
Figure 6-4. Aldis Gun 6-10
Figure 6-5. Procedure Signs 6-11
Figure 7-1. Jumping in Water.. 7-3
Figure 7-2. Resting Position.. 7-4
Figure 7-3. Preparing to Exhale 7-4
Figure 7-4. Exhalation 7-5
Figure 7-5. Inhalation 7-5
Figure 7-6. Return to Rest Position.. 7-6
Figure 7-7. Break the Surface.. 7-7
Figure 7-8. Fully Inhale 7-7
Figure 7-9. Swimming Through Thick Oil Fire 7-8
Figure 7-10. Swimming Through Thin Oil Fire... 7-8
Figure 7-11. Double-Up.. 7-12
Figure 7-12. Buddy-Up.. 7-12
Figure 7-13. Williamson Turn... 7-14
Figure 7-14. Bringing Ship into Wind 7-15
Figure 7-15. A Cradled Life Raft 7-16
Figure 7-16. A Typical Oval Inflated Life Raft 7-17
Figure 7-17. Throwing in Life Raft.. 7-18
Figure 7-18. Yanking on the Operating Cord 7-19
Figure 7-19. Boarding Life Raft.. 7-19
Figure 7-20. Container Rises to Surface... 7-20
Figure 7-21. Operating Cord Parting.. 7-20
Figure 7-22. Jumping into Canopy Opening 7-21

Figure 7-23. Boarding a Life Raft From the Sea 7-21
Figure 7-24. Bringing Aboard an Injured Crewman..7-22
Figure 7-25. Pushing Bottom Plate Through Hole..7-24
Figure 7-26. Sliding Top Clamp Over...7-24
Figure 7-27. Tightening Wing Bottom Clamp Nut...7-25
Figure 7-28. Signaling Mirror 7-25
Figure 7-29. EPIRB Secured to a Life Raft 7-26
Figure 7-30. Getting Aboard an Overturned Life Raft 7-27
Figure 7-31. Standing on Edge...7-27
Figure 7-32. Knees Bent...7-28
Figure 7-33. Several People Righting an Overturned Life Raft...7-28
Figure 7-34. Overhead Views of Round and Oval Rafts 7-29
Figure 7-35. Fishhooks Made From Wood 7-32
Figure 7-36. Fishhook Made From a Jackknife 7-32
Figure 7-37. Aircraft Signal...7-34
Figure 7-38. Aircraft Dismissal Signal 7-34
Figure 7-39. Rescue Basket Hoist..7-37
Figure 7-40. Rescue Basket..7-37
Figure 7-41. Rescue Sling 7-38
Figure 7-42. Stokes Litter 7-38
Figure 7-43. Standard NATO CBRNE Markers..7-45
Figure 8-1. Fabrication of Fiber Line 8-2
Figure 8-2. Opening View Coil Line...8-4
Figure 8-3. Coiling Line 8-5
Figure 8-4. Flemish Down a Line...8-5
Figure 8-5. Faking Down a Line 8-6
Figure 8-6. Plain or Temporary Whipping 8-6
Figure 8-7. Making a Permanent Whipping..8-7
Figure 8-8. Elements of the Knot, Bend, and Hitch 8-8
Figure 8-9. Overhand Knot...8-9
Figure 8-10. Figure Eight Knot 8-9
Figure 8-11. Square Knot 8-10
Figure 8-12. Tying the Single and Double Sheet or Becket Bend..8-10
Figure 8-13. Tying a Bowline..8-11
Figure 8-14. Tying a Bowline on a Bight 8-12
Figure 8-15. Tying a French Bowline 8-13
Figure 8-16. Tying the Double Carrick Bend 8-13
Figure 8-17. Half Hitch..8-14
Figure 8-18. Clove Hitch..8-14
Figure 8-19. Stopper Hitch 8-15
Figure 8-20. Stage Hitch..8-15
Figure 8-21. Monkey Fist...8-16

Figure 8-22. Making a Back Splice, Step 1..8-17

Figure 8-23. Making a Back Splice, Step 2..8-18

Figure 8-24. Making a Back Splice, Step 3..8-18

Figure 8-25. Making a Back Splice, Step 4..8-19

Figure 8-26. Making a Short Splice...8-20

Figure 8-27. Selecting the Middle 8-20

Figure 8-28. First Two Tucks in a Strand Eye Splice..8-21

Figure 8-29. Last Tucks of an Eye Splice 8-21

Figure 8-30. Makeup of Wire Rope..8-22

Figure 8-31. Strands and Wires 8-23

Figure 8-32. Measuring Wire Rope 8-24

Figure 8-33. Uncoiling Wire Rope..8-25

Figure 8-34. Seizing Wire Rope 8-26

Figure 8-35. Putting a Back Turn in Wire Rope 8-27

Figure 8-36. Selected Components of Rigger's Cargo Set...8-29

Figure 8-37. Correct and Incorrect Use of Wire Clips..8-29

Figure 8-38. Improved Type of Wire Rope Clip 8-30

Figure 8-39. Making a Hasty Eye (Molly Hogan) Splice, Step 1..8-31

Figure 8-40. Making a Hasty Eye (Molly Hogan) Splice, Step 2..8-31

Figure 8-41. Making a Hasty Eye (Molly Hogan) Splice, Step 3..8-32

Figure 8-42. Tucking Strands of a Liverpool Splice 8-33

Figure 8-43. How to Avoid a Kink Completing a Splice 8-34

Figure 8-44. Tubular Fid..8-35

Figure 8-45. Metal Wire Fid 8-35

Figure 8-46. Pusher 8-36

Figure 8-47. Minimum Lengths for Standard Eye Splice 8-37

Figure 8-48. Minimum Lengths for End-for-End Splice...8-38

Figure 8-49. Marking the Measurements (Step 1) 8-38

Figure 8-50. Extracting the Core (Step 2) 8-39

Figure 8-51. Marking the Core (Step 3) 8-39

Figure 8-52. Marking the Cover for Tapering...8-40

Figure 8-53. Putting the Cover Inside the Core 8-40

Figure 8-54. Performing the Taper...8-41

Figure 8-55. Reinserting the Core Into the Cover..8-41

Figure 8-56. Marking the Reduced Volume Tail Core 8-42

Figure 8-57. Burying the Exposed Core...8-43

Figure 8-58. Finishing the Eye Splice with Lockstitch..8-43

Figure 8-59. Continuing Lockstitching..8-44

Figure 8-60. Completing Lockstitching..8-44

Figure 8-61. Standard End-for-End Splice 8-44

Figure 8-62. Marking the Measurements 8-45

Figure 8-63. Extracting the Cores 8-45

Figure 8-64. Marking the Cores...8-46

Figure 8-65. Marking the Cover for Tapering 8-46

Figure 8-66. Performing the Taper..8-47

Figure 8-67. Repositioning the Lines...8-47

Figure 8-68. Putting the Cover Inside Core...8-48

Figure 8-69. Reinserting the Core Into Cover 8-48

Figure 8-70. Burying the Exposed Core 8-49

Figure 8-71. Finishing the Splice 8-49

Figure 9-1. Sharpening the Scraper..9-3

Figure 9-2. Sharpening a Chipping Hammer..9-3

Figure 9-3. Electric Portable Grinder...9-4

Figure 9-4. Power Scaling Tools and Wire Brushes 9-4

Figure 9-5. Steps in Boxing...9-10

Figure 9-6. Types of Brushes 9-11

Figure 9-7. Small Brush Keeper 9-12

Figure 9-8. Correct Way to Hold a Brush 9-13

Figure 9-9. Laying On and Laying Off 9-14

Figure 9-10. Painting Pipes and Stanchions 9-15

Figure 9-11. Applying Masking Tape..9-16

Figure 9-12. Removing Masking Tape 9-17

Figure 9-13. External-Mix Air Cap 9-18

Figure 9-14. Internal-Mix Air Cap 9-19

Figure 9-15. Suction-Feed Air Cap..9-19

Figure 9-16. Pressure-Feed Air Cap 9-20

Figure 9-17. Cross-Section of a Spray Gun 9-20

Figure 9-18. Principal Parts of the Spray Head..9-21

Figure 9-19. Pressure-Feed Cup..9-21

Figure 9-20. Pressure Tank..9-22

Figure 9-21. Air Transformer 9-23

Figure 9-22. Hold Spray Gun..9-24

Figure 9-23. Proper Spray Gun 9-24

Figure 9-24. Correct and Incorrect Methods of Spraying Corners 9-25

Figure 9-25. Steps in Cleaning a Pressure-Feed Gun 9-26

Figure 9-26. Steps in Cleaning a Container-Type Gun 9-27

Figure 9-27. Lubrication Points of a Spray Gun 9-28

Figure 9-28. Removing the Spray Head 9-28

Figure 9-29. Filter Respirator..9-29

Figure 9-30. Dust Respirator 9-29

Figure 9-31. Air Supply Respirator 9-30

Figure 9-32. Hood Respirator 9-30

Figure 10-1. Cross Section of the Surf...10-2

Figure 10-2. Crossing the Surf Line 10-3

Figure 10-3. Correct and Incorrect Angles for Towing Broached Boat Clear of Beach 10-8
Figure 11-1. Arm and Hand Signals 11-4
Figure 11-1. Arm and Hand Signals (continued) 11-5
Figure 11-2. Hydrographic and Beach Markers and Signs as Seen From Seaward 11-7
Figure 12-1. Retaining Body Heat 12-2
Figure 12-2. Sector Search Pattern—Ship 12-3
Figure 13-1. Display of Towlines 13-3
Figure 13-2. Chafing Gear 13-6
Figure 13-3. Small Plate Shackle 13-7
Figure 13-4. Doubling-Up Towlines 13-7
Figure 13-5. Hawser on Fantail 13-8
Figure 13-6. Leading the Hawser Over the "H" Bitt 13-8
Figure 13-7. Making Up a Hip Tow 13-9
Figure 13-8. Shifting a Tow From One Side to the Other 13-11
Figure 13-9. Towing Two Barges Alongside 13-12
Figure 13-10. Keeping a Tow in Step 13-14
Figure 13-11. Tandem Rig 13-15
Figure 13-12. Honolulu Rig 13-15
Figure 13-13. Christmas Tree Rig 13-15
Figure 13-14. Modified Christmas Tree Rig 13-16
Figure 15-1. Wire Rope Block 15-2
Figure 15-2. Common Cargo Blocks 15-3
Figure 15-3. Various Rigs and Fittings 15-4
Figure 15-4. Blocks and Tackles 15-5
Figure 15-5. Blocks at Right Angles 15-6
Figure 15-6. Reeving a Double 15-6
Figure 15-7. Reeving a Threefold Luff Tackle Purchase 15-7
Figure 15-8. Mechanical Advantages of Tackles 15-7
Figure 15-9. Threaded Rod on Turnbuckle 15-12
Figure 15-10. A Disassembled Block 15-13
Figure 15-11. Grounding a Shroud 15-15
Figure 16-1. Nomenclature of an Anchor 16-2
Figure 16-2. Types of Anchors 16-2
Figure 16-3. Detachable Link 16-3
Figure 16-4. Connecting Anchor to Anchor Cable 16-4
Figure 16-5. Standard Anchor Chain Markings 16-5
Figure 16-6. Horizontal Shaft Windlass 16-7
Figure 16-7. Vertical Shaft Windlass 16-7
Figure 16-8. Side View of Horizontal Shaft Anchor Windlass 16-8
Figure 16-9. Capstan/Wildcat and Brake 16-10
Figure 16-10. Turning the Drum Brake 16-10
Figure 16-11. Setting Up the Brake 16-13

Figure 16-12. Wildcat Engaged 16-13

Figure 16-13. Lifting Locking Ring 16-14

Figure 16-14. Inserting Anchor Bar 16-14

Figure 16-15. Wildcat Disengaged 16-15

Figure 16-16. Slacking Off Chain Stopper 16-15

Figure 16-17. Removing the Devil's Claw 16-16

Figure 16-18. Opening Riding Pawl 16-16

Figure 16-19. Setting Brake 16-17

Figure 16-20. Securing Devil's Claw 16-17

Figure 16-21. Closing Riding Pawl 16-18

Figure 16-22. Checking Controller 16-18

Figure 16-23. Lift Locking Ring Key 16-19

Figure 16-24. Engaging Wildcat 16-19

Figure 16-25. Wildcat Engaged 16-20

Figure 16-26. Sequence of Weighing Anchor 16-21

Figure 16-26. Sequence of Weighing Anchor (continued) 16-22

Figure 17-1. Flooding Effect Comparison: Unplugged Holes Versus Partially Plugged
Holes 17-2

Figure 17-2. Anchoring Fit 17-3

Figure 17-3. Correct Shoring Angles 17-4

Figure 17-4. Shoring for Bulging Plate 17-4

Figure 17-5. Shoring Against Horizontal Pressure 17-5

Figure 17-6. Driving Wedges 17-6

Figure 17-7. Adjustable Shore Measuring Batten 17-7

Figure 17-8. Shoring Spread to Adjacent Compartments 17-8

Figure 17-9. Measuring Length of Shore 17-9

Figure 17-10. Cutting the Angles of a Shore 17-10

Figure 17-11. Cracks with Holes 17-11

Figure 17-12. Shoring over Drilled Crack 17-12

Figure 17-13. Example of Hull Patching Using a Mattress 17-12

Figure 17-14. Hinged Plate Patch 17-13

Figure 17-15. Types of Hook Bolts 17-14

Figure 17-16. Patching Using Hook Bolts 17-14

Figure 17-17. Patching Using the Folding T-Type Hook Bolt 17-14

Figure 17-18. Installing a Soft Patch on a Pipe 17-15

Figure 17-19. Jubilee Pipe Patches 17-15

Figure 17-20. Three Types of Reinforced Clamps 17-16

Figure 17-21. Resin Temperature Versus Kick Over Time Graph 17-18

Figure 17-22. Relative Positions of Patch Materials 17-19

Figure 17-23. Simple Pipe Patch 17-21

Tables

Table 5-1. Index of Nautical Charts and Publications 5-10

Table 6-1. Standard International Phonetic Alphabet 6-6

Table 6-2. Pronunciation of Numbers 6-6

Table 7-1. Materials Effectiveness Against Gamma Radiation 7-42

Table 7-2. Recommended Personnel Action Against Nuclear Detonations 7-43

Table 8-1. Line Strengths 8-2

Table 8-2. Line Constants 8-2

Table 8-3. Line Strength Table (Safety Factor of 5) 8-4

Table 8-4. Size and Number of Wire Clips 8-30

Table 9-1. Types and Uses of Brushes 9-11

Table 9-2. Brush Cleaners for Different Finishes 9-13

Table 9-3. Spray Gun Troubleshooting Chart 9-27

Table 9-4. Paint Coverage per Gallon 9-31

Table 11-1. Day Signals 11-2

Table 11-2. Night Signals 11-2

Table 11-3. Maneuvering Signals 11-6

Table 12-1. Survival Times in the Water 12-1

Table 14.1 Elements of a Standard Fire Command 14-2

Table 14.2 Range Fire Commands 14-3

Table 14.3 Sample CSW Commands 14-4

Table 14.4 Reduced Fire Commands 14-4

Table 15-1. Safety Factor of Line 15-8

Table 15-2. Rule of Thumb for Computing the Safe Working Load and Breaking
Strength 15-8

Table 15-3. Line Strength Table 15-11

Table 15-4. Turnbuckle Rod SWL Table 15-13

Table 17-1. Ratio Conversion Table 17-9

Table 17-2. Table of Approximate Quantities of Materials Required for Pipe Patches 17-19

Preface

The United States (U.S.) Army watercraft fleet is made up of all types of vessels, including oceangoing vessels, tugs, landing craft, and barges. Although all of these vessels operate on water, their missions are different. The watercraft operator must have the skills and knowledge to perform the tasks required on any of these vessels. This training circular (TC) is for the 88K watercraft operator, skill levels 1 through 4. It will provide the subject matter that relates directly to the common technical tasks listed in STP 55-88K14-SM-TG.

The U.S. Army's environmental strategy into the 21st century defines the Army's leadership commitment and philosophy for meeting present and future environmental challenges. It provides a framework to ensure that environmental stewardship ethic governs all Army activities. The Army's environmental vision is to be a national leader in environmental and natural resource stewardship for present and future generations, as an integral part of all Army missions. The Army's environmental vision statement communicates the Army's commitment to the environment.

This TC applies to the Active Army, the Army National Guard/Army National Guard of the United States, and the United States Army Reserve.

The proponent of this publication is Headquarters, United States Army Training and Doctrine Command (USATRADOC). Submit comments and recommendations for improving this publication on DA Form 2028 (Recommended Changes to Publications and Blank Forms) directly to:

Commander, USACASCOM
ATTN: ATCL-TDM
2221 Adams Avenue
Fort Lee, VA 23801-2102

This page intentionally left blank.

Chapter 1

Introduction to Army Watercraft

INTRODUCTION

1-1.   U.S. Army watercraft is used in the following operations: harbor, coastal, intra-island, and logistics over-the-shore (LOTS). U.S. Army watercraft is also used for other operations such as ocean towing and security patrols. This chapter addresses these operations and the categories of watercraft.

WATERCRAFT OPERATIONS

1-2.   U.S. Army watercraft plays a major role in projecting and sustaining combat forces. From established ports to LOTS operations, Army watercraft provides a flexible means of moving troops and supplies. Through pre-positioning and self-deployable vessels, the Army's fleet of diverse watercraft is capable of playing a dynamic part in the Army Strategic Mobility Program.

MISSION

1-3.   The mission of Army watercraft is to support the Army's reception, staging, onward movement and integration (RSOI) movement plan. Army watercraft provides the vital link between the offshore arrival of combat power, loaded aboard strategic sealift ships, and placing that power ashore in a ready-to-fight configuration. The Army watercraft fleet must be prepared to do this mission anywhere in the world. This is accomplished by the following means:

- Transport of personnel and cargo between ship and shore and on inland waterways.
- Support of floating equipment support for terminal operations within a fixed-port or unimproved port facility complex.
- During riverine operations.
- Lighterage for cargo and personnel from ships lying offshore to transfer-segregation areas beyond the beach lines in LOTS operations.

HARBOR OPERATIONS

1-4.   Harbor operations include the movement of cargo and personnel within a harbor and the protected waters in the vicinity of the harbor. Tugs, barges, and floating cranes discharge and transfer cargo; small craft provide ferrying service and conduct security patrols. Tugs are used for providing berthing service for oceangoing vessels and for fire fighting in the port area.

INTRA-ISLAND AND COASTAL OPERATIONS

1-5.   Large watercraft carries cargo and personnel from central ports to smaller outlying installations (such as subports, radar installations, and other terminals). Where larger oceangoing freighters cannot navigate, logistic support vessels (LSV), and landing craft can safely transport cargo through shallow waters and narrow winding channels.

LOGISTICS-OVER-THE-SHORE OPERATIONS

1-6.   LOTS operations include ship-to-shore operations moving cargo and personnel onto a prepared beachhead from larger vessels anchored offshore or sea basing. This operation is the most difficult and time consuming. Planning, timing, and a skilled Weather Eye mean the difference in success or failure in this type of cargo operation. Landing craft and causeway ferry systems are normally used for such operations. Tugs with barges, floating cranes, and LSVs may also be used where causeway piers have been installed on the beach.

WORLD WIDE MISSIONS

1-7.   Army watercraft is capable of deployment to any theater of operation around the world. Vessels such as the LSV and the 128-foot tug are capable of self-deployment. The landing craft, utility (LCU) 2000, while capable of self-deployment, may also be transported aboard heavy lift ships. The remainder of the smaller vessels in the fleet also uses this method of transportation.

Chapter 2

Shipboard Life

INTRODUCTION

2-1.   The successful operation of a watercraft depends directly on the capabilities and knowledge of each crewmember.   The commanders and the watercraft operators are responsible for ensuring that the watercraft are operated according to federal, state, and DA regulations.   This chapter discusses the requirement for marine certification, customs and courtesies, watch standing procedures, shipboard sanitation and pollution control, and the marine logbook.

MARINE QUALIFICATION

2-2.   Qualification for marine personnel incorporates a twofold process of certification and licensing. This twofold process is as follows:

- Certification.   Verifies an individual's knowledge of common marine tasks by military occupational specialty (MOS) and skill level (SL).
- Licensing.   Verifies that an individual has the knowledge and ability to perform vessel-specific tasks on a designated vessel.   Watercraft operators will operate only those vessels for which they are licensed.

ARMY REGULATION 56-9

2-3.   Army Regulation (AR) 56-9 prescribes the responsibilities, policies, and procedures for the authorization, assignment, operation, maintenance, sanitation, and safety of Army watercraft.   It also defines the procedures for verifying the qualifications of Army marine personnel and for SL and/or vessel type.

MARINE QUALIFICATION BOARD (MQB)

2-4.   The MQB is responsible for evaluating and recommending approval of actions on the issuing, denying, suspending, or revoking of a United States Army Marine License (USAML).   The MQB also prepares, administers, and grades the appropriate examinations for the required USAML.

MARINE SERVICE RECORD (DA FORM 3068-1)

2-5.   This form is the method that a Soldier's sea service is tracked and reported to the Marine Qualification Officer (MQO). This document becomes a permanent part of each Soldier's sea service record and in accordance with AR 25-400-2 is maintained by the MQO for up to 40 years.

SKILL LEVELS OF WATERCRAFT OPERATORS

2-6.   Marine certification is an essential requirement for promotion in the marine field.   Advancing from one SL to another requires demonstrated improvement of skills and knowledge.   It also requires the recommendation of the commander.   The following paragraphs give an overview of deck duties according to SL.

- SL 1, USAML Annotated "Seaman (88K10)," Pay Grades E-1 through E-4.   A seaman assigned to harbor craft or landing craft will be required (under supervision) to demonstrate the following general seamanship duties:

- Perform marlinespike seamanship.
- Handle mooring lines and hawsers when docking and undocking.
- Perform deck maintenance by using hand and power tools to prepare metal surfaces for painting; maintain standing rigging, run rigging, and deck machinery. Also follow preventive maintenance procedures.
- Stand helm watch, lookout watch, and towing watch.
- Participate in shipboard emergency drills (such as fire fighting, abandon ship, man overboard, and chemical, biological, radiological, nuclear, high yield explosives (CBRNE) operations).
- Perform fire and emergency rescue procedures.
- Recognize international distress signals.
- Interpret single-flown international code flags having a special meaning.
- Communicate with other vessels and shore stations using correct radiotelephone procedures.
- Serve as a relief operator.
- Demonstrate a sound understanding of the Army's environmental ethics.
- Demonstrate the proper response to a fuel spill.

- SL 2, USAML Annotated "Watercraft Operator of Class B and Class C Vessels (88K20)," Pay Grade E-5. The coxswain is responsible for the operation, maintenance, and welfare of his/her vessel and crew. Small craft are versatile and are assigned to various types of operations. The craft requires constant maintenance and must be available for dispatch on short notice. The coxswain is required to work side-by-side with his/her seamen and engineers in maintaining his/her craft and stowing equipment and gear. The coxswain performs the following:
  - Exercises complete charge of his/her vessel, passengers, and cargo while underway and ensures compliance with safety regulations.
  - Handles his/her craft skillfully when maneuvering or mooring and exercises good seamanship practices at all times.
  - Knows the principles of advanced piloting and dead reckoning as well as the use of charts, compass, pelorus, and other navigational instruments.
  - Supervises first aid according to the procedures outlined in FM 4-25.11.
  - Maintains an accurate log and keeps fuel and supplies at authorized maximum levels.
  - Manipulates the helm and engine controls to dock, undock, beach, and retract during LOTS operations.
  - Communicates with other vessels and shore stations using correct radiotelephone procedures.
  - Operates the vessel at all times according to COMDTINST M16672.2C (Navigation Rules, International-Inland).

- SL 3, USAML Annotated "Watercraft NCO/Boatswain (88K30)," Pay Grade E-6. The watercraft boatswain is the senior NCO of the deck crew and will serve as the coxswain aboard the landing craft, mechanized (LCM) – 8 MOD 2. Aboard class A vessels the boatswain will—
  - Check all deck machinery and equipment for operating condition, reporting all discrepancies to the mate.
  - Ensure that the vessel is secured for sea before it leaves its moorings.
  - Supervise the stowage of mooring lines and fenders when leaving port.
  - Prepare the anchor for use when arriving or departing.
  - Be responsible to the mate for maintenance of the gear and equipment of the deck department as well as the conduct, discipline, and direct supervision of deck department personnel.
  - Assign, under the supervision of the mate, deck department personnel to watches and details.
  - Perform the first mate's duties on craft not authorized a mate.

- SL 4, USAML Annotated "Mate, Class A-1 Freight and Towing Vessels upon Coastal and Inland Waters; Radar Observer (88K40)," Pay Grade E-7. The mate at this SL can function as either a Detachment Sergeant aboard an LSV or be assigned as mate aboard an LCU. The mate acts as assistant to the master and assumes the master's responsibilities during his/her absence. The mate is specifically responsible for the following:
    - Ensuring that all of the master's instructions are obeyed.
    - Supervising the deck department. When in port, the mate will supervise deck maintenance, cargo operations, and general ship's business. At sea, the mate will take charge of the navigation of the ship during his/her watch, inform the master of any unusual circumstances that may arise, know the ship's position, and ensure that all watch standers are alert and attentive to the details of their duties. Before relieving the watch, the mate will read and initial the remarks in the master's night order log.
    - Keeping up and ordering charts and publications required aboard ship.
    - Assisting the master in the pilothouse in adverse weather, in confined waters, or as required.
    - Notifying the master when any unusual obstructions to navigation are discovered, when the vessel appears to be approaching danger, and when unusual changes in the weather or other unexpected occurrences are observed.
    - Maintaining the prescribed course. When necessary to avoid sudden danger, the mate will take action without awaiting the master's instructions.
    - Being familiar with and complying with federal and local pollution laws.

SHIPBOARD CUSTOMS AND COURTESIES

2-7.   The military has many customs and courtesies they follow. Watercraft personnel must also follow certain rules of customs and courtesies required aboard ship.

FLYING THE NATIONAL ENSIGN

2-8.   There are certain situations and times that the national ensign (Figure 2-1) is flown. The national ensign will be flown 24 hours a day during war or when sailing in unfriendly waters.

Figure 2-1. National Ensign

SHIP UNDERWAY

2-9.   When an Army ship is underway, the ensign is flown from sunrise to sunset. When underway, Class A ships fly the National Ensign from the main mast or the after mast. On ships fitted with only one mast (such as an LCM-8), the ensign is flown from the outboard halyard on the starboard yardarm.

ANCHORED OR MOORED

2-10.   When the ship is at anchor or tied up to the pier, the ensign is flown from the flagstaff. The ensign will be flown from 0800 hours until sunset.

HALF-MASTING THE ENSIGN

2-11.   The custom of flying the national ensign at half-mast is observed as a tribute to the dead. The Army follows the half-mast custom carefully and according to specific regulations. Whenever the ensign is to be flown at half-mast, it is first raised to the closed up position (the top-most position) on the gaff and then lowered to half-mast position. The ensign is flown half-mast during the following times.

- Memorial Day. Memorial Day is a time to remember the U.S. men and women who lost their lives serving their country. On this day, the national ensign is flown at half-mast from 0800 to sunset.
- President's Death. When official word is received of the death of the President of the United States, all U.S. vessels display the ensign at half-mast, starting at 0800 of the following day and for 29 days thereafter. The same ceremony is observed upon the death of an ex-President or President-elect. Upon the death of the Vice President or certain other high government officials, the ensign remains at half-mast for 14 days.
- Death aboard Vessel. If a death occurs aboard an Army vessel, the national ensign is flown half-mast until the remains are transferred from the ship. The ceremonies appropriate at Army installations are conducted upon the death of an officer, warrant officer, or enlisted.

CHURCH PENNANT

2-12.   The church pennant is unique in that it is the only flag or pennant flown on the same halyard as and above the national ensign. It flies only during divine services onboard Army vessels.

UNION JACK

2-13.   The blue star-studded field in the corner of the U.S. flag is the canton of the national ensign. Since each star represents a state, the canton symbolizes the union of the states of the United States. The union jack (Figure 2-2) must be the same size as the canton of the national ensign flown from that particular vessel. The union jack is flown only on those Army ships commanded by commissioned officers or warrant officers. The union jack is flown only when the ship is at anchor or moored to a pier. It is flown from 0800 to sunset from the jack staff at the bow of the vessel. The union jack is raised after the national ensign and lowered before the national ensign at evening colors. When the national ensign is flown half-mast, the union jack is also flown half-mast. The union jack is never dipped as a salute.

Figure 2-2. Union Jack

TRANSPORTATION CORPS FLAG

2-14.   The Transportation Corps (TC) flag (Figure 2-3) is flown from sunrise until sunset on the forward mast. On vessels not fitted with a forward mast, the TC flag is flown from the outboard halyard on the port yardarm unless it interferes with signal flag communications.

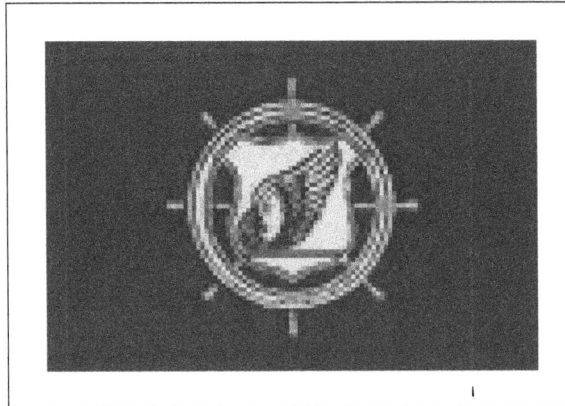

Figure 2-3. TC Flag

DRESSING AND FULL DRESSING SHIP

2-15.   Ships may be either dressed or full dressed during our national holidays or while in a foreign port during that nation's holidays. Dressing a ship, in honor of a person or event, consists of displaying flags and pennants on various halyards and stays. Usually the port or terminal commander specifies whether to dress or full dress the vessels. The latter is usually ordered when ceremonies are held at the port or terminal. In determining whether to dress or full dress vessels while in foreign ports, masters may be governed by the actions of the foreign nationals.

DRESSING SHIP

2-16.   Dressing ship is much less elaborate than full dressing. The national ensign is displayed at each masthead when a vessel is dressed. If the masts are all the same height, the ensigns at the mastheads must all be the same size. The largest ensign on board must be hoisted at the flagstaff and a union jack of corresponding size raised on the jack staff. If the occasion is one honoring a foreign nation, that nation's ensign is displayed at the main mast instead of the U.S. ensign. The U.S. ensign must be hoisted on the main mast and other mastheads during all U.S. celebrations.

FULL DRESSED SHIP

2-17.   The same procedure with the ensigns is followed when a ship is full dressed. A rainbow of the international code flags is also arranged as follows:

- From the jack staff to the fore masthead.
- Between the fore and main mastheads.
- From the main masthead to the flagstaff.

2-18.   If possible, all Army vessels should have their flags on the rainbow dressing lines in the same order. The flag order starts at the foot of the jack staff and extends to the foot of the flagstaff. This sequence should be repeated if one set of flags does not complete the rainbow. A ship is usually left dressed from 0800 to sunset during these celebrations. Dressing ship is never done when underway. If a ship enters port after 0800 or leaves before sunset on a dress occasion, she dresses or full dresses, as the occasion may be, upon anchoring and undresses upon getting underway.

SALUTES

2-19.   Army personnel, whether officer or enlisted, salute the national ensign when boarding a vessel and salute the mate on watch. When leaving the vessel, they give the two salutes in reverse order. The mate on watch returns salutes given him/her. This courtesy is required only when a vessel flies both the national ensign and union jack.

DECK WATCHES

2-20.   At sea, the day is divided into watches of 4 hours each (0001 to 0400, 0400 to 0800, 0800 to 1200, 1200 to 1600, 1600 to 2000, and 2000 to 2400). The 1600 to 2000 watch is sometimes divided into two watches (1600 to 1800 and 1800 to 2000). These are called "dog watches." Changing the watch at 1800 (dogging the watch) breaks the sequence, divides the evening recreational period, and allows the evening meal to be eaten without furnishing a relief section. Generally, a late mess is held for the 0400 to 0800 watch. The watch is referred to according to location or type of duty, such as the gangway watch, towing watch, and bow lookout.

WATCH LIST AND QUARTERS LIST

2-21.   The watch list specifies the hours and location for each crewman standing watch. Each crewman must check the watch list daily for the time and location of his/her duty. The quarter's list specifies the compartment and location of each crewman's berth. The watch list and quarter's list are posted at any two of the following places aboard ship: crew's quarters, passageways, crew's mess, and wheelhouse.

WATCH DUTIES: CLASS A VESSELS

2-22.   The dictionary describes the word "watch" (in nautical terms) as the periods of time into which the day aboard ship is divided and also during which a part of the crew is assigned to duty. The watches aboard Class A vessels are usually broken up into four-hour tours. The following guidelines describe how the watch routine functions.

RELIEVING THE WATCH

2-23.   Watches must be relieved 15 minutes before the hour. This allows time for the relief to receive instructions from the man on watch and permits the night relief to accustom his/her eyes to darkness. The quartermaster of the watch generally assigns watch stations. Once assigned the relief reports directly to the Soldier to be relieved and receives any instructions about the watch (such as targets being tracked and so forth). When he/she understands the instructions, he/she first requests permission to relieve the watch from the officer of the deck. Once permission is given, the relief will state loudly, "I relieve you," and then becomes completely responsible for the watch. The relieved watch then reports to the officer of the watch, informing him/her that he/she has been relieved. The watch relieving the helmsman follows this same procedure for relieving the off going watch. The helmsman in turn reports to the officer of the watch, informing him/her that he/she has been relieved and reporting the course on which the vessel was being steered when he/she was relieved.

HELM WATCH

2-24.   The helmsman may be a seaman or a quartermaster. He/She is responsible for the safe steering, either by compass or by terrestrial objects, as ordered by the master or watch officer. His/Her tour of duty normally consists of a 2-hour watch at the helm. He/She must know the degree and full point markings of a compass card and the vessel handling characteristics at various engine speeds.

LOOKOUT WATCH

2-25.   A crewman is stationed in the bow or bridge where he/she acts as a lookout, reporting anything he/she sees or hears to the bridge. This information includes ships, land, obstructions, lights, buoys, beacons, discolored water, reefs, fog signals, and anything that could pertain to the navigation of the vessel.

When reporting, the lookout names the object and gives the direction to the target using the point system (Figure 2-4) for example, "lighthouse two points on the port bow." If the officer of the watch asks for further information on the object sighted, the lookout describes it as briefly and clearly as possible. When port and starboard lookouts are posted, each lookout keeps watch only on his/her side of the bridge. Each notes the running lights on his/her side and reports immediately if the lights are dim or go out. In general, the orders given to the lookout are as follows:

- Remain alert, giving your attention only to your own special duty.
- Remain at your station until you are relieved.
- Keep on your feet; do not sit or lounge.
- Do not talk to others except as required by your duty.
- Speak loudly and distinctly when making a report.
- Repeat a report until it is acknowledged by the officer of the deck.
- When stationed, be sure that you understand your duties.
- Report everything, even if it was reported on the previous watch.

Note: 1 point is equal to 11.25 degrees.

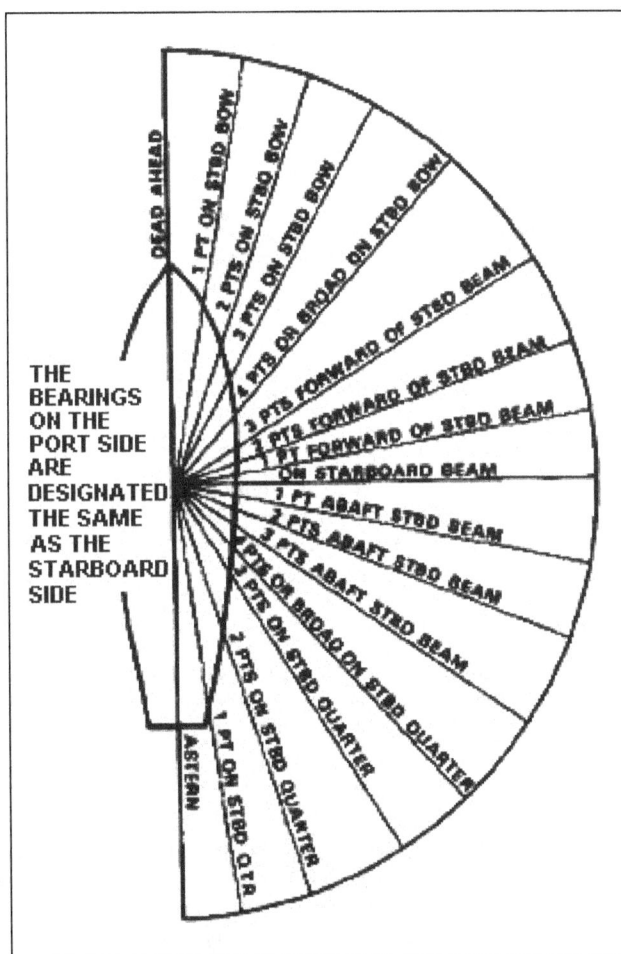

Figure 2-4. Point System

TOWING WATCH

2-26.   Aboard a tug, a member of the deck department is detailed to a towing watch. Normally stationed on the after section of the towing vessel, the crewmember—

- Observes how the tow is riding (Figure 2-5) and reports any unusual conditions to the bridge.
- Checks and makes adjustments as required to the towing engine, towing cable chafing gear, and bridles.
- Reports equipment failures immediately to the bridge.

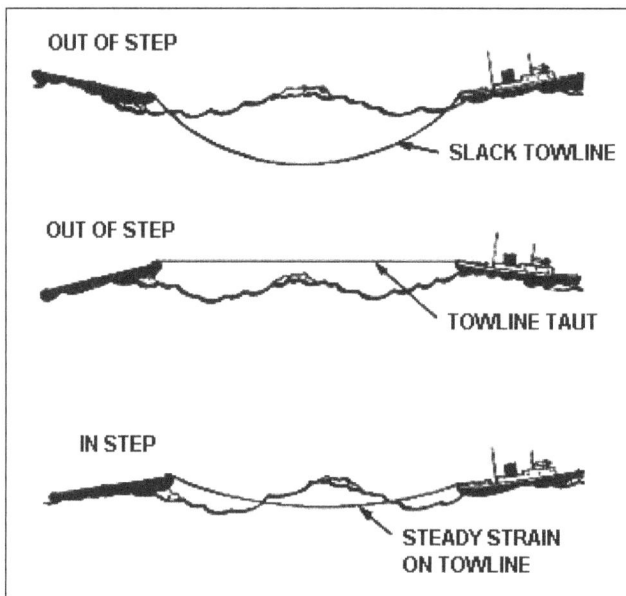

Figure 2-5. Tow Riding In/Out of Step

STANDING NIGHT WATCH

2-27.   Night watches are the most critical periods of responsibility for the lookout. Knowing how to stand night watches is necessary to ensure the safety of the vessel. During a night watch, the officer of the watch frequently enters the chart room or other lighted areas. He/She depends primarily on the lookout to spot objects.

EYE REACTION AT NIGHT

2-28.   Night vision differs from day vision to a much greater extent than is generally realized. Eyes respond slowly in the dark, picking out a moving object more easily than a stationary object. An object can more readily be seen at night by looking a little to the right or left of it rather than at it directly.

DARK ADAPTATION

2-29.   Developing the ability to see and to recognize distant objects at night is known as "dark adaptation." To compensate for the slower eye reaction at night, the lookout should scan the sky and sea slowly because he/she may not notice an object until he/she has looked near it several times. Since an object cannot always be located at night by looking straight at it, the lookout should look above the object. Moving the head from side to side will give an object the appearance of movement, making it easier to locate. To learn to see things at night requires considerable and continual practice.

RULES FOR DEVELOPING AND MAINTAINING DARK ADAPTATION

2-30.   A competent night lookout should do the following:

- Take advantage of the 10-minute interval before the hour of his/her watch to adjust his/her eyes to darkness.
- Use only a dim red light when a light is necessary.
- Look out of the corner of the eyes when scanning the horizon.
- Scan the region under observation slowly and regularly.
- Wear red dark adaptation goggles that permit vision without disturbing dark adaptation if it is necessary to enter a lighted place.
- Avoid looking at instrument panels, even if they are illuminated by a red light.
- Use light binoculars if available.
- Keep optical equipment clean.
- Refrain from looking at an object already spotted and reported.
- Keep in good physical condition. Fatigue, alcohol, and tobacco reduce dark adaptation.

ANCHOR WATCH AND FIRE WATCH

2-31.   A member of the deck crew is detailed as anchor watch and fire watch when the vessel is at anchor or moored. He/She performs the following:

- Frequently checks the lead and strain on the anchor cable.
- Sounds the fog bell when fog closes in and notifies the watch officer of the weather condition.
- Checks the position of the vessel by taking compass bearings of known objects and checks the drift lead.
- Notifies the watch officer when there is a change in bearings or a strain on the drift lead (which indicates the vessel may be dragging anchor).
- Checks for fire and fire hazards; sounds the alarm in case of fire.

GANGWAY WATCH IN PORT

2-32.   A member of the deck department is detailed as security or gangway watch in port to assist the watch officer in maintaining the security of the ship. The crewmember on watch takes a position near the gangway to prevent unauthorized persons and contraband from coming aboard. He/She is also responsible for—

- Tending the mooring lines and gangway during the rise or fall of the tide.
- Keeping the gangway log, that is, recording the activities of the watch, such as noting persons coming aboard or going ashore, the weather, and other information designated by the officer of the watch.
- Notifying the watch officer in case of emergency.
- Maintaining discipline aboard ship and notifying the watch officer of any disorder or unusual circumstances.
- Inspecting the vessel periodically for evidence of fire hazards.

WATCH DUTIES

CLASS B VESSELS

2-33.   Department of the Army (DA) regulations require small craft certification of operating personnel for Class B vessels. On these vessels, the duty of the watch consists of performing all the duties and responsibilities of a coxswain or seaman while in command or on duty. The watch (not to exceed a 12-hour shift) on a Class B vessel lasts for the duration of the task assigned the vessel.

CLASS C VESSELS

2-34. Watch duties consist of carrying out all the duties and responsibilities necessary for the safe operation and maintenance of the vessel and its machinery. The general duties assigned are comparable to similar positions on Class A or B vessels, although modified to meet the requirements of the assigned mission.

LOGBOOKS

DECK AND ENGINE LOGBOOKS

2-35. Deck and engine department logbooks are maintained according to AR 56-9. The logbooks provide a permanent legal record of the operations and conditions of the vessel and the status of its cargo, crew, and/or passengers. All occurrences of importance, interest, or historical value concerning the crew, passengers, operation, and safety of Army watercraft will be recorded daily, by watches, in three types of deck logbooks.

- The Deck Department Log (DA Form 4640, Harbor Boat Deck Department Log for Class A&B Vessels) (Available through normal forms supply channels.). This log is required for use on Class A and Class B vessels.
- The Deck and Engine Log (DA Form 5273, Harbor Boat Deck and Engine Log for Class B Vessels) (Available through normal forms supply channels.). This log is required for use on all Class B vessels. DA Form 5273 may also be used on the deck or liquid barge, design BG 231B, and the refrigerator barges (BR 7010 and BR 7016).
- The Engine Department Log (DA Form 4993, Harbor Boat Engine Department Log for Class A and C-1 Vessels) (Available through normal forms supply channels.). This log is required on Class A and Class C-1 vessels.

Note: Logbooks will be prepared according to instructions in AR 56-9. Logbooks and other pertinent records must be preserved for use in claims against the United States for damage caused by an Army watercraft and for affirmative claim by the United States for damage to Army property caused by other vessels or floating objects. Requirements for preserving deck and engine logs as well as other records are in AR 27-series regulations (Legal Services). Headquarters, Department of the Army (HQDA) Department of the Army Judge Advocate (DAJAZA) will be notified when a log (or any portion of the log) is to be used in litigation or is to be withheld for any other legal proceedings. When no longer required for the legal proceedings, the log will be returned to the installation having command over the Army watercraft that was involved. Commanders having assigned watercraft will periodically review log requirements to ensure that logs are maintained according to provisions in AR 56-9. Watercraft less than 30 feet do not require maintenance of logs—provided adequate records are maintained by the unit.

MAINTENANCE AND RETENTION OF LOGBOOKS

2-36. The ship's deck log will be presented to the master or coxswain each day. Should any inaccuracies or omissions be noticed, the master or coxswain will have the necessary corrections made. After corrections have been made, they will approve the log by signing the page. After the log has been approved, no change or addition will be made without the master's or coxswain's permission or direction. The mate, on whose watch the matter under consideration occurred, must make any change or addition. When the master or coxswain calls attention to an inaccuracy or omission, the mate will not decline to make a change in or an addition to, the log unless he/she believes the proposed change or addition to be incorrect. He/She will, if required, explain in writing to the master or coxswain his/her reason for the change or addition. The master or coxswain may then make any appropriate remarks concerning the inaccuracy or omission, entering them at the bottom of the page of the log over his/her own signature. When a correction is necessary, a single line will be drawn through the original entry (in red ink) so that the entry remains legible. The correct entry will then be made clearly and legibly. Corrections will be initialed by the person making the original entry and also initialed by the master or coxswain. Entries will also be made of all drills and inspections prescribed in the Code of Federal Regulations (CFR) 46, paragraph 97.35-1. These entries will be made or underlined in red ink. For reserve components' (RC), non-drill dates will be noted in the log, together with the vessel location, and annotated "nonduty days." Logs will be made available to the area maintenance support activity personnel for entries when applicable. Night order books are used aboard class A vessels where sea watches are maintained on a 24-hour basis. Each day the master reviews, updates, and prepares the general standing orders, special orders, and specific instructions for the night watches. Each mate coming on watch must read and sign the night order book. Bell logbooks will be maintained on every vessel except those capable of complete engine control and operation from the pilothouse. The time and all changes in engine speed and/or direction must be recorded. Vessels with pilothouse control are also included when using the bell system. A logbook will be retained for 5 years as the onboard record of the deck and engine departments. At the end of this period, it will be destroyed according to the AR 25-series of regulations.

OIL RECORD LOGBOOK

2-37.   In addition to required deck and engine logbooks, Class A and Class B vessels will maintain a record of ballasting or cleaning of bunker fuel tanks and disposal of oily residues from bunker fuel tanks. Other exceptional discharges of oil will also be recorded. All masters, coxswains, and chief engineers of Class A, Class B, and Class C-1 vessels and assigned crews will comply with oil pollution regulations. Oil record books will also be kept for 5 years.

MODIFICATION OF LOGBOOKS

2-38.   Make changes to individual log sheets by drawing ruled lines in ink and then making appropriate entries on them. Changes are required for floating cranes to show the number and weight of heavy lifts made, as well as any other entries appropriate to the type of service in which employed.

COMMUNICATION LOGBOOKS

2-39.   The vessel master or coxswain will ensure that the following logs are maintained.

- Bridge-to-Bridge (VHF). For vessels equipped with bridge-to-bridge VHF radio/telephone, this record may be kept on the logbook. Each page shall be dated and identified by the vessel name or number. The log of the bridge-to-bridge station (channel 13, 156-650 MHz) shall include, as a minimum, the following entries:
    - All radio/telephone distress and alarm signals.
    - All communications transmitted or intercepted, and any information heard which might be of importance to maritime safety.

Note: Text should be as complete as possible, including the time, frequencies used, and position of vessel in distress.

    - The times when watch is begun, interrupted, and ended.
    - A daily entry concerning the operating condition of the radio.
- Military Tactical Communications. For vessels equipped with military tactical communication capability, records and procedures shall be according to existing Army regulations.
- High Frequency (HF) and Low Frequency (LF) Communications. On vessels equipped with HF or LF communication ability, a record of the following shall be kept as a minimum:
    - Name of the operator on watch. The entry "On Watch" is made by the operator going to watch. The entry "Off Watch" is made when an operator is relieved or the station is closed down. The operator's signature must accompany both entries.
    - All calls and replies to calls, the call sign of station called, the time that traffic is handled, and the frequency and mode used. The time that traffic is handled shall be noted as "Time In" to note when a communication begins and "Time Out" to note when a communication is completed. Times shall be suffixed for the applicable time zone.

- Cases of unlawful interference and failure of equipment.
- The full text of distress, urgent, and safety messages.
- Results of tests of auto-alarm receivers, including the times that the auto-alarm is in operation.

RADIO LOGBOOKS

2-40.   Radio logbooks will be kept by calendar year. They will be kept for a period of 1 year after the last entry. Station logs involving communications, incident to or involved in distress, disaster, or accidents will be kept for a period of 3 years after the last entry is made.

SHIPBOARD SANITATION

2-41.   Cleanliness and sanitary conditions are essential since personnel aboard ship live and work in restricted quarters. The health of each crewmember is the concern of all aboard, as any infection or unhealthy habit can affect the overall health or efficiency of the crew.

2-42.   High sanitary standards must be set to protect the crew from infection and illness. Conditions aboard ship should include—

- Adequate cleaning and laundry facilities.
- Adequate locker space for each member of the crew.
- Clean and orderly quarters.
- Recreation facilities separate from the crew's sleeping quarters.
- Adequate ventilation and temperature control in the crew's quarters.
- Enough storage space for refrigerated foods, dry storage, vegetables, and dairy products.
- Proper food handling and storage.
- A daily balanced diet for the crew.
- Rodent control. Rat guards should be used on the mooring lines and traps (if required) inside the vessel.
- Insect control. Cleanliness and the use of powders, insecticides, and fumigation.
- Water purification. Water will be chlorinated if there is any doubt of its purity.

PERSONAL HYGIENE

2-43.   This is something that must be done by each crewmember. Aboard ship, crewmembers can do this by—

- Wearing clean and dry clothing.
- Bathing at least once a day.
- Washing their hands after using the toilet facilities.
- Keeping their fingernails and toenails clean and clipped.
- Brushing their teeth after each meal or at least once a day.
- Getting the proper sleep and rest.
- Doing some type of physical exercise on a daily basis.

TAKING POTABLE WATER ABOARD SHIP

2-44.   Using improper or careless procedures when taking potable water (water suitable for drinking) aboard ship can result in contaminated water being introduced into the drinking water system. The following describes the operational procedures that will be followed to provide for the safe and sanitary intake of water aboard ship.

POTABLE WATER HOSES/RISERS

2-45.   Hoses will be labeled "POTABLE WATER ONLY" at 10-foot intervals and used only for that purpose. Potable water risers will be labeled "POTABLE WATER" and color-coded light blue. Also ensure the following:

- The end couplings of the hoses will be color-coded light blue.
- When not in use, potable water hoses will be rolled, coupled, or otherwise protected from contamination and stored.
- Hoses will be stored in a vermin-proof, closed locker specifically designated for potable water hose storage. Preferably, the locker will be located off the weather deck, installed 18 inches above the deck and labeled "POTABLE WATER HOSE."
- All risers will be equipped with screw caps and keeper chains.

MAKING THE SHIP-TO-SHORE POTABLE WATER HOSE CONNECTION

2-46.   The following procedure will be used when making ship-to-shore potable water connections.

- Before making the potable water connection, disinfect the potable water risers on the vessel and the shore facility. Do this by preparing a chlorine solution and immersing or swabbing the risers with the solution.
- To disinfect the insides of the hoses, fill them with a chlorine solution and allow to stand for 2 minutes. To do this, elevate both ends of the hose, pour required amount of chlorine agent into one end, and fill remainder of hose with water.
- Open the valve on the shore supply and flush for 15 to 30 seconds to remove any debris which may be present in the piping.
- Connect the potable water hose to the shore facility riser.

2-47.   After disinfecting of the vessel riser, the ship connection can be made and transfer of water initiated. If during the transfer or connection procedures the hose is contaminated by hanging into or dropping into the harbor water, pumping operations will be stopped and the hoses disinfected.

Note: All water supplied by public or private systems outside the United States should be considered of doubtful quality. If doubt exists as to the quality of the water, medical authorities ashore should be requested to evaluate the source and provide recommendations to the vessel commander.

POLLUTION CONTROL

WATER POLLUTION

2-48.   The U.S. government has passed many laws to protect our country's natural resources starting with the River and Harbor Act of 1899 (which is still in use today) to the Federal Water Pollution Control Act of 1970 and later amendments. These federal laws are concerned with the dumping of sewage and garbage and oil pollution of our navigable waters. New Army vessels and older vessels going into the shipyard for overhaul and rehabilitation are now fitted with a filtering system and holding tanks. There will be no more overboard discharge of vessel-generated waste (sewage, laundry drains, galley waste, slop oil, and so on).

Vessels at Sea

2-49.   When sailing in any waters, Army vessels will not discharge vessel-generated waste, throw garbage overboard, or create oil pollution.

Vessels in Port

2-50.   When in port, Army vessels will discharge their holding tanks, slop oil, and sludge either into fixed shore disposal facilities or make arrangements with the shore authorities to bring alongside tank trucks or barges fitted for this task. Some ports have a disposal system where oil can be returned for reprocessing while others may provide only storage barrels. Make use of whatever is provided.

OIL POLLUTION

2-51.   Any substance that will create a visible sheen on the surface of the water or create an emulsion on or below the surface will constitute as an oil spill. This may consist of only one cup of oil or oil product. Vessel personnel will not dump the following over the side:

- Soapy water.
- Galley water.
- Garbage.
- Paint, thinner, kerosene, or other oil-base products.
- Sanitary waste.

OIL TRANSFER PROCEDURES

TITLE 33, CODE OF FEDERAL REGULATIONS

2-52.   Title 33, Code of Federal Regulations, Parts 154, 155, and 156, were written by the U.S. Coast Guard based on the Federal Water Pollution Control Act. Their purpose is to prevent, as much as possible, any accidental oil spills. Part 156 of the regulation concerns the actual oil transfer operations. The regulation states: "No person may transfer oil to or from a vessel unless—

- The vessel's moorings are strong enough to hold in all expected weather conditions.
- Hoses or arms are long enough to allow the vessel to move at its mooring without strain on the hose or arm.
- Hoses are supported, so that couplings have no strain on them.
- All parts of the transfer system are lined up before beginning the transfer.
- All parts are blanked or shut off.
- The transfer system is connected to a fixed piping system on the receiving end.
- Overboard discharge or sea suction valves connected to the transfer system are sealed shut during oil transfer.
- Transfer hoses are in good shape--no cuts, slashes, or soft spots.
- Flange couplings are properly bolted.
- Discharge containment equipment (such as drip pans) is in place.
- Scuppers and drains are plugged.
- Communications are available between the vessel and facility.
- An emergency shutdown system is available.
- Enough people to do the job are on duty.
- The person in charge on the vessel is able to speak to the person in charge at the facility (translators must be available if there is a language difference).
- The person in charge on the vessel and the person in charge at the facility hold a meeting (to discuss the transfer operation) before starting the transfer.
- Both persons in charge agree to begin the transfer before it is started.
- Both persons in charge are present during the transfer.
- Required lighting is available at night.

2-53.   Part 156 also describes what is meant by proper connections in oil transfer systems. The regulation states that the materials, in joints, must make a tight seal. If a coupling is a standard ANSI coupling, at least four bolts (one in every other hole) must be used. If it is not a standard ANSI coupling, a bolt must be used in each hole. Bolts must be of the same size in each coupling and be evenly tightened. Bolts must not be strained or deteriorated. Unless authorized by the Coast Guard, no quick-connect coupling may be used.

ACTION TO TAKE IN EVENT OF AN OIL SPILL

2-54.   U.S. Code, 40 CFR, 110.10 Notification, states that when an oil spill occurs, the person in charge of the vessel, an onshore facility, or an offshore facility must immediately notify the appropriate federal agency. The maximum fine for failing to report an oil spill is 1 year in prison and a fine of up to $10,000. When in a military port area, notify the harbor master immediately. The harbor master's office will in turn notify the U.S. Coast Guard National Response Center. If finding an oil spill in an area outside of military control (within CONUS), notify the National Response Center on their TOLL FREE number 1-800-424-8802. Provide them with the following information:

- Identify yourself.
- Give the area of pollution.
- If possible, state how much and what type of pollution is involved.

Note: Small oil spills, especially those with heavy oils, tend to be cumulative. This means that every time oil is spilled, it tends to join with oil already in the water and increases the problem.

DISCHARGE CONTAINMENT AND CLEANUP

2-55.   When oil is accidentally discharged into water, there are two things that should be done after reporting the discharge. The oil should be contained or fenced in to prevent the slick from spreading. Containment is usually more desirable in the case of heavy oil spills. Because of possible dangers of fire from light oils, such as gasoline and kerosene, it is sometimes better not to contain these spills. The proper people who will consider all the problems involved must make the decision. As much oil as possible, must be removed from the water. Since oil floats on water, much of it can be contained up to a point. However, weather conditions, tides, and currents can make it more difficult to contain the oil on the surface and can cause the oil and water to mix or emulsify. This also makes it harder to clean up. The most common type of containment device is a boom. Booms are basically floating fences used to surround a patch of spilled oil. There are many types of booms, made of many different materials like wood, cork, or plastic. The simplest boom is a chain of logs or lumber. Some booms have skirts that hang down in the water and some are blown up with air like long balloons. Some are towed out to the area of the spill and others are permanently mounted in areas where oil is transferred. Oil can also be contained with a special system that allows a stream of air bubbles to escape from submerged permanent piping. Removal of oil from water is done by different methods. Special suction equipment can be used to suck up the oil like a big vacuum cleaner. Skimmers (small boats that skim across the surface of the water to pick up the top layer of oil and water) are used in some places. Dockside or barge-mounted skimmers are also used to skim the surface oil.

This page intentionally left blank.

Chapter 3

Vessel Terms and Definitions

INTRODUCTION

3-1.   The watercraft operator must know and use the correct terms for following commands and instructions. He/She must also know the general layout of his/her vessel. Some terms, which are necessary to understand structural and operational nomenclature, are not explained in this chapter. These terms are explained in their appropriate chapter or in the glossary. The terms used in this chapter are the same on all types and sizes of vessels.

NAUTICAL TERMINOLOGY

3-2.   The floors of a ship are called decks, the walls are called bulkheads, and the stairs are called ladders. There are no halls or corridors in a ship, only passageways. There are no ceilings in a room, only the overhead in the compartment. Openings in the outside of the ship are ports, not windows. Entrances from one compartment to another are called doors. Openings from one deck to another are called hatches. The handles on the watertight hatch or door are called dogs. When you close a door or watertight hatch, you secure it. If you close down the dogs on the door or hatch, you dog it down. You never scrub the floor or wash the walls; rather you swab the deck and scrub the bulkheads. When you get up to go to work, turn to. You never go downstairs, you lay below, and if you are going up from one deck to another, you lay topside. If you are going up the mast or into the rigging, you are going aloft.

STRUCTURAL PARTS OF THE HULL

3-3.   The hull (Figure 3-1) is the main body of the ship below the main outside deck. The hull consists of an outside covering (or skin) and an inside framework to which the skin is secured. The skin and framework are usually made of steel and secured by welding. However, there may still be some areas where rivets are used. The steel skin may also be called shell plating. The main centerline structural part of the hull is the keel, which runs from the stem at the bow to the sternpost at the stern. The keel is the backbone of the ship. To the keel are fastened the frames, which run athwart ship. These are the ribs of the ship and give shape and strength to the hull. Deck beams and bulkheads support the decks and give added strength to resist the pressure of the water on the sides of the hull.

SKIN

3-4.   The skin, or shell plating, provides water-tightness. The plates, the principal strength members of a ship, have various thicknesses. The heaviest plates are put on amidships. The others are put on so that they taper toward both ends of the ship (from the keel toward the bilge and from the bilge toward the upper row of plates). Using plates of various thicknesses reduces the weight of the metal used and gives the vessel additional strength at its broadest part. The plates, put on in rows from bow to stern, are called strakes. They are lettered consecutively, beginning at the keel and going upward.

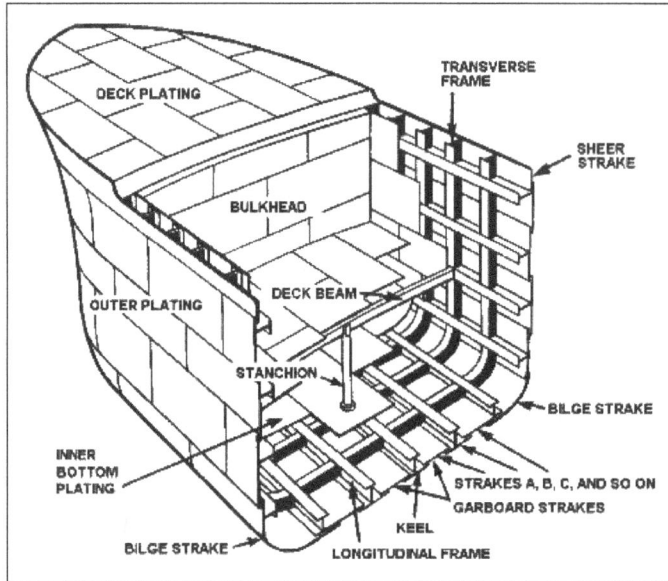

Figure 3-1. Construction of a Hull

STRAKE NAMES

3-5.   The bottom row of strakes on either side of the keel, are called garboard strakes. The strakes at the turn of the hull, running in the bilge, are bilge strakes. The strakes running between the garboard and bilge strakes are called bottom strakes and the topmost strakes of the hull are sheer strakes. The upper edge of the sheer strake is the gunwale.

BULKHEADS

3-6.   The interior of the ship is divided by the bulkheads and decks into watertight compartments (Figure 3-2). A vessel could be made virtually unsinkable if it were divided into enough small compartments. However, too many compartments would interfere with the arrangement of mechanical equipment and the operation of the ship. Engine rooms must be large enough to accommodate bulky machinery. Cargo spaces must be large enough to hold large equipment and containers.

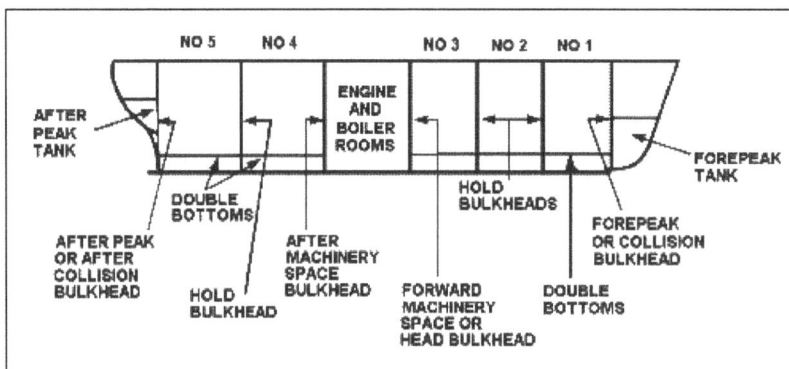

Figure 3-2. Bulkheads and Decks

ENGINE ROOM

3-7. The engine room is a separate compartment containing the propulsion machinery of the vessel. Depending on the size and type of propulsion machinery, other vessel machinery may be located there (such as generators, pumping systems, evaporators, and condensers for making fresh water). The propulsion unit for Army vessels is a diesel engine. The "shaft" or rod that transmits power from the engine to the propeller leads from the aft end of the engine to the propeller.

EXTERNAL PARTS OF THE HULL

3-8. Figure 3-3 shows the external parts of the hull. The waterline is the water-level line on the hull when afloat. The vertical distance from the waterline to the edge of the lowest outside deck is called the freeboard. The vertical distance from the waterline to the bottom of the keel is called the draft. The waterline, draft, and freeboard will change with the weight of the cargo and provisions carried by the ship. The draft of the ship is measured in feet and inches. Numbered scales are painted on the side of the ship at the bow and stern. The relationship between the drafts at the bow and stern is the trim. When a ship is properly balanced fore and aft, she is in trim. When a ship is drawing more water forward than aft, she is down by the head. If the stern is too far down in the water, she is down by the stern. If the vessel is out of balance laterally or athwart ship (leaning to one side) she has a list. She may be listing to starboard or listing to port. Both trim and list can be adjusted by shifting the weight of the cargo or transferring the ship's fuel and water from one tank to another in various parts of the hull. The part of the bow structure above the waterline is the prow. The general area in the forward part of the ship is the forecastle. Along the edges of the weather deck from bow to stern are removable stanchions and light wire ropes, called life lines. Extensions of the shell plating above the deck are called bulwarks. The small drains on the deck are scuppers. The uppermost deck running from the bow to the stern is called the weather deck. The main deck area over the stern is called the fantail or poop deck. The flat part of the bottom of the ship is called the bilge. The curved section where the bottom meets the side is called the turn of the bilge. Below the waterline are the propellers or screws which drive the ship through the water. The propellers are attached to and are turned by the propeller shafts. A ship with only one propeller is called a single-screw ship. Ships with two propellers are called twin-screw ships. On some ships (especially landing craft) there may be metal frames built around the propellers (called propeller guards) to protect them from damage. The rudder is used to steer the ship.

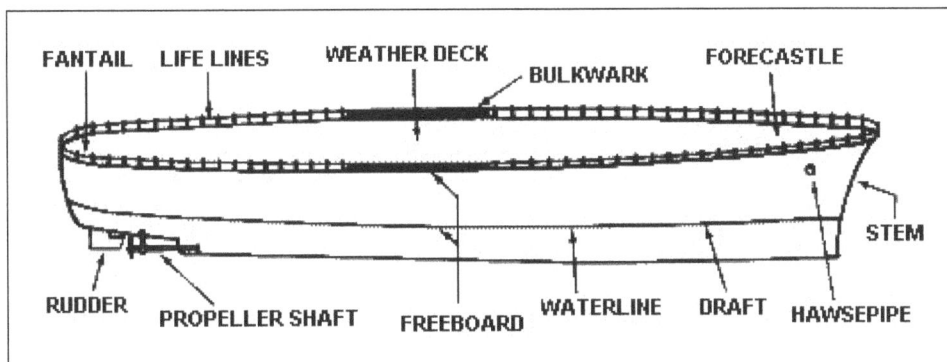

Figure 3-3. External Parts of the Hull

NAMES OF DECKS

3-9. The decks aboard ship (Figure 3-4) are the same as the floors in a house. The main deck is the first continuous watertight deck that runs from the bow to the stern. In many instances, the weather deck and the main deck may be one and the same. Any partial deck above the main deck is named according to its location on the ship. At the bow, it is called a forecastle deck, amidships it is an upper deck, and at the stern it is called the poop deck. The term weather deck includes all parts of the forecastle, main, upper, and poop decks exposed to the weather. Any structure built above the weather deck is called superstructure.

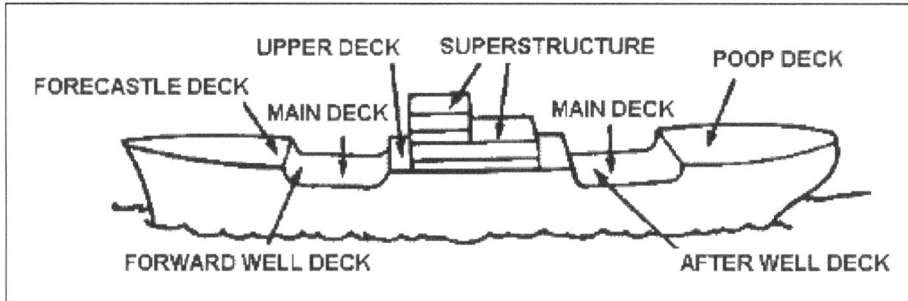

Figure 3-4. Weather Decks

SHIPBOARD DIRECTIONS AND LOCATIONS

STOWAGE AREAS

3-10. You must be able to identify and locate stowage areas when involved in operations aboard ship. Refer to Figure 3-5 to locate the following:

- Bow. The front end of the ship is the bow. When you move toward the bow, you are going forward, when the vessel is moving forward, it is going ahead. When facing toward the bow, the front-right side is the starboard bow and the front-left side is the port bow.
- Amidships (Center). The central or middle area of a ship is amidships. The right center side is the starboard beam and the left center side is the port beam.
- Stern (Back). The rear of a vessel is the stern. When you move in that direction you are going aft, when the ship moves in that direction it is going astern. When looking forward, the right-rear section is called the starboard quarter and the left-rear section is called the port quarter.

OTHER TERMS OF LOCATION AND DIRECTION

3-11. The entire right side of a vessel from bow to stern is the starboard side and the left side is the port side. A line or anything else running parallel to the longitudinal axis or centerline of the vessel is said to be fore and aft and its counterpart, running from side to side, is athwart ships. From the centerline of the ship toward either port or starboard side is outboard and from either side toward the centerline is inboard. However, there is a variation in the use of outboard and inboard when a ship is on berth (moored to a pier). The side against the pier is referred to as being inboard; the side away from the pier as outboard.

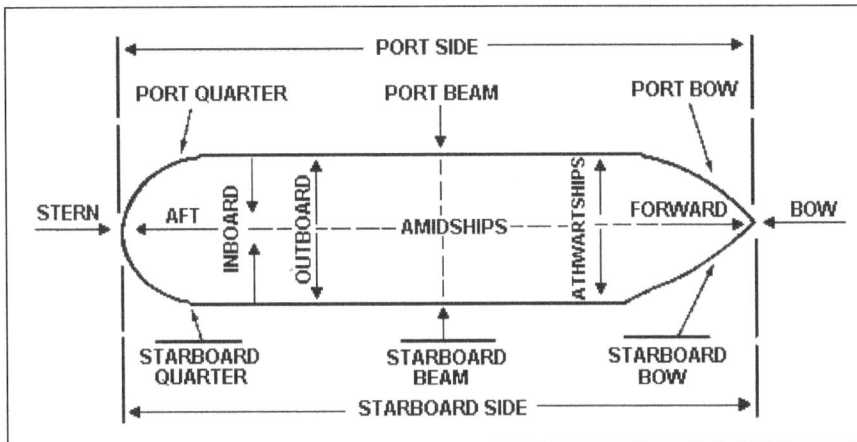

Figure 3-5. Locations and Directions Aboard Ship

SHIPBOARD MEASUREMENTS

3-12.   A ship's size and capacity can be described in two ways—linear dimensions or tonnages. Each is completely different yet interrelated. A ship's measurement is expressed in feet and inches—linear dimensions. A ship is a three dimensional structure having length, width, and depth (Figure 3-6).

LENGTH

3-13.   A ship's length is measured in different ways for ship's officers, for architects and designers, and for registry. Terms used for technical or registry purposes include registered length, tonnage length, floodable length, and length by the American Bureau of Shipping (ABS) rules. We mention these terms for familiarization only. The more commonly used length measurements—length overall, length between perpendiculars, and length on load waterline—are discussed as follows.

Length Overall (LOA)

3-14.   A ship's LOA is measured in feet and inches from the extreme forward end of the bow to the extreme aft end of the stern. The top portion of Figure 3-6 shows how the LOA is measured. Watercraft operators must be familiar with this and similar dimensions to safely maneuver the ship. The dimension is commonly found in lists of ship's data for each vessel.

Length Between Perpendiculars (LBP)

3-15.   A ship's length is sometimes given as LBP. It is measured in feet and inches from the forward surface of the stem, or main bow perpendicular member, to the after surface of the sternpost, or main stern perpendicular member. On some types of vessels this is, for all practical purposes, a waterline measurement.

Length on Load Waterline (LWL)

3-16.   A ship's LWL is an important dimension because length at the waterline is a key factor in the complex problem of speed, resistance, and friction. On vessels with a counter stern, the LWL and LBP can be the same or about the same. On a ship with a cruiser stern, the LWL is greater than the LBP, as shown in the top portion of Figure 3-6.

WIDTH

3-17.   A ship's width or, more properly, a ship's breadth is expressed in a number of ways and, like length, for a number of reasons. A ship's extreme breadth, commonly called beam, is measured in feet and inches from the most outboard point on one side to the most outboard point on the other at the widest point on the ship, as shown in the bottom portion of Figure 3-6. This dimension must include any projections on either side of the vessel. Like length overall, this measurement is important to a ship's officer in handling the vessel.

DEPTH

3-18.   The depth of a vessel involves several very important vertical dimensions. They involve terms like freeboard, draft, draft marks, and load lines. The vessel's depth is measured vertically from the lowest point of the hull, ordinarily from the bottom of the keel, to the side of any deck that you may choose as a reference point. Therefore, it has to be stated in specific terms such as depth to upper deck amidships. It is impractical to measure depth in any other way, since it varies considerably from one point to another on many ships. For example, the depth is greater at the stern than amidships. The term "depth" is where the measurement is taken from the bottom--from the keel upward. Ordinarily, if such a measurement were being made in a room of a building, taken from the floor to the ceiling, it would be called height.

Note: You must know a ship's draft or maximum allowable draft when selecting a berth for loading or discharging operations.

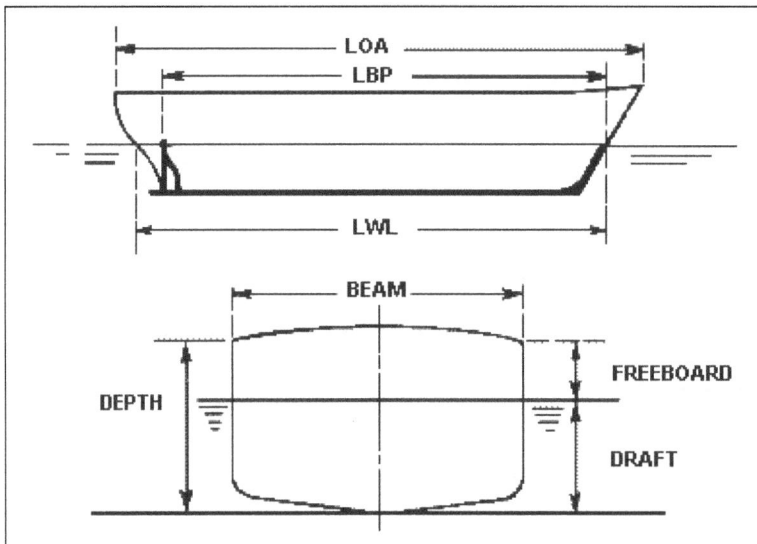

Figure 3-6. A Ship's Dimensions

PROCEDURE FOR READING DRAFT MARKS

3-19.   Draft marks are numbers marked on each side of the bow and stern of the vessel (Figure 3-7). Draft marks show the distance from the bottom of the keel to the waterline.

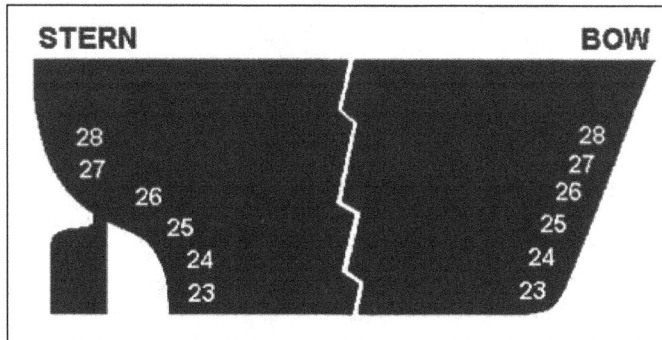

Figure 3-7. Draft Marks on Bow and Stern of Vessel

3-20.   The draft numbers shown in Figure 3-8 are 6 inches high and 6 inches apart. The bottom of each number shows the foot draft mark.

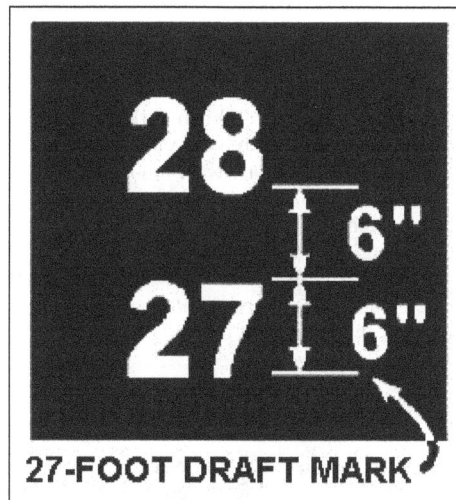

Figure 3-8. Draft Numbers Showing Foot Draft Mark

3-21.   Figure 3-9 shows four different draft readings.

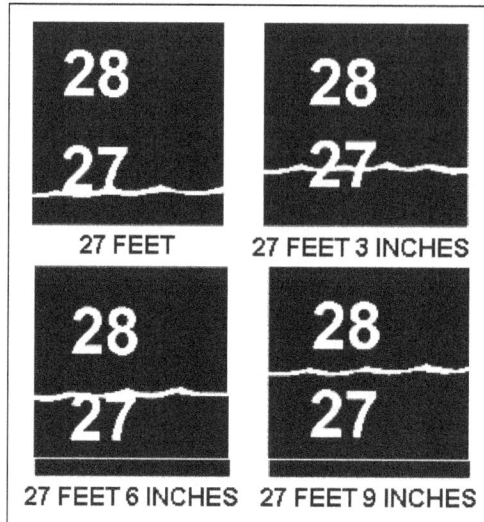

Figure 3-9. Various Draft Readings

WEIGHT TONNAGE TERMS

3-22.   The word "ton" comes from the English "tun" meaning cask or barrel. To the English, it meant a wine barrel with a capacity of about 252 gallons. When Parliament imposed duties on the wine entering England in these barrels, the duty imposed on each tun eventually led to the use of tunnage in describing a ship's capacity to carry such barrels. The original use of tun meant a barrel of a particular size, the space that such a barrel would occupy, and a ship's capacity to carry a given number of such barrels. Tun was originally a figure for space--not weight. By law, Parliament fixed the tun at 252 gallons. Since this fixed tun weighed an average of 2,240 pounds, it brought into existence the weight term "long ton." A long ton is used throughout the shipping business. It is not to be confused with the familiar ton of 2,000 pounds, the short ton, used so widely in the United States in relation to so many things other than ships and shipping. The metric ton is 1,000 kilograms, the equivalent of 2,204.6 pounds. Tonnages throughout this manual refer to the long ton of 2,240 pounds.

CATEGORIES OF SHIP'S DECK GEAR

3-23.   Watercraft operators must be familiar with ship's gear. The term "ship's gear" is used to describe the gear and equipment aboard ship that is used for cargo transfer activities and deck operations. Ship's gear can be divided into four categories:

- Standing rigging.
- Running rigging.
- Deck fittings.
- Deck machinery.

STANDING RIGGING

3-24.   Standing rigging gear (Figure 3-10) includes the rigging that supports masts or king posts. This gear includes the following:

- **Shrouds**. These are heavy wire ropes that provide athwart ship support for the mast or king posts. Two or more shrouds are used on either side of a mast or king post. They are secured to the outboard side of the deck or to the bulwark to provide maximum support.

- **Turnbuckles**. These are internally threaded collars turning on two screws threaded in opposite directions. They are used to secure and to take up the slack in the shrouds and stays.
- **Stays and Backstays**. These are heavy wires similar to shrouds. The difference is that they will lead in a forward or aft direction. They are found at the mast where the jumbo boom (heavy lift boom) is located. When they support the mast from a forward direction, they are called stays. When they support the mast from an aft (back) direction, they are called backstays.

RUNNING RIGGING

3-25. This gear includes the moving or movable parts that are used to hoist or operate gear (such as cargo runners, topping lifts, and guy tackles).

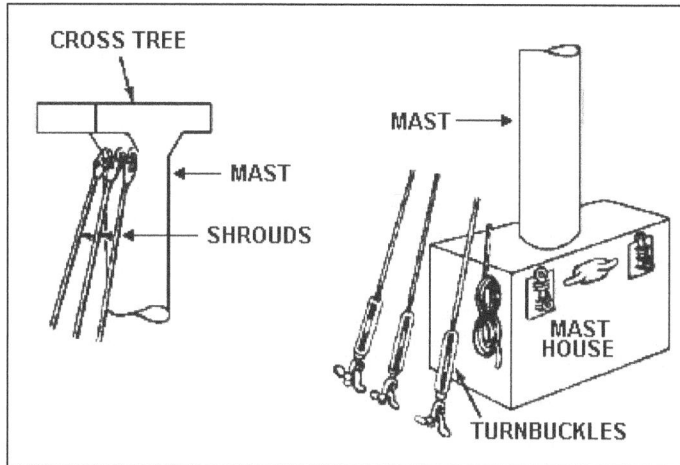

Figure 3-10. Standing Rigging Gear

DECK FITTINGS

3-26. These are the devices that are used to secure standing rigging, running rigging, and mooring lines. These devices (Figure 3-11) are described as follows:

- **Bitts**. These are heavy metal bed plates with two iron or steel posts. They are used on ships for securing mooring or towing lines. Usually there is a set forward and after each chock.
- **Chocks**. These are heavy fittings secured to the deck. Lines are passed through them to bollards on the pier. The types of chocks used are closed, open, roller, and double roller.
- **Cleats**. These are metal fittings having two projecting horns. They are used for securing lines.
- **Pad Eyes**. These are fixtures welded to a deck or bulkhead. They have an eye to which lines or tackle are fastened and are used for securing or handling cargo. A bulwark is the wall around any deck exposed to the elements. This includes the weather deck, the poop deck, the foredeck, and any deck on the superstructure. On top of the bulwark is a flat rail (or plate) called the rail. Pad eyes and cleats are often welded to the rail.

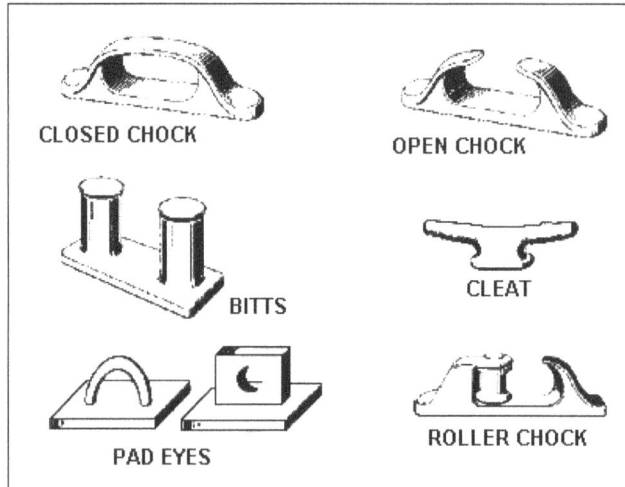

Figure 3-11. Deck Fittings

DECK MACHINERY

3-27. This includes the standard machinery that is found on the decks of Army watercraft. The size and shape of the deck machinery may vary depending upon type of vessel, but the operating principles remain the same.

Cargo Winches

3-28. These are power-driven machines used to lift, lower, or move cargo. Winches are classified according to their source of power. Electric winches are standard equipment on most vessels. An electric winch (Figure 3-12) has a steel base on which the winch drum, motor, gears, shafts, and brakes are mounted. The drum, which has cable wound on it, is usually smooth with flanged ends. It revolves on a horizontal axis and is driven through single or double reduction gears by an electric motor (usually direct current). A solenoid brake and a mechanical brake are fitted to the motor shaft. The winch is located on deck or on a deckhouse. The winch controls consist of a master controller or switchbox located on a pedestal at the end of the hatch square and a group of relays, contactors switches, and resistors located near the winch motor.

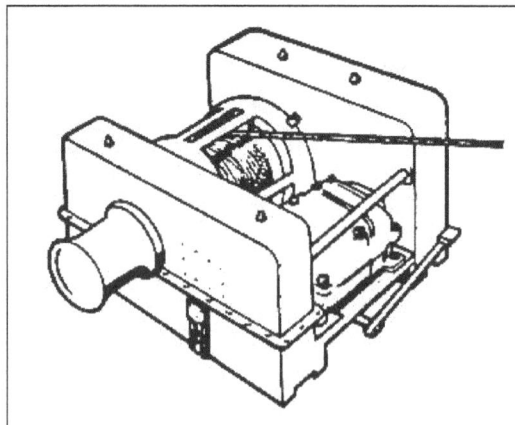

Figure 3-12. Electric Winch

Windlass

3-29.   The windlass (Figure 3-13) is a special type of winch used to raise and lower the anchors and to handle the forward mooring lines. It consists of a wildcat (a steel casting in the form of a deeply grooved drum with projecting ribs [whelps]) used to grip the anchor chain, controls for connecting or disconnecting the wildcat from the engine, and a friction brake which can be set to stop the wildcat when disconnected. There are horizontal drums at each end of the windlass for warping.

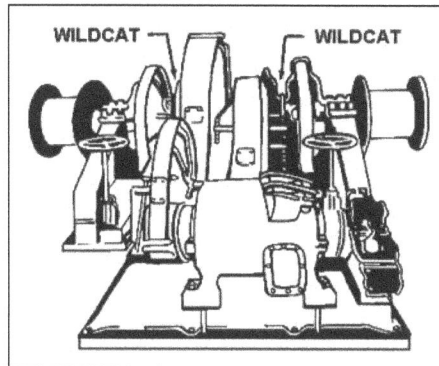

Figure 3-13. Windlass

Capstan

3-30.   The capstan (Figure 3-14) is a vertically mounted winch head used aboard ship when mechanical power is required for raising anchor, lifting heavy weights, or for any similar work. It is a cast steel drum mounted on a vertical spindle with the largest diameters at top and bottom and the smallest in the middle to allow the rope around it to surge up or down as the number of turns are increased. The drum is fixed to the spindle by keys.

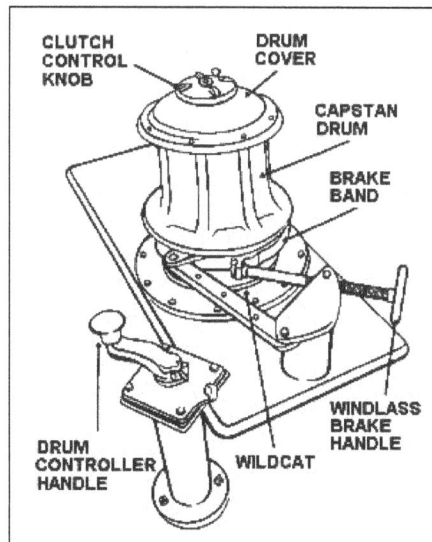

Figure 3-14. Capstan

This page intentionally left blank.

Chapter 4

Boat Handling

INTRODUCTION

4-1.　Boat handling requires an understanding of the many variable and complex problems of seamanship. The basic principles involved in handling small boats are essentially the same as those used in handling larger craft.

FORCES AFFECTING BOAT HANDLING

4-2.　Before attempting to handle a boat, it is important to understand the forces that affect a boat under various conditions. A watercraft operator who thoroughly understands these forces can use them to maneuver his/her boat. Therefore, he/she will not have to fall back on the often painful, trial-and-error method of learning boat handling. The following vessel characteristics influence the control of single-screw boats having right-hand propellers.

DESIGN OF THE VESSEL

4-3.　The design of a ship includes the size and shape of the hull, draft, trim, weight, and amount of superstructure. Ships with shallow draft, low superstructure, and slim design normally handle more easily than ships with high superstructure, deep draft, and wide beam because they are less affected by wind and current and respond more rapidly to the rudder.

POWER

4-4.　Each phase of motive force as it reacts on the vessel has its own peculiarities. No set of rules can be devised to cover all types. Every vessel has its own power characteristics, which the operator must learn to determine their effect upon handling of the vessel.

PROPELLER ACTION

4-5.　A propeller draws its supply of water from every direction forward and around the blades, forcing it in a powerful stream toward the stern. This moving current which provides the power for propulsion is called "screw current." The water flowing into the propeller is called "suction screw current," that being ejected is called "discharge current." Figure 4-1 shows this water pressure effect of the suction current vaporizing off the tips of the blades and spiraling back in a helical pattern. The factors that affect propeller action are described below.

PITCH

4-6.　The pitch of a propeller (Figure 4-2) is the distance the propeller would advance in one revolution if the water was a solid medium.

SLIP

4-7.　The difference between the speed of the ship and the speed of the propeller is known as the "slip". Slip is caused by the yield of the water against the propeller thrust. In other words, it is the percentage of distance lost because water is a yielding substance.

CAVITATIONS

4-8.   When the blade-tip speed is excessive for the size and shape of the propeller, the vessel rides high in the water. There is also an unequal pressure on the lower and upper blade surfaces. This condition produces cavities or bubbles around the propeller known as "cavitation." The result is an increase in revolutions per minute without an equivalent increase in thrust. This results in loss of efficiency. When cavitation is fully developed, it limits a vessel's speed regardless of the available engine power.

RUDDER ACTION

4-9.   The rudder acts the same on a large vessel as on a small craft. The rudder is placed directly behind the propeller to use the powerful discharge current to turn the boat. Moving the rudder to the right deflects the discharge current to the right, which forces the stern to the left. This action is reversed when the left rudder is applied. At very slow propeller speed and with very little way on, there may not be enough control over a boat to maneuver it, especially if other forces are acting upon it at the same time. When this condition prevails, the propeller may be speeded up enough to give it a more powerful thrust against the rudder. Using sudden thrusts of power to kick (move) the stern in this manner is one of the fundamental principles of vessel handling. A vessel can often be turned in twice its length by kicking the stern.

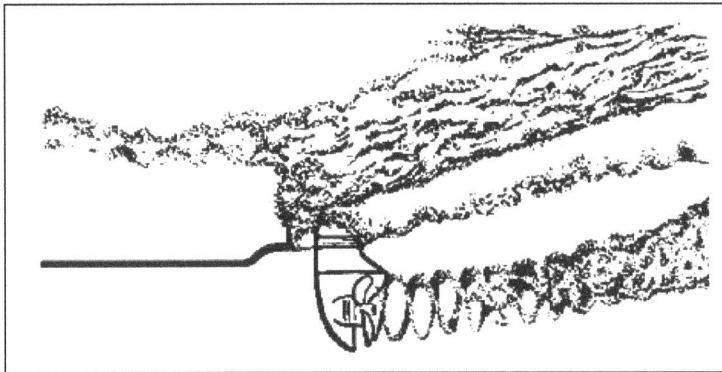

Figure 4-1. Propeller Action and Resulting Screw Action

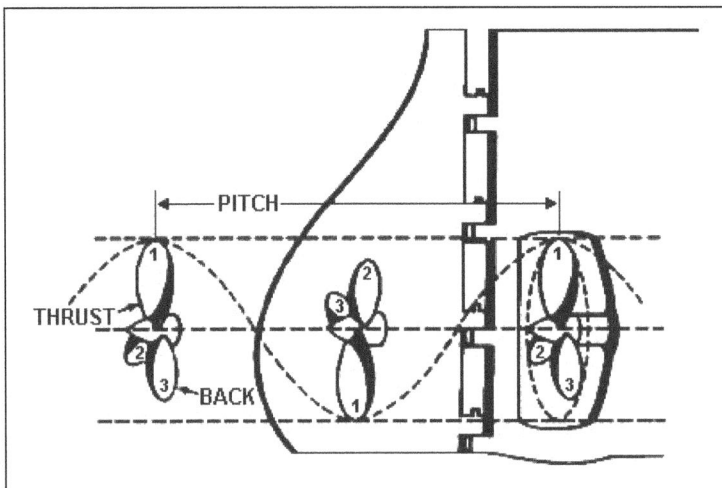

Figure 4-2. Propeller Pitch

OTHER FACTORS AFFECTING CONTROL

4-10.   Wind, tidal, ocean currents (waves or sea), and depth of water must be considered when handling a vessel. Shallow water particularly affects deep draft vessels because of the cushion effect similar to that encountered when navigating in narrow channels.

STANDARD STEERING COMMANDS

4-11.   There are some standard steering commands used to give orders to the helmsman. These are described below.

RIGHT RUDDER AND LEFT RUDDER

4-12.   "Right rudder" or "left rudder" are orders given for the wheel to be turned to the right or left. When the wheel is turned to the right or left, the rudder and rudder angle indicator must turn to the same side; that is, they must not be rigged reversely.

COMMONLY USED COMMANDS

4-13.   When a command is given to the helmsman, the first part of the order indicates the direction (right or left) for the helmsman to turn the wheel. The second part of the command states the amount of angle. The following are some commonly used steering commands.

- **"Right (or left) full rudder."** Full rudder designates a 30-degree rudder. When the rudder is turned past 30 degrees (usually designated hard right or left), care must be exercised to avoid jamming it against the stops.
- **"Right (or left) 5, 10, 15 degrees, and so on."** This indicates the angle, in degrees, that the rudder is to be offset.
- **"Right (or left) easy."** Usually indicates 2 or 3 degrees of rudder angle in the direction indicated. Some Masters may prefer 5 degrees of rudder angle for this command. This should be understood in the vessels Standard Operating Procedures (SOP).
- **"Give her more rudder."** To increase the rudder angle already on when it is desired to turn the ship more rapidly in the direction in which she is already turning.
- **"Ease the rudder."** To decrease the rudder angle that is already on. The order may also be: "Ease to (state number) degrees."
- **"Rudder amidships."** To place the rudder on the centerline.
- **"Meet her."** To check, but not stop, the swing by putting the rudder in the opposite direction. Usually this order is used when it is desired to keep the ship from swinging past her new course.
- **"Steady" or "steady as you go."** To steer the present course while the ship is swinging. The course should be noted at the time the order is given and the ship steadied on that course.
- **"Shift the rudder."** To change from right to left (or left to right) rudder. Usually given when a ship loses her headway and begins to gather sternway and it is desired to keep her turning in the same direction.
- **"Mind your rudder."** To steer more carefully or stand by for an order.
- **"Keep her so."** To steer the course just reported, following the quest for that course.

REPEATING COMMANDS

4-14.   To assure the watch officer that his/her orders have been correctly received, the helmsman must always repeat, word for word, any command received. As soon as the command has been executed, the helmsman must also report it to the watch officer. The watch officer confirms that the order is understood by replying, "Very well."

HANDLING CHARACTERISTICS OF SINGLE- AND TWIN-SCREW VESSELS

4-15.   Characteristics or factors, such as the power, propeller, rudder, and design of a ship affect handling in various ways. For illustrating the effects of these factors, it will be assumed that the sea is calm, there is neither wind nor current, and the ship has a right-handed propeller.

HANDLING SINGLE-SCREW VESSELS

4-16.   The single-screw vessel has only one propeller. The operation of this vessel is described below.

VESSEL AND PROPELLER GOING FORWARD

4-17.   With the ship and propeller going forward and the rudder amidships, the ship tends to move on a straight course. The sidewise pressure of the propeller is offset by the canting of the engine and shaft. When the rudder is put over (either to the right or left) the water through which the ship is moving strikes the rudder face, forcing the stern in the opposite direction. At the same time, discharge current strikes the rudder face and pushes the stern over farther. As a result of these forces, the bow moves in the direction in which the rudder has been thrown.

VESSEL WITH STERNWAY, PROPELLER BACKING

4-18.   When backing, the sidewise pressure is opposite to that exerted when the ship is moving forward. The discharge current from the propeller reacts against the hull. This current is rotary; therefore, when the propeller is backing, the current strikes the hull high on the starboard side and low on the port side. This current exerts a greater force on the starboard side and tends to throw the stern of the vessel to port (Figure 4-3).

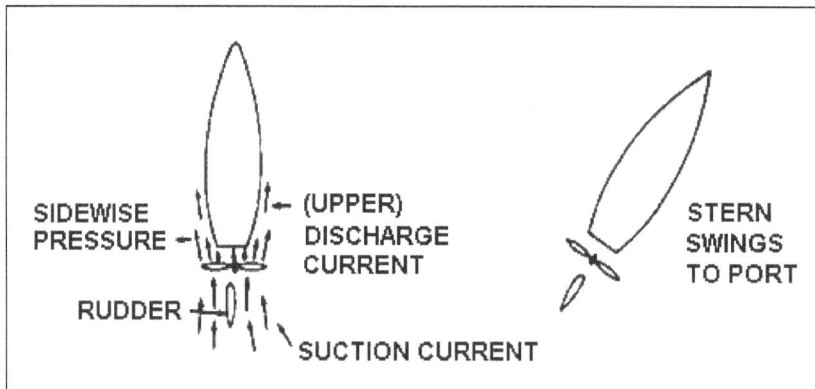

Figure 4-3. Vessel with Sternway, Propeller Backing, Rudder Amidships

4-19.   With rudder amidships, the vessel will back to port from the force of the sidewise pressure and the discharge current. When the rudder is put over to starboard (Figure 4-4), the action of the suction current against the face of the rudder will tend to throw the stern to starboard. Unless the ship is making sternway, this force will not be strong enough to overcome the effect of the sidewise pressure and the discharge current, and the stern will back to port.

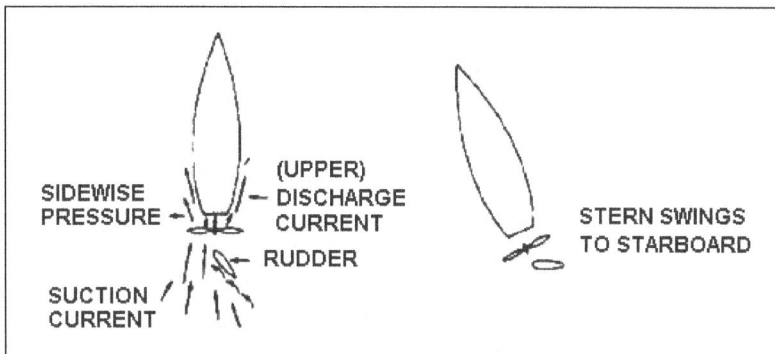

Figure 4-4. Vessel with Sternway, Propeller Backing, Rudder to Starboard

4-20.   When the rudder is put over to port (Figure 4-5), the force of the suction current on the face of the rudder intensifies the effect of the sidewise pressure of the propeller and the discharge current and will force the stern rapidly to port. Because of these forces, all right-handed, single-screw vessels tend to back to port.

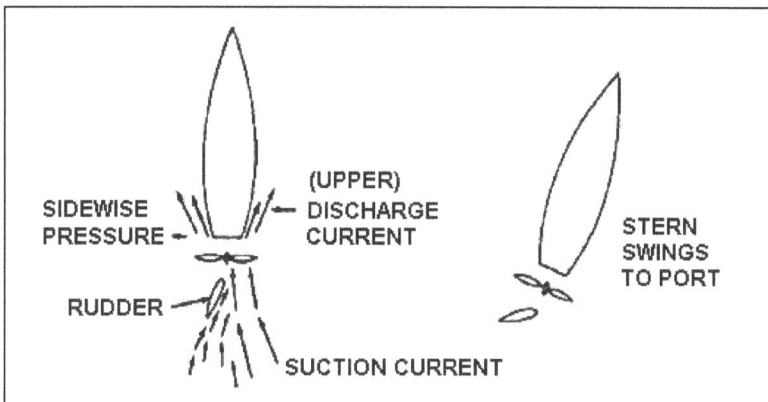

Figure 4-5. Vessel with Sternway, Propeller Backing, Rudder to Port Vessel with Headway Propeller Backing

4-21.   With the rudder amidships (Figure 4-6), the stern will go to port because the only active forces are the sidewise pressure of the propeller and the discharge current.

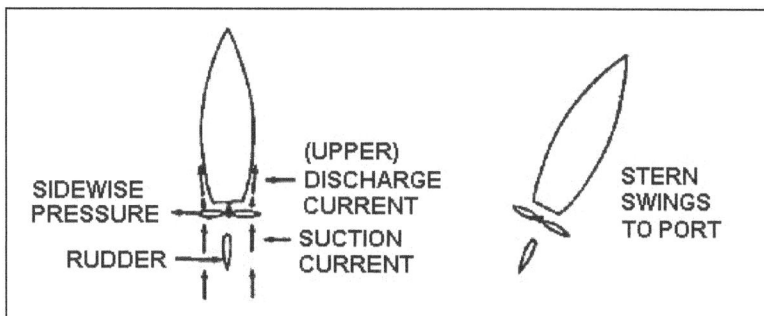

Figure 4-6. Vessel with Headway, Propeller Backing, Rudder Amidships

4-22.   With the rudder to starboard (Figure 4-7), the stern rapidly goes to port but, as headway is lost and the vessel begins to go astern, the effect of the suction current on the face of the rudder slows the swing. However, since a single-screw vessel tends to back to port when moving astern, the stern will tend to port unless the vessel gathers considerable speed astern.

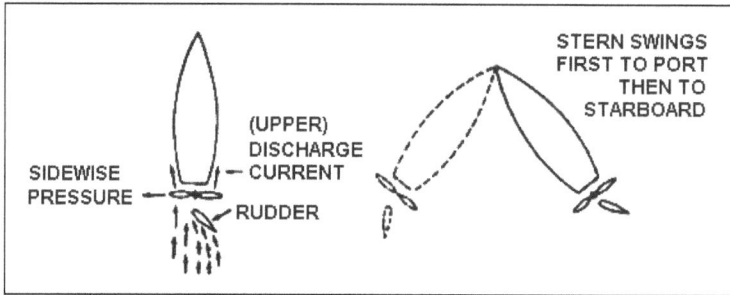

Figure 4-7. Vessel with Headway, Propeller Backing, Rudder to Starboard

4-23.   With the rudder left (Figure 4-8), the normal steering tendency of the rudder will throw the stern to starboard. This starboard motion will occur when the vessel has considerable headway, but as headway is lost, the effect of the sidewise pressure of the propeller and the discharge current, in conjunction with increasing forces of the suction current against the face of the rudder, swings the stern rapidly to port.

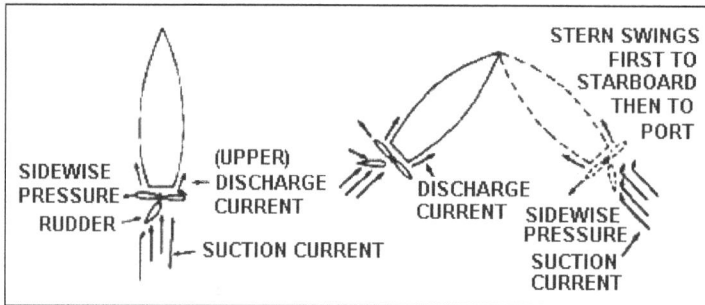

Figure 4-8. Vessel with Headway, Propeller Backing, Rudder Left Vessel with Sternway, Propeller Going Ahead

4-24.   In this situation, the sidewise pressure of the propeller and the discharge current are persistent factors and may offset each other. Therefore, if the rudder is amidships with no forces acting against it (Figure 4-9), the vessel will tend to follow a straight course.

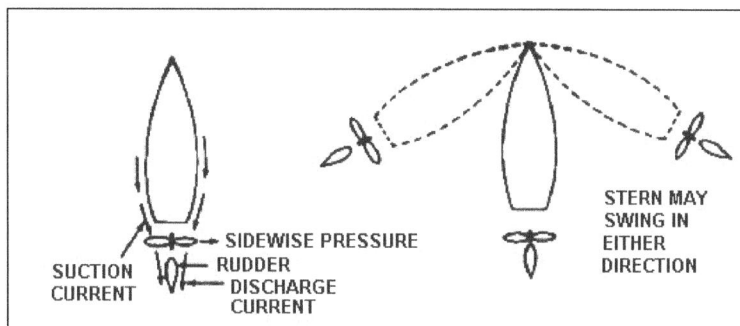

Figure 4-9. Vessel with Sternway, Propeller Going Ahead, Rudder Amidships

4-25.   With the rudder to the right (Figure 4-10), the action of the water on the back face of the rudder as it moves astern will tend to throw the stern to the starboard. The action of the discharge current against the forward face of the rudder tends to throw the stern to port. Direction is determined by the stronger force. However, as the vessel loses sternway, the direct steering effect of the rudder takes over and the stern swings to port.

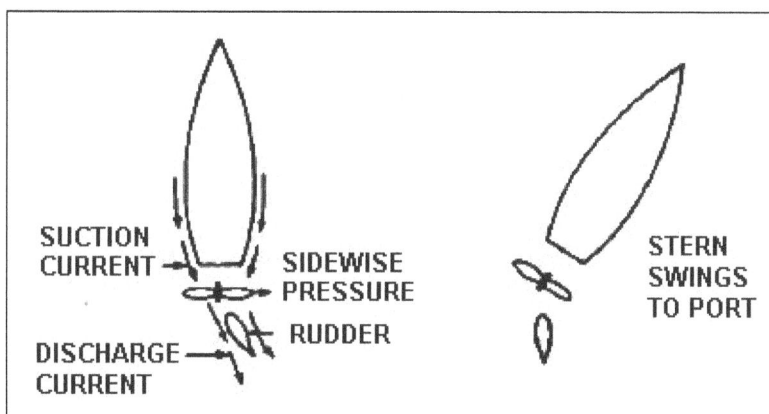

Figure 4-10. Vessel with Sternway, Propeller Going Ahead, Rudder to Right

4-26. With the rudder left (Figure 4-11), the action is the same as with the rudder right. In either case, the rudder action is determined by the strength of the forces, and as the rudder loses sternway, the direct steering effect of the rudder takes over and the stern swings rapidly to starboard.

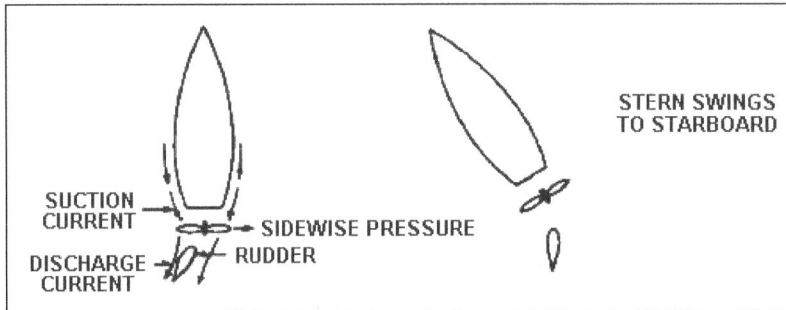

Figure 4-11. Vessel with Sternway, Propeller Going Ahead, Rudder to Left Handling Twin-Screw Vessels

HANDLING TWIN-SCREW VESSELS

4-27. The twin-screw vessel has two propellers—one on each side of the centerline. These propellers are maneuvered by separate throttle controls. Generally the propellers are outturning; that is, the starboard propeller is right-handed and the port, left-handed. This balances the sidewise pressure of the propellers and makes it possible to keep the ship on a straight course with no rudder. Discounting outside influences, the twin-screw vessel backs with equal facility to port or starboard. The various forces affecting the action of the single-screw ship are still present, but normally a twin-screw vessel is not affected by these forces as much as a single-screw vessel. This is because the forces from one screw balance the forces from the other screw. One powerful force is the momentum of the ship, ahead or astern, acting through the center of gravity. When a twin-screw ship is going ahead and one screw is backed, two opposing forces are set in motion; namely, the force of the backing screw acting in one direction and the weight of the ship acting in the opposite direction. This is in addition to the forces from the action of the pressure on the rudder if it is put over. Other than this force and the turning action accomplished by one engine ahead and the other astern, the vessel handling characteristics of a twin-screw vessel are similar to those explained in the preceding paragraphs for the single-screw vessel.

TURNING IN A LIMITED SPACE

4-28. Single-screw vessels can be turned easily in restricted waters. To start the swing, the engine speed is set at full ahead and the rudder is put full right; then the engine is reversed to full astern until way is lost. When way is lost, the rudder is shifted; after sternway has started, the rudder is again shifted and the engine put full ahead. This procedure is repeated until the vessel is on the desired heading. This maneuver makes use of the tendency of right-handed propellers back to port. In strong winds, it is wise to turn in such a way that the tendency to back into the wind can be used to increase the turn. A twin-screw vessel with a single rudder can be turned by going ahead on one engine and astern on the other, using the rudder only when headway or sternway has been gained. When the vessel is fitted with twin rudders that are directly behind the propellers, the rudder is placed hard over in the direction of the turn before the maneuver is begun and one engine is backed at the speed necessary to prevent headway.

DOCKING AND UNDOCKING

4-29. In some ports, particularly on the East and Gulf Coasts, individuals frequently referred to as docking pilots or docking masters direct the docking and undocking of vessels. In most cases, these individuals are employees of tug boat companies.

MOORING LINES

4-30.   The lines used to secure the ship to a wharf or to another ship are called "mooring lines." They must be as light as possible for easy handling and, at the same time, be strong enough to take considerable strain when coming alongside and holding a ship in place. Five-inch to six-inch nylon line is the customary material for mooring lines on freight supply and other large vessels. Figure 4-12 shows the locations and names of the lines. Lines should be neatly coiled or arranged to prevent fouling, to eliminate hazards, and to keep the working area clear.

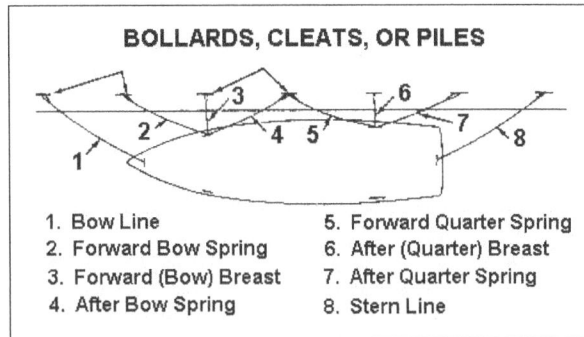

Figure 4-12. Names and Locations of Mooring Lines

4-31.   The bow line and the stern line lead well up the wharf to reduce the fore and aft motion of the ship. Breast lines are run at right angles to the keel to prevent a ship from moving away from the wharf. Spring lines leading forward or aft prevent a vessel from moving aft or forward respectively. Two spring lines placed close together and leading in opposite directions act as a breast line from wharf to ship.

USING THE LINES

4-32.   Lines assist in coming alongside or clearing a wharf. Before a ship comes alongside, the required lines with eye splices in the ends should be led outboard through the chocks, up and over the lifelines and/or rails. Heaving lines (light lines with weighted ends) are used on larger vessels to carry heavier lines to the wharf. With small boats, there is rarely any need to use a heaving line. Generally, a seaman can either step ashore with the mooring line or throw it the short distance required. Heaving lines should be made fast near the splice--not at the end of the bight where they may become jammed when the eye is placed over the bollard. Heaving lines should be passed ashore as soon as possible.

DIPPING THE EYE

4-33.   If two bights or eye splices are to be placed over the same bollard, the second one must be led up through the eye of the first and then placed over the bollard. This makes it possible for either line to be cast off independently of the other and is called dipping the eye.

STOPPING OFF A MOORING LINE

4-34.   When a mooring line is taut, it is stopped off with a stopper line (Figure 4-13). The stopper line is secured to the bitts and applied to the mooring line with a half hitch and three or four turns taken in a direction opposite to the one in which the hitch is taken. When the stopper takes the strain, the turns are thrown off the mooring line and it is made fast to the bitts.

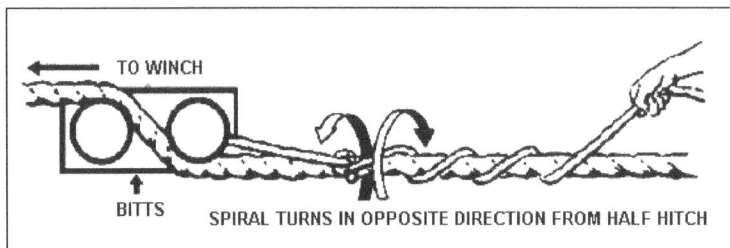

Figure 4-13. Stopping Off a Mooring Line

DOCKING (SINGLE-SCREW VESSEL)

4-35.   In securing alongside a wharf, attention must be paid to the tide. When securing at high water, enough slack must be left in the lines to ensure that at low tide they will not part, carry away bollards or in extreme cases, list the ship to a dangerous degree or capsize a small vessel.

MAKING LANDINGS

4-36.   Wharves and piers may be built on piles that allow a fairly free flow of water under them and in the slips between them. Their underwater construction may be solid, in which case there will be no current inside the slips, but eddies may whirl around them. Warehouses or other buildings may be built on piles, which vary the effect of the wind on the upper works of a vessel when making a landing. Making a landing is more dangerous when the wind and current are at right angles to the wharf than when blowing or running along its face. In coming alongside, as in all ship handling, the wind and current should be observed and if possible, used as an advantage. Making a landing usually involves backing down. For this reason, procedures for landing port-side-to differ from those for starboard-side-to landing. Let us first consider a port-side-to landing.

Note: A coxswain should remember that boats do not always respond exactly as theory predicts and that there is no substitute for actual experience.

PORT-SIDE-TO LANDING

4-37.   Making a port-side-to landing is easier than a starboard-side-to landing because of the factors already discussed. With no wind, tide, or current to contend with, the approach normally should be at an angle of about 20 degrees with the pier. The boat should be headed for a spot slightly forward of the position where you intend to stop. Several feet from that point (to allow for advance), put your rudder to starboard-to, bring your boat parallel to the pier, and simultaneously begin backing. Quickly throw the bow line over. Then, with the line around a cleat to hold the bow in, you can back down until the stern is forced in against the pier. If wind and current are setting the boat off the pier, make the approach at a greater angle and speed. The turn is made closer to the pier. In this situation, it is easier to get the stern alongside by using hard right rudder, kicking ahead, and using the bow line as a spring line. To allow the stern to swing in to the pier, the line must not be snubbed too short. If wind or current is setting the boat down on the pier, make the approach at about the same angle as when being set off the pier. Speed should be about the same or slightly less than when there is no wind or current. The turn must begin farther from the pier because the advance is greater. In this circumstance, the stern can be brought alongside by either of the methods described, or the centerline of the boat can be brought parallel to the pier and the boat will drift down alongside.

STARBOARD-SIDE-TO LANDING

4-38.   Making a starboard-side-to landing is a bit more difficult than landing-to port. The angle of approach should always approximate that of a port-side-to landing. Speed however, should be slower to avoid having to back down fast to kill headway, with the resultant swing of the stern to port. Use a spring line when working the stern in alongside the pier. Get the line over, use hard left rudder, and kick ahead. If you cannot use a spring line, time your turn so that when alongside the spot where you intend to swing, your bow is swinging out and your stern is swinging in. When it looks as though the stern will make contact, back down; as you lose way, shift to hard right rudder.

MAKING USE OF THE CURRENT

4-39.   If there is a fairly strong current from ahead, get the bow line to the pier, and the current will bring the boat alongside as shown in Figure 4-14 (View 1). If the current is from aft, the same result can be achieved by securing the boat with the stern fast as shown in Figure 4-14 (View 2). Care must be exercised during the approach because an oncoming following current decreases rudder efficiency and steering may be slightly erratic.

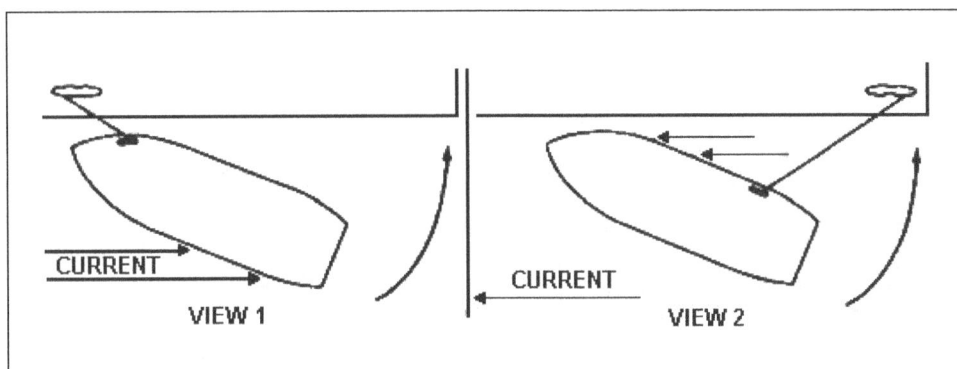

Figure 4-14. Making Use of the Current

TYING UP TO A PIER (HEAVY WEATHER PROCEDURES)

4-40.   If heavy winds are forecast (less than 50 knots), make sure storm lines are out fore and aft and additional breast lines are set. The greatest damage to the ship will result from the ship banging against the pier or other nested ships. Make sure all lines are properly set and that adequate fenders are rigged between the ships nested alongside.

GETTING UNDERWAY FROM A PIER

4-41.   As when coming alongside, procedures for getting underway depend upon which side of the pier the boat is located, as well as the state of current, wind, and so on.

STARBOARD-SIDE-TO

4-42.   The easiest way to get underway, when starboard-side-to a pier, is to cast off the stern first, hold the bow line, give the boat hard left rudder, and begin backing. When the stern is clear of the pier and any boat or other object astern, cast off the bow line and back out of the slip.

PORT-SIDE-TO

4-43.   The easiest way to clear a port-side-to landing is to use the bow line as a spring line. Cast off the stern line, give the boat left full rudder, and kick ahead until the stern is well clear. Then castoff the spring line and back out of the slip.

HANDLING GROUNDED HARBOR CRAFT

4-44.   Grounded vessels can cause physical damage to fragile reefs. They also pose serious pollution threats to the marine environment because fuel and waste oil tanks can be damaged as a result of grounding.

4-45.   If a vessel runs aground, an intelligent and careful estimate of the situation must be made before attempting to move the vessel. The following procedures should be observed:

- Notify unit headquarters (HQ) to send assistance; establish communication procedure; furnish a relief party; and provide for another vessel or vessels to take off fuel, water, stores, and/or cargo.
- Stop engines and inspect the double bottoms and bilges to find out the extent of leakage resulting from the initial impact of grounding.
- Take soundings around the entire vessel to determine the depth of the water and character of the bottom. Send out a boat to take soundings to determine if engines can be maneuvered.
- Inform the engineer about the condition of the bottom so that he/she may take precautions to protect the pumps, pipelines, and engine-cooling spaces from damage by sand and mud.
- Examine the hull to determine the extent of damage. If the shell has been pierced, IMMEDIATELY close watertight doors.
- Determine whether backing vessel off would be an advantage or would increase damage and whether or not pumps could control flooding compartments if the vessel were floated free.
- Study tide tables, sailing directions, and charts to determine the time of high tide and tidal current, depths, and type of bottom in the vicinity of the vessel's position. A leadsman should be stationed to take frequent soundings to note any change in tides. If it appears that some time will be required to ready the vessel for floating or to secure outside aid, weather reports should be obtained before planning action.

SALLYING

4-46.   One of the first observations the master should make is whether the vessel is lively, that is, affected by the swells. If so, it may be possible to refloat her at once by sallying ship. Sallying a ship is accomplished by securing a line from another vessel at right angles to the keel and alternately pulling the vessel back and forth in an effort to free the bottom. At the same time, if the propellers are clear, the engines should be backed at full speed and, if another vessel is in the area, a line should be secured to the vessel to exert pulling force.

FLOODING THE VESSEL

4-47.   When the vessel is held fast and is in danger of being pounded by heavy seas, it may be best to flood the compartments with water to prevent the vessel from moving over the bottom and smashing the hull. The water can be pumped out later after the heavy seas have abated.

REMOVING BALLAST

4-48.   The vessel may be lightened and its draft reduced by discharging some of its liquid ballast. This weight may be enough to decrease the draft and free the vessel. The tanks should not be emptied completely because a certain amount of fuel and water is needed to keep machinery running to deliver the vessel to safety. If the vessel is aground at the bow, the tanks should be pumped from bow to stern; if the stern is aground, the tanks should be pumped from stern to bow. You must keep in mind that intentional dumping of fuel is a criminal offense. If it is necessary to save the ship, a means of transferring the fuel to another vessel should be attempted.

KEDGING

4-49.   When a vessel grounds on a bar or a river bank in quiet water and the engine either is of no assistance or cannot be used, the vessel may be cleared by kedging. Kedging consists of carrying out kedge anchors with sufficient scope and taking a strain on them. If in conjunction with any possible lightening or shifting, a steady strain often will work a vessel free. If the vessel has gone hard on a bar so that she is aground amidships, it is possible that the best means of freeing her may be to go ahead on the engines and shove full across the bar.

SHIFTING CARGO

4-50.   If the vessel is not aground along the full length of her keel, all weight should be shifted to the part of the vessel still afloat. On a vessel aground by the bow, ballast, fuel, and water may be shifted to the aftermost tanks available and the cargo may be shifted from the forward to the after hatches. When there is no room in the after hatches, as much cargo as is deemed safe may be deck-loaded aft. Such shifting of weight and cargo should not be attempted if it would merely put the full length of the keel on the bottom. However, when there is enough depth (as there often is when grounded on a bar) such operations may be the quickest and simplest means of working free.

TOWING OFF

4-51.   The aid of another vessel should be obtained immediately when the following occurs:
- When a vessel is aground and the master is not sure that he/she can get it off quickly without damage.
- When a vessel strands on a beach in open surf.

The master of the assisting vessel must acquaint himself as fully as possible with the whole situation. This includes the following:
- Knowing the nature of the bottom.
- Prevailing winds.
- Current and tidal data.
- Any damage to grounded vessel (such as possibility of pierced hull or compartment).

He/She must also confer with the master of the stranded vessel and use the following procedures.

IN QUIET SEAS

4-52.   In quiet seas an assisting vessel may anchor to seaward with a good scope of cable. The lines should be secured to the stern of the stranded vessel and kept taut until the assisting vessel tails in as near the stranded vessel as wind and tide permit. The anchor windlass should be used with full power to keep these lines taut and pick up every inch of slack until the vessel is pulled off. Engines should be used and a good strain kept on anchor cable.

IN HEAVY SEAS

4-53.   If seas are heavy, it is often necessary to pass a light line (messenger line) between the assisting vessel and the stranded vessel. The messenger line is secured to one end of a heavy towing line and is hauled to the stranded vessel by pulling in the messenger line.

IN STRONG CURRENT

4-54.   The assisting vessel can use two anchors, but if they drag and the vessel is being set down on the beach, the lines should be castoff immediately. The vessel can then maneuver clear of difficulty, heave up anchors if necessary, and make a fresh attempt.

TOWING VESSEL—NOT ANCHORED

4-55.   When the towing vessel is not anchored, caution should be used to prevent grounding (Figure 4-15). A stern line must not be used, especially in a cross current or wind, as it would make the rudder useless. It is best to secure a line to a bit farther forward and then head the vessel slightly forward and across the current, gradually adding strain to the line and using it with the helm to pivot the towing vessel (Figure 4-16).

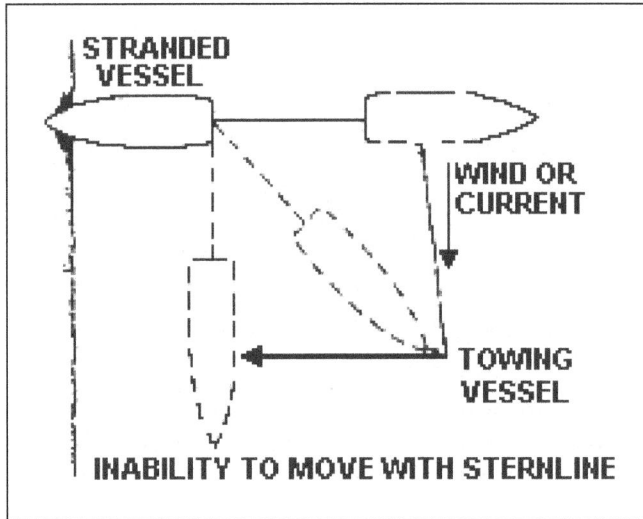

Figure 4-15. Maneuverability of Towing Vessel (Not Anchored)

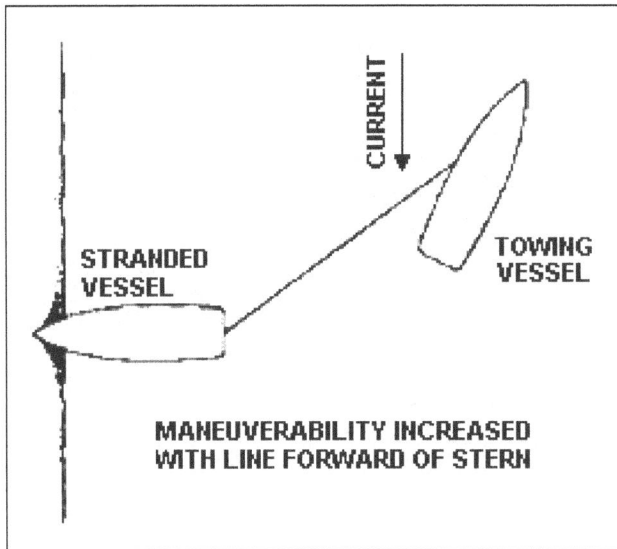

Figure 4-16. Maneuverability of Towing Vessel (with Secured Line)

APPROACHING BOW ON

4-56.   When the wind is offshore, it is possible for the towing vessel to approach bow on and pass a line forward (Figure 4-17). After taking the line in through a forward bow chock, the towing vessel can back her engines to pull off the stranded vessel, thereby saving time which would have been lost in maneuvering to take a line aft. However, pivoting power will be lost if the line is taken exactly at the bow. Instead, it should be taken through a chock a little farther aft. This procedure may assist in slowing the stranded vessel's stern and thereby causing it to break free.

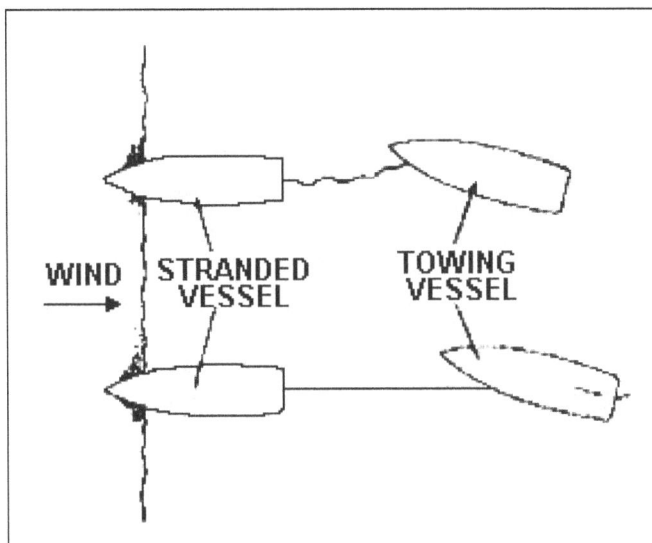

Figure 4-17. Receiving Line Forward

4-57.   A towing vessel which approaches bow on should come in a little to the windward, drift toward the stern of the stranded vessel, and receive the line in that position (Figure 4-18).

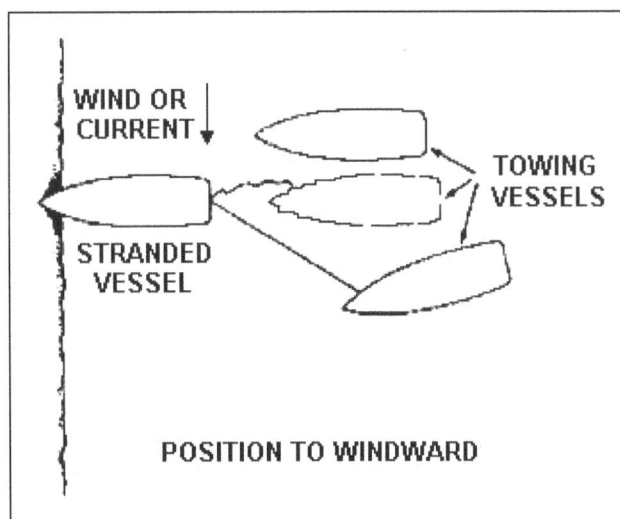

Figure 4-18. Towing Vessel Approaching Bow On in Packed Sand

4-58.   When a vessel is grounded by the bow on a sandy beach, sand will frequently become packed around the stern. Soundings taken by lead line will show the exact location of this sand. Water jets may be rigged over the side and connected to the fire mains. At the beginning of the high tide during which the vessel is to be pulled off, these pumps and jets should be started. The force of the water will create live sand, which will move away from the bottom and sides of the vessel.

FOR JUMPING A LINE

4-59.   Small vessels may be pulled off the beach by a sudden pull under full power. This method is never used on heavy vessels, but is sometimes useful on smaller craft at high tide. The hawser must be securely fastened to towing bitts because no other part of the vessel will stand so great a strain. Care must be taken to keep the hawser away from the propellers and personnel must stand clear of the hawser.

HEAVY WEATHER MEASURES

4-60.   Since vessels vary in design and size and weather conditions vary in severity, so do the measures that need to be performed. The following are some measures you should know in heavy weather.

- Meet with the crew to explain the situation and reassure them. Make sure that they know what to do and what not to do when the extreme weather arrives. Explain such things as keeping low in the boat, not moving around excessively and not going out on deck unless necessary. Give them all an assignment to keep them occupied and keep their minds off the situation.
- Determine position of storm, wind direction, speed, and estimate time to your location.
- Secure all hatches and close all ports and windows to keep the water on the outside.
- Pump bilges dry (into holding tank) and repeat as required. This helps eliminate "free water affect." (Sloshing of water in the bilge as the boat rolls which can effect stability)
- Secure all loose gear above and below decks. Put away small items and lash down larger ones.
- Break out personal floatation devices (PFD) and foul weather gear.
- Ready emergency equipment that you may need such as hand pumps, bailers, first aid kit, sound signaling device, and so on.
- Get a good fix of your position and plot it on your chart. Make note of the time, your heading, and speed.
- Make plans to alter course to sheltered waters if possible.
- Monitor all means of receiving weather for updates to severe forecasts.
- Review abandon ship procedures.
- Make sure the life raft is ready to be deployed.
- Make sure emergency food and water are in the life raft.
- Rig jack lines and/or life lines. Require anyone who must go on deck to wear a safety harness.
- Make ready your sea anchor or drogue if needed.
- Turn on navigation lights.
- Keep away from metal objects.
- Change to a full fuel tank if possible.
- Keep a sharp lookout for floating debris and other boats.
- If you have a choice, do not operate the boat from the fly bridge.

STANDARD PRECAUTIONS

4-61.   Before a vessel leaves port and passes the sea buoy, standard precautions are taken to make her secure. All booms are lowered and stowed, movable gear on deck is lashed down, and covers are placed over machinery that may be damaged by saltwater. When a vessel enters a storm area, a check should be made to see that these standard precautions have been taken. Extra lashings should be added where needed to avoid damage to gear or cargo. Hatch coverings should be checked and the battens secured. Ventilators should be trimmed away from the wind and spray or taken down entirely and plugs or tarpaulins should be fitted over the openings. Boat gripes should be inspected and tightened. Watertight doors should be closed securely and dogged, skylights battened, deadlights closed, and, if necessary, lifelines rigged.

SECURING CARGO AND GEAR

4-62.   When stowing or supervising the stowing of cargo, keep in mind that the vessel will be at sea and the cargo will be subjected to the forces constantly generated by the roll and pitch of the vessel. A stiff roll or continuous pitching has an element of impact that tends to loosen cargo. Once the stowage has become loose, it creates an impact of its own. After the damage is done, it is usually too late and too dangerous to attempt to correct. Gear used to handle cargo should always be stowed securely. Booms should be cradled and bolted and guys and pendants should be coiled and lashed. There should be no loose lines on deck while at sea because they jeopardize life and property.

PROTECTING DECK CARGO

4-63.   The chief advantage of deck cargo is that it is always visible and can be easily checked. All deck cargo should be well lashed and secured. In foul weather, turnbuckles should be tightened and tarpaulins rigged. When they sweep the deck, waves exert an immense hydraulic force, which the deck cargo must withstand.

USE OF SEA ANCHOR

4-64.   Before it becomes impossible to steam either with or against the seas, the vessel must be hove to, that is, headed so she will take the seas most comfortably. It must be remembered that each vessel will heave to in a manner dependent on her design and trim. Some vessels will lay their quarters into the wind and others, their bows. The master, bearing in mind that the most comfortable and safe position for a vessel is with a small angle to the seas, should estimate what position his/her vessel will assume when lying powerless. Steaming slowly ahead or astern, depending on whether the vessel is laying its bow or quarters into the wind and whether the storm is of average strength, will preserve the desired angle. However, if the storm is so violent that the vessel is unable to proceed at all, a sea anchor may be rigged. A sea anchor (Figure 4-19) is used to create a drag through the water and hold either the bow or the stern into the sea. Small vessels carry a sea anchor, which is a canvas bag that is dropped over the bow or stern and secured with a heavy line. Large vessels can improvise a sea anchor by rigging one from hatch covers or other available material.

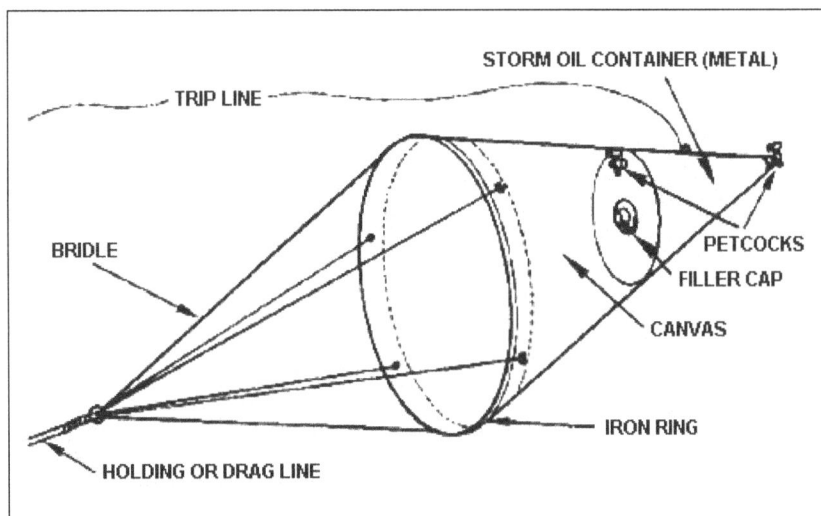

Figure 4-19. Sea Anchor

This page intentionally left blank.

Chapter 5

Charts and Publications

INTRODUCTION

5-1.   The mariner must be able to identify and describe the following:
- Navigation aids used aboard ship.
- Chart numbers.
- Correct charts.
- Basic navigation publications.
- Agencies that are responsible for updating, publishing, and issuing charts and publications. The mariner must also be able to interpret chart numbers, use basic navigation publications, and know how to requisition charts and publications.

CARE OF CHARTS

5-2.   Charts are one of the most important aids to the navigator, and they must be treated as such. Charts should be kept dry and clean. Permanent chart corrections should be made in ink so that they cannot be erased. All other lines and markings should be made in pencil so that they can be erased. To avoid confusion, lines drawn on a chart should be drawn no longer than necessary and they should be labeled. After you have finished using the chart, all lines should be erased. The chart should be inspected for damage and stored flat with the least amount of folding. Charts are stored in a drawer or kept in a portfolio. They should be properly indexed so that any desired chart can be found when needed.

CHART PORTFOLIOS

5-3.   The chart portfolio system divides the world into 52 geographical areas. This system assigns a two-digit designator that represents a portfolio number to each area (see Figure 5-1). An "A" and/or "B" prefix is also used. The "A" series of portfolios contains all the general charts and the principal harbor and approach charts for each of the 52 geographical areas. The "B" series of charts supplements the "A" coverage. To determine the chart portfolio number, locate the Defense Mapping Agency (DMA) stock number in the lower right-hand corner of the chart. Figure 5-2 shows the system used to establish the portfolio and sequence of chart numbers within the portfolio. The last three digits of the chart number show the chart number within the sub area.

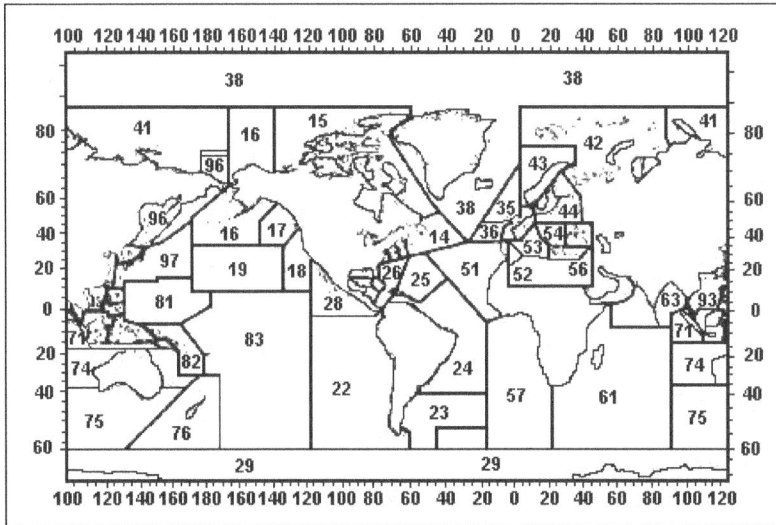

Figure 5-1. World Sub Regions

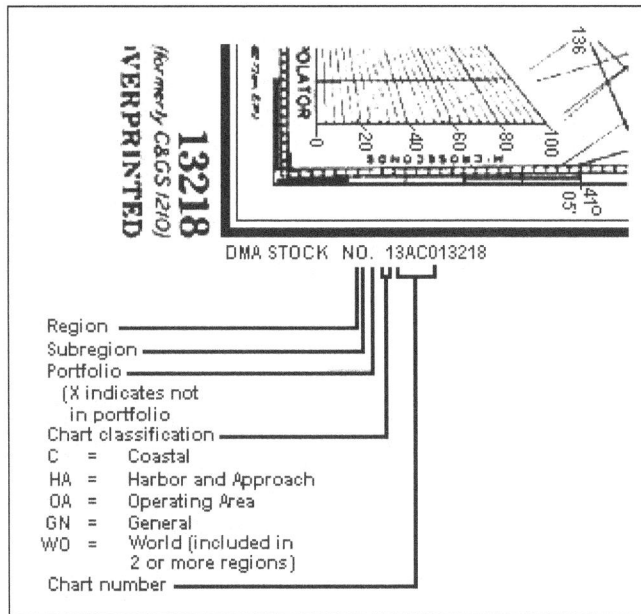

Figure 5-2. Lower Right-Hand Corner of Chart

CORRECTING A CHART

5-4.   The date printed on the lower left-hand corner of the chart is the date of the latest Notice to Mariners used to update the chart. After this date, the responsibility for updating the chart belongs to the user. The weekly Notice to Mariners will list corrections to be posted on charts as they occur. The USCG Local Notice to Mariners is updated weekly and is available for download in Portable Document File (PDF) Format at; http://www.navcen.uscg.gov/lnm/default.htm. The weekly Notice to Mariners presents corrective information affecting charts as well as coast pilots, sailing directions, fleet guides, catalogs of nautical charts, light lists (USCG and DMAHTC), radio navigational aids, and other publications as may from time to time require updating. Chart corrections are listed numerically by chart number, beginning with the lowest and progressing through all charts affected. Each correction pertains to a particular chart and that chart only. Related charts, if any, have their own corrections which, in turn, pertain to a single chart only. The following paragraphs and the example in Figure 5-3 explain the individual elements of a typical correction. A correction preceded by an asterisk (*) indicates it is based on original U.S. source information, the letter "T" indicates it is temporary, and the letter "P" indicates it is preliminary. Courses and bearings are given in degrees clockwise from 000 true. Bearings of light sectors are toward the light. The visibility of lights is usually the distance that a light can be seen in clear weather and is expressed in nautical miles. Visibilities listed are values received from foreign sources. The visibility of lights maintained by the USCG is given as "nominal range."

This section contains corrections to federally and privately maintained Aids to Navigation, as well as NOS corrections.

This section contains corrective actions affecting chart(s). Corrections appear numerically by chart number, and pertain to that chart only. It is up to the mariner to decide which chart(s) are to be corrected. The following example explains individual elements of a typical chart correction.

Chart Number	Chart Edition	Edition Date	Last Local Notice to Mariners	Horizontal Datum Reference	Source of Correction	Current Local Notice to Mariners
12327	91st Ed.	19-APR-97	Last LNM: 26/97	NAD 83		27/97

Chart Title: NY-NJ-NEW YORK HARBOR - RARITAN RIVER
   Main Panel 2245 NEW YORK HARBOR   CGD01

(Temp) ADD   NATIONAL DOCK CHANNEL BUOY 3   at 40-41-09.001N 074-02-48.001W
                Green can
Corrective Action   Object of Corrective Action   Position

(Temp) indicates that the chart correction action is temporary in nature. Courses and bearings are given in degrees clockwise from 000 true. Bearings of light sectors are toward the light from seaward. The nominal range of lights is expressed in nautical miles (NM) unless otherwise noted

Figure 5-3. Explanation of Format

CHART/PUBLICATION CORRECTION RECORD CARD

5-5. Before changing any chart, you should go through the Chart/Publication Correction Record (DMAHC-8660/9) cards (Figure 5-4) and remove those affected by that particular notice. After withdrawing the cards corresponding to the number entered on the chart correction list, you are ready to enter the necessary data on the cards. Prepare a card for each chart/publication by inserting the following information:

- Chart/publication number.
- Portfolio.
- Edition number/date.
- Classification.
- Title of chart/publication (if title is too long, use an abbreviated descriptive title).

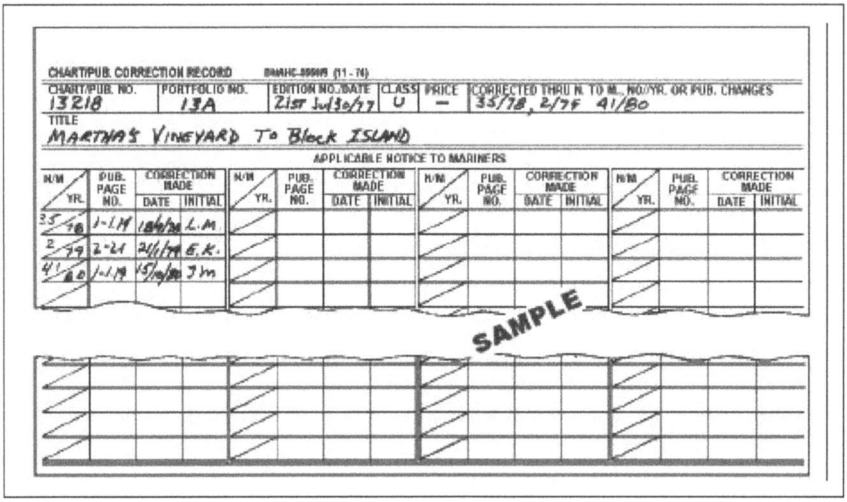

Figure 5-4. Chart/Publication Correction Record (DMAHC-86609)

5-6.   Perform the following steps when updating the record card:

- Review section I of the current Notice to Mariners and determine which charts on board are to be corrected.
- Get the chart/publication correction record card for each chart to be corrected.
- Make corrections in ink on chart.
- Post chart/publication correction record card, date, and initial.
- Make notation on lower left-hand side of chart, showing date of correction and Notice to Mariners number.

5-7.   Each chart on board will have a chart correction card on file. With the chart correction card system, only the charts in current use in the operating area of your ship must be kept up-to-date at all times. Corrections are not made to other charts until the charts are needed. If a chart is not corrected, a notation is made on the card. The entry gives the Notice to Mariners number and page number. When a chart is corrected, the date and the initials of the person who corrected the chart are entered in the prescribed spaces on the card. The weekly Notice to Mariners features a new format for presenting corrective information affecting charts, Sailing Directions, and U.S. Coast Pilots. In section I, chart corrections are listed by chart number, beginning with the lowest and progressing in sequence through each chart affected. The chart corrections are followed by publication corrections, which are also listed in numerical sequence. Since each correction pertains to a single chart or publication, the action specified applies to that particular chart or publication only. Related charts and publications, if any, are listed separately.

Note: In correcting charts that have accumulated many corrections, it is more practical to make the latest correction first and work backwards. The reason for this is because late corrections may cancel or alter earlier corrections.

5-8.   Upon receipt of a new chart and/or a new edition, a new card should be made so that the card will show only those corrections (including temporary changes) which have been published since the date to which the chart was corrected by the DMAHTC. Temporary changes are not incorporated in new editions of charts and must be carried forward from old editions. Consult Notice to Mariners 13, 26, 39, and 52 for outstanding temporary corrections. At the end of each quarter, the DMAHTC will include in the Notice to Mariners a chart correction list for that quarter containing all effective Notice to Mariners corrections to charts. The list for Navy Notice to Mariners corrections will be published annually. These lists should be checked against the cards to make sure that all corrections have been entered.

CHART CORRECTION TECHNIQUES

5-9. The first step in correcting a chart is usually to erase that part of the charted information that will be changed. Lay the chart on a smooth area. Erase in a back and forth motion with an abrasive eraser (this should remove most of the ink). To preserve the quality of the paper's surface, remove the remainder of the ink with a nonabrasive eraser. Most typewriter erasers are very abrasive and general erasers are mildly abrasive. A sharp penknife or razor blade can be a valuable tool in the hands of an experienced draftsman, but poor handling of a knife can quickly ruin a chart. Rubbing an erased surface with an ivory or bone burnishing tool, or even the thumbnail, may improve its inking qualities. The ink from a conventional lettering or drafting pen tends to feather to a degree, depending on the condition of the erased surface. Normally, a ballpoint pen with a fine point feathers less and reduces the possibility of ink smears. Another advantage of the ballpoint pen is the variety of colored inks available. Although black is the principal color used in chart correction work, other colors such as green, purple, magenta, or blue may be occasionally needed. Corrections in writing should be kept clear of water areas as much as possible, unless the objects referred to are on the water. When inserting written corrections, care must be taken not to obliterate any of the information already on the chart. When "notes" are to be inserted (such as cautionary, tidal, and so forth) they should be written in a convenient but conspicuous place where they will not interfere with any other details. The year and number of every Notice to Mariners from which corrections have been made are to be entered in ink at the lower left-hand corner of the charts (for example, "1968-6, 9, 18"). Temporary changes should be made in pencil.

REQUISITIONING PROCEDURES FOR CHARTS

5-10. DMAHTC is responsible for providing navigational charts and publications to the DOD and civilian users. The DMA requisitioning requirements are designed primarily for DOD activities as a means of simplifying the control of inventories and to reduce order processing time.

5-11. The requisitioning procedures outlined below are an integral part of the DMA automated distribution management system (DADMS) and cover requisitions by both DOD and non-DOD activities. DOD activities authorized to request DMA products must obtain a DOD activity address designator code through their respective service or agency.

REQUISITION FORMS

5-12. Use any one of the following forms, when ordering DMA products:
- DD Form 1348 (*DOD Single Line Item Requisition System Document [Manual]*) (Available through normal forms supply channels.).
- DD Form 1348M (*DOD Single Line Item Requisition System Document [Mechanical]*) (Available through normal forms supply channels.).
- SF 344 (*Multiuse Standard Requisitioning/Issue System Document*).

Note: DOD activities should use any one of the above forms that best fit their individual needs. Use SF 344 to requisition items not identified in DMA publications (1-N-A and 1-N-L) or items where the stock number is not known. When used in this manner, the item description may be written across the entire line or lines under requisition data without regard to columnar headings. Such data as the quantity, serial number, supplementary address, and signal and advice codes will be entered directly below the item description in the appropriate blocks.

PREPARATION OF REQUISITIONS

5-13. Requisition documents (DD Form 1348, DD Form 1348M, and SF 344) will be prepared as required.

REQUISITIONING OF NAUTICAL CHART PORTFOLIOS

5-14.   To preclude the submission of many requisition documents when an entire nautical chart portfolio is desired, the requestor may prepare a single requisition identifying the portfolio in the first five positions of the stock number followed by the word "ALL" in the next three positions. The DADMS will generate a requisition for each nautical chart included in the portfolio applying the basic requisition information to each item including document number and quantity requested. Validation, status, manifesting, and issue will be accomplished on a line item basis. Requisitions of this type can be submitted only to the Department of Defense Control Point (DDCP), DMA Office, Pacific, and Defense Mapping Agency Office of Distribution Services (DMAODS) branch offices. Examples of portfolios include the following:

- Standard Nautical Chart Portfolios:
    - Portfolio 11A—order as 11AXXALL.
    - Portfolio 37B—order as 37BXXALL.
    - Portfolio 97A—order as 97AXXALL.

Note: This same sequencing will be used in requisitioning any of the designated nautical chart portfolios included in this catalog.

- World Portfolios:
    - Portfolio WOA—General Charts of the Atlantic—order as WOAXXALL.
    - Portfolio WOP—General Charts of the Pacific—order as WOPXXALL.
    - Portfolio WOB—General Charts of the World—order as WOBXXALL.

SUBMISSION OF REQUISITIONS DOD ACTIVITIES

5-15.   All DOD activities will submit requisitions for DMA products by Automatic Digital Network (AUTODIN), message, or mail to the designated source of supply.

Method of Transmitting Requisitions

5-16.   DOD activities having punch card facilities will submit the DD Form 1348M to the designated supply source. Those DOD activities having transceiver facilities may receive the DD Form 1348M via the AUTODIN. The DD Form 1348M may be mailed if AUTODIN is not readily available. Those activities that do not have punch card equipment or transceiver facilities may use DD Form 1348 or SF 344. These forms will be mailed to the designated supply source. Requisitions forwarded by mail will be identified by the word "MILSTRIP" printed in the lower left corner of the envelope. Message requisitions are acceptable from authorized requisitioners, provided that they are in the format shown. The term "MILSTRIP REQUISITION" will precede the text of the message. Telephone requests from DOD activities are acceptable when the urgency of the requirement dictates. When the telephone is used, the requester will contact the DDCP, overseas depot, DMA field office, or other issuing activity, as appropriate, and provide pertinent data. The issuing activity will record the data on a machine able requisition document and complete processing of the telephone request. Telephone requests from non-DOD activities must be confirmed by letter or message as such deliveries normally involve reimbursements.

Designated Supply Source

5-17.   The mail and message address, routing identifier and telephone number of the designated source of supply for nautical products is as follows:

Mailing Address	Message Address:
DMA Office of Distribution Services	DMAODS WASHINGTON, DCC//DDCP//
ATTN: DDCP	Routing Identifier: HM8
Washington, D.C. 20315	DODAAD: HM0028
	Telephone: DSN 287-2495 or 287-2496
	COML: (202) 227-2495/227-2496

PUBLICATIONS

5-18. Every ship must carry the charts and publications required for its safe operation. These include the COMDTINST M16672.2C (Navigation Rules—International-Inland), all charts applicable for the vessel's navigational area of operation, and all TMs, TCs, and ARs that apply to the class of vessel. The following discusses the various publications and the agencies responsible for publishing them.

THE NATIONAL OCEAN SURVEY

5-19. This agency is charged with the survey of the coast, harbors, and tidal estuaries of the United States and its insular possessions and is a part of the National Oceanic and Atmospheric Administration, Department of Commerce. It is responsible for issuing the following publications.

United States Coast Pilot

5-20. The United States Coast Pilot publication consists of nine volumes. These volumes include the following:
- Atlantic Coast—
  - Eastport to Cape Cod.
  - Cape Cod to Sandy Hook.
  - Sandy Hook to Cape Henry.
  - Cape Henry to Key West.
  - Gulf of Mexico, Puerto Rico, and Virgin Islands.
- Great Lakes—
  - The Lakes and Connecting Waterways.
- Pacific Coast—
  - California, Oregon, Washington, and Hawaii.
  - Alaska-Dixon Entrance to Cape Spencer.
  - Alaska-Cape Spencer to Beaufort Sea.

Coast Pilot

5-21. Coast Pilot, besides its standard information on U.S. ports and waterways, contains the following:
- Descriptions of ports and harbors.
- Pilot information.
- Quarantine and marine hospital information.
- Coast Guard stations.
- Radio services, distances, and bearings.
- Time signals.
- Atmospheric pressure, temperature, and wind tables.
- Rules of the road.
- Instructions in case of shipwreck.

- General harbor regulations.

Tide Tables

5-22.   These are published each year for various parts of the world and are published in four volumes. Each volume consists of the following:

- **Table 1.** A list of reference stations for which the tide has been predicted. The time and heights of high and low tides are tabulated for each day of the year for each of these reference stations.
- **Table 2.** A list of subordinate stations for which the tidal differences have been predicted with respect to a reference station having nearly the same tidal cycle.
- **Table 3.** A convenient means of interpolation which allows for the characteristics of the tidal cycle. While Tables 1 and 2 provide times and heights of high and low tides, the state of the tide may be desired for a given time in between.
- **Table 4.** A sunrise-sunset table at 5-day intervals for various latitudes.
- **Table 5.** A table that provides an adjustment to change the local mean time from Table 4 to zone or standard time.
- **Table 6** (two volumes only). A table that gives the zone time of moonrise and moonset for each day of the year at selected places.

Tidal Current Tables

5-23.   These tables are prepared annually in two volumes for various areas to provide predictions of the state of the current. Each volume consists of the following tables:

- **Table 1.** A list of reference stations in geographical sequence for which the current has been predicted.
- **Table 2.** A list of subordinate stations for which the difference between local current and current at a reference station has been predicted. Above the groups of subordinate stations, the appropriate reference station is listed.
- **Table 3.** A table that provides a means of interpolation for the state of the current at any time between tabulated times.
- **Table 4.** A table that gives the number of minutes that the current does not exceed the stated amounts for various maximum speeds.
- **Table 5** (Atlantic Coast of North America only). A table that gives information on rotary tidal currents.

Tidal Current Charts

5-24.   The tidal current tables are supplemented by 11 sets of tidal current charts. These charts present a comprehensive view of the hourly speed and direction of the current in 11 bodies of water. These bodies of water include the following:

- Boston Harbor.
- Narragansett Bay to Nantucket Sound.
- Narragansett Bay.
- Long Island Sound and Block Island Sound.
- New York Harbor.
- Delaware Bay and River.
- Upper Chesapeake Bay.
- Charleston Harbor.
- San Francisco Bay.
- Puget Sound (northern part).
- Puget Sound (southern part).

5-25.   They also provide a means for determining the speed and direction of the current at various localities throughout these bodies of water.

Chart Catalog

5-26.   National Ocean Survey (NOS) publishes four catalogs. Each catalog covers the following areas:

- Catalog No. 1. The waters of the Atlantic and gulf coasts, including Puerto Rico and the Virgin Islands.
- Catalog No. 2. The Pacific coast, including Hawaii, and Pacific islands such as Guam and Samoa.
- Catalog No. 3. The Alaskan waters, including the Aleutian Islands.
- Catalog No. 4. The Great Lakes and the adjacent waterways.

5-27.   Each catalog contains several small-scale outline charts with lines showing the limits of each nautical chart in that region.

DEFENSE MAPPING AGENCY HYDROGRAPHIC AND TOPOGRAPHIC CENTER PUBLICATIONS

5-28.   The identification of charts, publications, and other products of DMAHTC is based on a numbering system of one to five digits without a prefix. The number will be based on the scale and on the basis of region and sub region. Publications will continue to carry the "HO" prefix until they are reprinted by DMAHTC. At that time, DMAHTC will assign a number and give the prefix "pub."

Portfolio Chart List

5-29.   The portfolio chart list (Publication No. 1-PCL) is a publication furnished by the DMAHTC to United States ships. It is issued in two volumes: one for the Atlantic side of the world and one for the Pacific side. This publication contains a complete list of charts, by portfolio, arranged according to their numbers. The portfolio chart list is intended as a guide for selecting and storing nautical charts aboard ship. It also provides a ready reference to the grouping, by geographical sequence, of the charts in the various portfolios. Most of the necessary information concerning charts, such as chart number, edition number and date, and title required by a mariner in establishing a chart correction card system, is included within this publication.

Index of Nautical Charts and Publications

5-30.   Nautical charts and publications may be found indexed in one of the catalogs shown in Table 5-1. These catalogs provide additional information of interest to the navigator that may not be found in Publication No. 1-PCL. Suppose you are to sail from Norfolk, Virginia to Sao Luis, Brazil, in South America. For general planning and for sailing the open sea between the two ports, you would refer to Publication No. 1-N-A to find the appropriate small-scale charts, Sailing Directions, light lists, any charts needed for Loran navigation, and position plotting sheets. For the large-scale charts needed when you navigate in ports, channels, and so forth, you would refer to Region 2.

Note: Additions and changes to the catalogs may be obtained from the Notice to Mariners.

5-31.   Only the latest editions of charts are issued. All charts issued by the depots will be corrected through NM 26/75, after which corrections will no longer be applied. All charts are corrected as of the print date shown in the lower left-hand corner of the chart. Corrections affecting charts after the issue date are published in the weekly Notice to Mariners received by all ships.

Table 5-1. Index of Nautical Charts and Publications

Catalog Number	Contents
Pub 1-N-A	General information on the catalogs, graphics showing regions covered by Lists of Lights and Sailing Directions, a listing of Loran and Omega charts and plotting display charts, and instructions and forms for ordering charts and publications.
Pub 1-N-L	Complete numerical listing of charts issued by DMA Hydrographic and Topographic Center. Standard terminology for the DMA Hydrographic and Topographic Center and National Ocean Survey charts. Instructions and forms for ordering charts and publications.
Pub 1-N, Region 1	United States and Canada
Pub 1-N, Region 2	Central and South America and Antarctica
Pub 1-N, Region 3	Western Europe, Iceland, Greenland, and the Arctic
Pub 1-N, Region 4	Scandinavia, Baltic, and Russia
Pub 1-N, Region 5	West Africa and the Mediterranean
Pub 1-N, Region 6	Indian Ocean
Pub 1-N, Region 7	Australia, Indonesia, and New Zealand
Pub 1-N, Region 8	Oceania
Pub 1-N, Region 9	East Africa

Publication 9–American Practical Navigator

5-32. This is commonly referred to as "Bowditch." This is an extensive text on piloting, celestial navigation, and other nautical matters. Publication 9 also contains tables, data, equations, and instructions needed to perform navigational calculations.

List of Lights–Publications 111A and 111B, 112 through 116

5-33. These publications contain detailed information on the location and characteristics of every light in the world not located in the United States or possessions. Brief descriptions of lighthouses and fog signals are included. List of Lights is published in seven volumes, and it is corrected through Notice to Mariners.

Sailing Directions

5-34. The Sailing Directions provide the same type of information as the Coast Pilots, except the Sailing Directions pertain to foreign coasts and coastal waters. Typical information includes the following:
- Pilotage.
- Appearance of coastline (mountains, landmarks, visible foliage, and so forth).
- Navigational aids in general.
- Local weather conditions.
- Tides and currents.
- Local rules of the road, if any.
- Bridges—type and clearance.
- Anchorage facilities.
- Repair facilities.
- Availability of fuel and provisions.
- Transportation service ashore.
- Local industries.

Note: Under a new concept begun in 1971, the 70 volumes of existing Sailing Directions are being replaced by 43 publications—namely 35 new graphic sailing directions (en route) and 8 new sailing directions (planning guide).

Publication 102 (HO 102) - International Code of Signals

5-35.   This publication includes both general procedures that apply to all forms of signaling and the specific rules for signaling by flags, blinking lights, sound, radio, and radio telephone.

Publication 117A-B - Radio Navigational Aids

5-36.   These two volumes are based on geographic location and contain information such as—
- Time and frequency of hydrographic broadcasts.
- Radio time signals.
- Distress traffic.
- Radio beacons.
- Direction-finding stations.
- Loran coverage and stations.
- Radio regulations for territorial waters.

Notice to Mariners

5-37.   The DMAHTC publications mentioned so far are published at more or less widely separated intervals. As a result, provisions must be made for keeping mariners informed of changes in hydrographic conditions as soon as possible after they occur. The principal medium for distributing corrections to charts, light lists, and other DMAHTC publications is Notice to Mariners. One or more copies are distributed to each vessel. Each notice is divided into three sections:
- Section I, chart corrections.
- Section II, light list corrections.
- Section III, broadcast warnings and miscellaneous information.

Summary of Corrections

5-38.   Every six months, DMAHTC publishes a Summary of Corrections, in two volumes. Volume I covers the Atlantic, Arctic, and Mediterranean areas. Volume II covers the Pacific and Indian Oceans and the Antarctic. These volumes cover the full list of all changes to charts, Coast Pilots, and Sailing Directions.

Pilot Charts

5-39.   These are small-scale charts of large areas. They are of little use in actual piloting but are still valuable to navigators. They present, in graphic form, a complete review of the hydrographic, navigational, and meteorological situation in a given area. Included is information concerning the following:
- Average winds, tides, currents, and barometer readings.
- Frequency of storms, calms, or fog.
- Possibility of the presence of ice, wrecks, or other dangers.
- Location of ocean station vessels.

5-40.   Lines of equal magnetic variations are given for each full degree of variation. Indicated also are the shortest and safest routes between principal ports. Pilot charts of the North Atlantic Ocean and North Pacific Ocean are issued monthly by the DMAHTC. Other pilot charts are published in atlas form for the Northern North Atlantic Ocean, South Atlantic Ocean and Central American Waters, and the South Pacific Indian Oceans.

COAST GUARD PUBLICATIONS

5-41.   The USCG prepares one major publication, marine safety manuals, and also extracts from CFRs.

List of Lights and Other Marine Aids

5-42.   This publication is often referred to as the "light list" and comes in five volumes.

- Volume I. Atlantic Coast--St. Croix River, Maine to Little River, South Carolina.
- Volume II. Atlantic Coast--Little River, South Carolina to Rio Grande, Texas.
- Volume III. Pacific Coast and Pacific Islands.
- Volume IV. Great Lakes.
- Volume V. Mississippi River System.

5-43.   The light list describes the lighthouses, lesser lights, buoys, and day beacons maintained in all navigable waters of the United States by the USCG and various private agencies. The information shown will include the following:

- The official name of the aid.
- Characteristics of its lights, sounds, and radio signals.
- Its structural appearance and position.
- Any other important information.

5-44.   The List of Lights is published about every 12 months, and any corrections or changes are posted in the Notice to Mariners.

Navigation Rules

5-45.   This is the law for all mariners operating in either inland or international waters. Commandant Instructions M16672.2C is published and issued by the USCG. It is an extract from the Code of Federal Regulations, Title 33, Navigation and Navigable Waters.

Chapter 6

Shipboard Communications

INTRODUCTION

6-1.   Ship-to-ship and ship-to-shore communications are vital for shipboard operations.   These communications can be by satellite, radio, radiotelephone, flag hoists, and flashing lights. The watercraft operator and mate aboard ship must be thoroughly familiar with their communication equipment. Shipboard communications are essential in normal operations, distress situations, and/or sea-air rescue missions. This chapter covers communications by satellite, radiotelephone, flashing lights, flag hoists, and the international distress signals.

SHIPBOARD TACTICAL AND MARINE RADIOS

6-2.   All Army vessels are equipped with some kind of radio which is easy to operate.

- **Command, Control, Communications, Computers and Intelligence-Surveillance and Reconnaissance (C4ISR).** The C4I system consists of the International Maritime Satellite Organization (INMARSAT) system, Ultra High Frequency (UHF), frequency hopping Very High Frequency/Ultra High Frequency (VHF/UHF) transceiver, High Frequency (HF) tactical communications system, secure telephone and secures fax machine cryptographic devices. The C4I communications suite also includes an INMARSAT/SALTS computer and a V-mail computer used to transmit and receive V-mail messages. These systems integrate with the Identification Friend or Foe (IFF) and Global Maritime Distress and Safety System (GMDSS).
- **The UHF SATCOM Systems** - consist of the transceiver, amplifier power supply, handset and a dual angle satellite communications antenna. The transceiver is an all weather tactical Amplitude Modulation/Frequency Modulation/Phase Shift Keying (AM/FM/PSK) UHF radio operating in a frequency band of 225 – 399.995 Mega Hertz (MHz). The transceiver operates in a Line of Sight (LOS), satellite secure or non secure, plain text, voice or data communications mode. It also operates in half-duplex mode, where it transmits on one frequency and receives on another, but not at the same time.
- **VHF/UHF LOS System** - The radio set consists of a Receiver-Transmitter (R-T), an Adapter Unit and Power Amplifier. The R-T provides tactical AM and FM communications in an operating frequency range of 30 – 512 MHz; VHF low 30 – 89.99999MHz, VHF high 90 – 224.99999 MHz and UHF 225 – 512 MHz. It provides communication in the LOS, SATCOM, and Electronic Counter-countermeasures (ECCM) frequency hopping mode. The R-T is connected to three antennas, a VHF low band whip, a VHF high band whip and a UHF whip, through and adapter unit for transmitting and receiving signals. It can operate in Communications Security (COMSEC) mode and in a simplex/half duplex mode. The simplex allows transmitting and receiving on one frequency and half duplex allows transmitting on one frequency and receiving on another.
- **The HF tactical communication system** - consists of the Radio Receiver-Transmitter, Radio Amplifier, two power supplies and HF antenna coupler. The R-T operates in a frequency range of 1.5 – 29.99999 MHz. It operates in four different modes, Single Side Band (SSB), Adaptive Frequency Control (AFC), Automatic Link Establishment (ALE) and Frequency Hopping.
- **INMARSAT Satellite Communications System** – The International Maritime Satellite Organization (INMARSAT), a key player within GMDSS, is an international consortium

comprising over 75 international partners who provide maritime safety communications for ships at sea. In accordance with its convention, INMARSAT provides the space segment necessary for improving distress communications, efficiency and management of ships, as well as maritime correspondence services. The basic components of the INMARSAT system include the INMARSAT space segment, Land Earth Stations (LES), also referred to as Coast Earth Stations (CES), and mobile Ship Earth Stations (SES). The INMARSAT space segment consists of 11 geostationary satellites. Four operational INMARSAT satellites provide primary coverage, four additional satellites (including satellites leased from the European Space Agency (ESA) and the International Telecommunications Satellite Organization (INTELSAT)) serve as spares and three remaining satellites (leased from COMSAT Corporation, the U.S. signatory to INMARSAT) serve as back-ups. The Polar Regions are not visible to the operational satellites and coverage is available from 70°N to 70°S. Satellite coverage is divided into four regions, which are—

- Atlantic Ocean Region East (AOR-E)
- Atlantic Ocean Region West (AOR-W)
- Pacific Ocean Region (POR)
- Indian Ocean Region (IOR)

6-3.   The LESs provide the link between the Space Segment and the land-based National/International fixed communications networks. These communications networks are funded and operated by the authorized communications authorities of a participating nation. This network links registered information providers to the LES. The data then travels from the LES to the INMARSAT Network Control Station (NCS) and then down to the SESs on ships at sea. The SESs provide two-way communications between ship and shore. INMARSAT A, the original INMARSAT system, operates at a transfer rate of up to 9600 bits per second and is telephone, telex and facsimile (fax) capable. It is being replaced by a similarly sized INMARSAT B system that uses digital technology to give better quality fax and higher data transmission rates. INMARSAT C provides a store and forward data messaging capability (but no voice) at 600 bits per second and was designed specifically to meet the GMDSS requirements for receiving Maritime Safety Information (MSI) data on board ship. These units are small, lightweight and use an omnidirectional antenna (Publication 9, 2002).

- **Standard Automated Logistics Tool Set (SALTS)** – With SALTS, the user can electronically submit support requests including (but not limited to); APL/AEL parts, technical manuals, training, warranty information, technical assistance, technical documentation, shipping information, feedback reports, payroll data from disbursing, e-mail from the crew, and so on. These can be submitted using the Readiness Support System (RSS) link. RSS will automatically pass the above support requests to the appropriate Support Provider based on information loaded to the Support Provider Matrix. WinSALTS is the client/server application used to "prepare" files for transmission, from various sources within your command (for example, requisitions from the SNAP system, payroll data from disbursing, or e-mail from the crew) to the SALTS project at NAVICP, Philadelphia. WinSALTS Central then routes the information to its final destination, based on the file suffix (which is assigned by the WinSALTS Client), or by the addresses provided (for e-mail and messages). SALTS can be accessed via the SALTS website, over regular telephone lines or through any Department of Defense network. In order to use SALTS, however, you must be running the latest version of WinSALTS. To get the latest version of WinSALTS, go to the SALTS website and download the software from the Download area. You may also receive the WinSALTS software by contacting a SALTS customer service representative at (215) 697-1112 (DSN prefix is 442). The SALTS customer service email address is help@salts.navy.mil. Outside help on installing the application is available by calling the Global Distance Support Center (GDSC) at 1-877-4-1-TOUCH (86824). (Naval Sea Logistics Center, 2007)
- **Secure Telephone Unit (STU)** -Third Generation (STU-III) is a low-cost, user-friendly, secure telephone device. The terminals are designed to operate reliably, with high voice quality, as both ordinary telephones and secure instruments over the dial-up public switch telephone network. STU-III operates in full-duplex over a single telephone circuit using echo canceling modem technology. STU-IIIs come equipped with 2.4 and 4.8 Kbps code-excited linear

prediction (CELP) secure voice. Secure data can be transmitted at speeds of 2.4, 4.8, and 9.6 Kbps. There are many manufacturers each having different maximum throughput rates. The data throughput between two STU-IIIs can only be as great as the slowest STU-III connected.

- **The Secure Terminal Equipment (STE)** is the evolutionary successor to the STU-III. The STE program will improve shore secure voice communications as well as shipboard communications by changing out the analog STU-III products with digital-based STE products. The STE cryptographic engine is on a removable Fortezza Plus KRYPTON ™ Personal Computer Memory Card International Association (PCMCIA) Card, which is provided separately. The STE Data Terminal provides a reliable, secure, high rate digital data modem for applications where only data transfer (FAX, PC files, Video Teleconferencing, and so on) is required. All STE products will be STU-III secure mode compatible with the following enhanced capabilities:
- Voice-recognition quality secure voice communication.
- High-speed secure data transfers (up to 38.4 Kbps for asynchronous or 128 Kbps for synchronous).

6-4.   STE terminal products can use Integrated Services Digital Network (ISDN), analog PSTN, TRI-TAC, or direct connection to Radio Frequency (RF) assets via RS-530A/232E ports. Maximum STE performance may be attained only by those commands employing ISDN service with two Bearer Channels (2B+D ISDN Service). When connected to a PSTN (Analog Telephone) service, the STE/Office units will only support current STU-III voice and data capabilities. A tactical version, STE/Tactical is a replacement for MMT 1500 with a Digital Non-secure Voice Terminal (DNVT) adapter. Though not a direct replacement for the KY-68, the STE/Tactical can serve as a DNVT replacement with secure voice communication capabilities in STU-III modes over TRI-TAC/Mobile Subscriber Equipment (MSE). STE/Tactical is not secure mode compatible with the Digital Secure Voice Terminal DSVT KY-68. A STE Direct Dial capability comprised of the STE/C2 Tactical terminal and/or associated STE/Interworking Function(s) will improve on the existing Navy "Direct Dial" secure voice ship to shore dial-up operations. STE Direct Dial improves secure mode connectivity, provides operational flexibility support for both plain text and cipher text voice modes, and provides a standardized secure ship digital telephone system solution and Joint CINC interoperability with forces at sea and ashore. (Navy INFOSEC, 2008)

MARINE RADIOS

6-5.   Marine radio sets, often called "bridge-to-bridge" radiotelephones, are designed for vessel control and more informal communications. With them you can contact other vessels, even if they are not military, and receive civilian and Coast Guard emergency information. The requirement for carrying these radios is a U.S. law and you cannot operate your vessel unless they work. The DSC 500 is a 25-watt (bridge-to-bridge) transceiver designed to communicate with other ships and shore-based radio stations. All international and United States VHF/TC marine channels in the frequency range of 156.025 to 163.775 MHz are accessible. The unit can also work 10 weather channels and 42 programmable channels. The DSC 500 has the following capabilities:

- Directory entry of 200 ship to ship stations.
- Directory entry of 50 coastal stations.
- Can store up to 100 call waiting events.
- Can store up to 50 group calls.
- Can distress log up to 20 calls.

6-6.   The DSC is designed to cut down on excessive radio traffic and make radio calling more efficient. Typical Digital Selective Calling offers a better way of calling other vessels. If you connect your DSC to a navigation receiver (such as loran or GPS) your position can be given quickly to another vessel. Another vessel can request your position, in an emergency situation, through your radio.

Note: In order to provide the best fidelity at high volume settings, your DSC 500 uses a speaker with a strong magnet. If any rearranging of equipment in your wheelhouse becomes necessary, be sure to place the DSC 500 well away from your compass.

6-7.   Observe your compass during remounting to ensure that the radio is not affecting your compass heading. Remember that you will have to swing ship after any equipment in your wheelhouse is moved to determine the effects on your compass.

GLOBAL MARINE DISTRESS AND SAFETY SYSTEM

6-8.   The GMDSS was developed by the maritime nations in the IMO. GMDSS was implemented on 1 February 1992 and has become mandatory for all new ships built after 1 February 1995. GMDSS should be installed on all ships by 1 February 1999 (unless this deadline is extended by the IMO). GMDSS is designed to ensure maximum coverage of safety communications for all passenger vessels and cargo vessels of 300 GT or more engaged in international voyages. The major reason for the GMDSS is to guarantee that complying vessels will be able to communicate at anytime (in case of distress or to exchange safety information) with a shore station or a ship. The GMDSS describes four sea areas based on the location and capability of shore-base communications facilities. These are described as follows:

- Sea Area A1. An area within the coverage of at least one VHF coast station in which continuous DSC alerting is available (normally 20 to 30 NM).
- Sea Area A2. An area, excluding Sea Area A1, within the coverage of at least one MF coast station in which continuous DSC alerting is available (normally within 150 NM).
- Sea Area A3. An area, excluding Sea Areas A1 and A2, within the coverage of an INMARSAT Satellite in which continuous alerting is available (normally everywhere on the globe except the Polar Regions).
- Sea Area A4. An area outside Sea Areas A1, A2, and A3 which is in the Polar Regions.

6-9.   GMDSS vessels carry the communications equipment appropriate to the Sea Area in which they are operating. GMDSS vessels also carry standard equipment that operates on the same frequencies and mode to ensure communication between other vessels.

HOW TO TALK ON A RADIO

6-10.   It is important that you use the correct radio procedures when using the radio. Radio messages should be short and to the point. Speak slowly and distinctly and do not try to impress the other station with your knowledge of current slang terms or CB talk. When talking on the bridge-to-bridge set (DSC 500), use plain language that can be understood. There are no requirements for special codes or words when using channel 13. Make sure the other person (whether civilian or military) can understand you. Speak clearly and use short sentences. The following practices are forbidden:

- Violation of radio silence.
- Unofficial conversation between operators.
- Transmitting on a directed net without permission (except flash and immediate traffic).
- Excessive tuning and testing.
- Transmitting the operator's personal sign or name.
- Unauthorized use of plain language.
- Use of other than authorized pro-words.
- Unauthorized use of plain language in place of applicable pro-words or operating signals.
- Linkage or compromise of classified call signs and address groups by plain language disclosures or association with unclassified call signs.
- Profane, indecent, or obscene language.

6-11.   BE ALERT while transmitting by radiotelephone. Release your PUSH TO TALK BUTTON occasionally (usually after each phrase or two) to allow another station to break in, if necessary, and to listen for a few seconds for possible distresses. Keep the receiver gain (volume control) turned high enough to hear weak signals through static and other interference.

HOW TO CALL AND REPLY

6-12.   The following are some important things to remember when calling or replying.

- Listen on the frequency before transmitting to make sure that you will not interfere with another transmitting station.
- Set your transmitter on the proper frequency.
- Speak clearly, in a normal voice, holding microphone about 1 to 3 inches from your lips.
- Reduce operating room noise level.
- Avoid excessive calling and unofficial transmissions. Transmit call signs only once when communication conditions are favorable and twice when unfavorable.

6-13.   When a station called does not reply to a call sent three times at intervals of 2 minutes, the calling will cease and will not be renewed until after an interval of 15 minutes. However, if there is no reason to believe that harmful interference will be caused to other communications in progress, the call sent three times at intervals of 2 minutes may be repeated after an interval of less than 15 minutes but not less than 3 minutes. The DSC 500 will do the calling for you automatically and let you know when called station answers. This will free up the operator for other bridge chores. End every transmission with either "OVER" or "OUT" (except when the sending operator wishes to pause a moment before continuing transmission). Use the pro-word "WAIT" in this instance. If you intend to pause for a longer period of time before resuming your transmission, use the pro-word "WAIT OUT". Never use OVER AND OUT together.

USEFUL OPERATING FREQUENCIES

6-14.   The following are the most important frequencies that are available.

- 2182 kHz. This is for international distress and calling voice frequency. It may be used for distress, urgent, and safety traffic. (Safety messages should be sent on 2670 kHz after a preliminary announcement on 2182 kHz.) Ship stations and shore stations will also establish initial contact on 2182 kHz and then shift to an appropriate working frequency for the passing of operational messages.
- 2670 kHz. This is a Coast Guard frequency. Use by non-Coast Guard stations will be restricted to communications with the Coast Guard. This is a normal working frequency for communications with nongovernment vessels after initial contact on 2182 kHz. Group stations also use this frequency for Coast Guard safety information broadcasts.
- 2638 and 2738 kHz. International ship-to-ship frequencies. Coast Guard ships may use these frequencies to communicate with non-Coast Guard ships. They are authorized for use by certain shore stations only for communicating with non-Coast Guard vessels that are in distress situations when no other common frequency is available.
- 3023.5 and 5680 kHz. International search and rescue (SAR) on-scene frequencies. Either of these frequencies may be used to conduct communications at the scene of an emergency or as the SAR control frequency.
- 156.3 MHz, Channel 6. International VHF-TC ship-to-ship frequency (nationally used by maritime mobile stations for SAR communications at the scene of the SAR incident).
- 156.6 MHz, Channel 12. Port operation's working frequency. Coast Guard use of this frequency shall be limited to shore station communications with non-Coast Guard ships involving port operations.
- 156.65 MHz, Channel 13. Vessel bridge-to-bridge VHF-TC frequency for navigational purposes.
- 156.7 MHz, Channel 14. Second choice port operation's working frequency. Coast Guard use of this frequency will be limited to shore station communications with non-Coast Guard ships involving port operations.
- 156.8 MHz, Channel 16. International VHF-TC calling and safety frequency (nationally used also as a distress frequency). It may be used for calling or answering messages preceded by the distress, urgency, and safety signals.

Note: There are no restrictions on obtaining radio checks from Coast Guard Stations on 156.8 MHz.

- 157.1 MHz. Primary liaison frequency for communications between nongovernment vessels and Coast Guard vessels and coast stations. Also used by the Coast Guard for the national VHF-TC radiotelephone safety and distress system and the Coast Guard Marine Information Broadcast Frequency.
- 157.05 MHz, Channel 21; 157.15 MHz Channel 23. Intra- Coast Guard VHF-TC working frequencies. These frequencies are authorized for communications between Coast Guard units engaged in maritime mobile operations and are common to all districts.
- 157.075 MHz, Channel 81. Joint command, control, and surveillance frequency. Used by U.S. and Canadian mobile units, that are operating according to Marine Pollution Contingency Plan for Spills of Oil and Other Noxious Substances. This frequency is also authorized for other Coast Guard maritime mobile command and control operations when not required for marine environmental purposes.
- 157.175 MHz, Channel 83. Coast Guard command and control frequency when required. Coast Guard auxiliary operational and training frequency in the VHF band. This frequency can also be used by Coast Guard Reserve training units (on a not to interfere basis) to Coast Guard operations.

PHONETIC ALPHABET

6-15.   You should be familiar with the standard International Phonetic Alphabet as shown in Table 6-1. It should be practiced and used for all transmissions. It is not a code; it is a means to better understand your radio transmission.

Table 6-1. Standard International Phonetic Alphabet

A -- Alfa	B -- Bravo	C -- Charlie	D -- Delta
E -- Echo	F -- Foxtrot	G -- Golf	H -- Hotel
I -- India	J -- Juliett	K -- Kilo	L -- Lima
M -- Mike	N -- November	O -- Oscar	P -- Papa
Q -- Quebec	R -- Romeo	S -- Sierra	T -- Tango
U -- Uniform	V -- Victor	W -- Whiskey	X -- X ray
Y -- Yankee	Z -- Zulu		

PRONUNCIATION OF NUMBERS

6-16.   You should also be familiar with the standard pronunciation of numbers as shown in Table 6-2. It should be practiced and used for all transmissions.

Table 6-2. Pronunciation of Numbers

0	Zero	1	Wun	2	Too
3	Thur-ree	4	Fo-wer	5	Fi-yiv
6	Six	7	Seven	8	Ate
9	Niner				

ARMY KEY MANAGEMENT SYSTEM (AKMS)

6-17.   AKMS integrates all functions of crypto management and engineering, signal operation instructions (SOI), electronic protection (EP), cryptographic key generation and distribution, key accounting, and key audit trail record keeping into a total system designated the Automated Communications Security Management and Engineering System (ACMES). ACMES provides commanders the necessary tools to work with the widely proliferating communications security

(COMSEC) systems associated with the mobile subscriber equipment (MSE), echelon above corps communications (EAC comms), Joint Tactical Information Distribution System (JTIDS), Enhanced Position Location Reporting System (EPLRS), Single Channel Ground to Air Radio System (SINCGARS) and other keying methods (electronic key generation, over-the-air rekey (OTAR) transfer, and electronic bulk encryption and transfer) being fielded by the Army. ACMES is a 2-phase program.

ACMES (Phase I)

6-18.   ACMES (Phase 1) focuses primarily on requirements for combat net radio (CNR) frequency management, common fill device (CFD), and electronic SOI. ACMES provides users with an enhanced SOI, frequency hopping (FH) data, and COMSEC key generation capability. The automated net control device (ANCD) provides the capability to electronically store and rapidly distribute SOI and key material. In addition, the ANCD provides radio operators the capability to load all FH and COMSEC data plus sync time into the SINCGARS radio in one simple procedure. Phase I consists of two functional elements:

ACMES Workstation

6-19.   The workstation generates SOI and FH data and integrates COMSEC cryptographic keys. The workstation consists of the AN/GYK-33A, lightweight computer unit (LCU), a rugged desktop computer (486 processor), and the AN/CSZ-9, random data generator (RDG). The LCU, in conjunction with the RDG, generates SOI and FH data (transmission security key (TSK), net identifiers (IDs), and hopset). The ACMES workstation replaces the AN/GYK-33 basic generation unit (BGU). Workstations with RDGs are organic to corps, divisions, and separate brigades. Workstations without RDGs are organic to subordinate brigades and separate battalions.

ANCD, System Designation AN/CYZ-10

6-20.   The ANCD is an electronic data storage and CFD procured by the National Security Agency (NSA) and configured by the Army with unique application revised battlefield electronics communications system (RBECS) communications-electronic operating instructions (CEOI), data transfer device (DTD) software (RDS), and keypad. The ANCD, in conjunction with the integrated communications security (ICOM) SINCGARS, performs the full range of CNR cryptonet support functions to include COMSEC key generation, transfer, and storage. In addition, the ANCD serves as an electronic SOI and replaces the need for most paper SOI products. The ANCD replaces the KYK-13, KYX-15, MX-18290, and MX-10579 in support of SINCGARS.

ACMES (Phase II)

6-21.   ACMES (Phase II) is a follow-on system with enhanced and expanded capabilities. Phase II consists of three functional elements:

ACMES Workstation

6-22.   The Phase-II workstation provides commanders with a fully automated capability to plan, control, and generate FH data and COMSEC keys and manage complex cryptonets. The Phase-II ACMES workstation provides cryptonet managers with the means to distribute cryptographic keys, SOI, and FH data; audit trail databases, design crypto nets; accomplish net configuration; accommodate key supersession; and manage all operational keys and SOI. This workstation is fully interoperable with all electronic key management system (EKMS) elements. A key processing equipment (KPE) will replace the RDG for FH data generation and SINCGARS and ANCD for COMSEC cryptographic key generation.

ANCD

6-23.   The Phase-II ANCD is a software-improved version of the Phase I.

Key Distribution Device (KDD)

6-24.   The KDD ANCD is a limited keypad version of the DTD. Its application software can perform the tasks performed by an ANCD without network control station (NCS) functions.

SIGNALING BY INTERNATIONAL CODE FLAGS

6-25.   Use visual signals when your radio goes out or radio silence is ordered. Also use them when you need to get a message to that ship you are unloading or to the next boat in your convoy. Sending someone over in a boat is a possibility, but not usually a good idea. There are different types of visual signals. This paragraph will give you the basic information you will need to know about the international signal flags and flag hoist methods. Although pyrotechnics (signal flares) are used less often, they are covered in another chapter.

FLAG HOISTS

6-26.   Signaling by flag hoist is a method of communication in which a set of flags of different patterns and colors is used. The set consists of 26 alphabetic flags, 10 numeral pennants, 3 substitutes, and 1 answering pennant. There are six single letter flag hoists that all crewmembers should immediately recognize (Figure 6-1). These signals warn the mariner of danger or are an urgent request for assistance. Any vessel seeing one of these signals will immediately take the proper action. Even though these six flags warn of danger, mariners should know all 26 signal flag hoist meanings from memory. Except for proper names, the international signal alphabetical flags are used only to send messages by code. Each flag has a meaning by itself in addition to the alphabetical meaning. Each flag will also have a different meaning when used with another flag. For example, the "A" flag by itself means "I have a diver down; keep well clear at slow speed." If you hoist two flags that read "AC," it means "I am abandoning my vessel." Two and three letter signals are described in Publication No. 102, International Code of Signals.

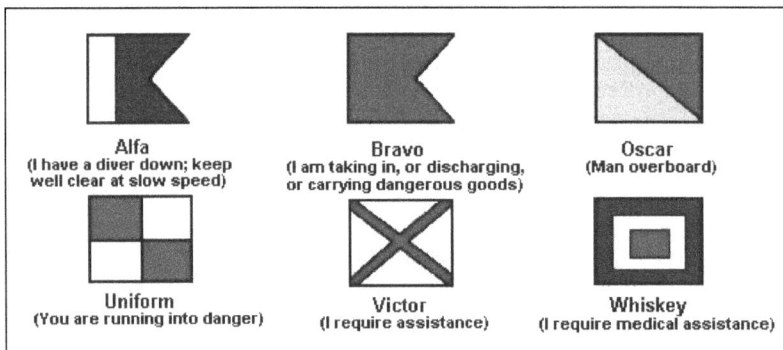

Figure 6-1. Urgent Code Flags

6-27.   It is possible to communicate with several ships at one time if the flags are visible to all. Simple signals for towing and other activities may be worked out between vessel masters to make routine movements easier. This method of signaling is slow and not suitable for the transmission of long messages. Flag hoists may be used only for short distances and cannot be seen in heavy weather or darkness.

USING THE FLAG HOIST SYSTEM

6-28.   Do the following to use the flag hoist system:
- As transmitting vessel, hoist signals where they can best be seen by the receiving ship and make sure they blow out clear and are free from smoke.
- Fly each hoist or hoists until answered.

Note: A signal is superior to another when hoisted first or in time or position.

- When more than one signal is shown on the same halyard, separate each one from the other by tack lines. Always read from the top down.
- When several hoists are displayed at the same time, read in this order: masthead, triatic stay, starboard yardarm, and port yardarm.
- When more than one hoist is shown on the same yardarm, read from outboard to inboard (Figure 6-2).

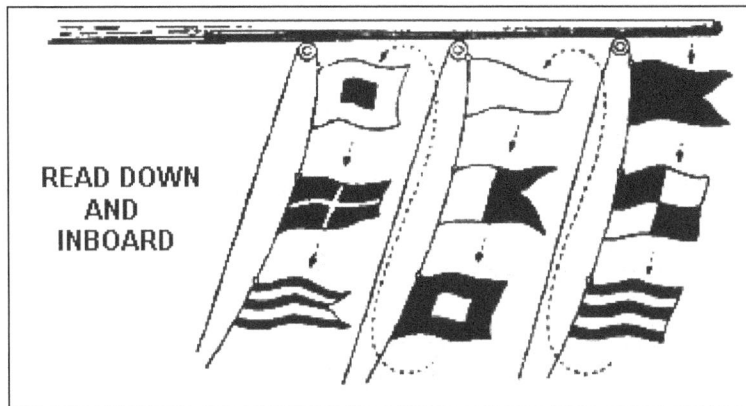

Figure 6-2. Order of Reading Flag Hoists on Yardarm

ANSWERING FLAG HOIST SIGNALS

6-29.   Do the following to answer flag hoist signals:
- Hoist the answering pennant at the dip as soon as you see each hoist. As soon as you understand the signal, immediately close up the pennant.
- Lower the answering pennant to the dip as soon as the hoist is hauled down; close up again when the next signal is understood.
- Keep the answering pennant at the dip if you do not clearly see the message. If you can see but cannot understand the message, hoist "ZQ -- Your signal appears incorrectly coded. You should check and repeat the whole," or hoist "ZL -- Your signal has been received but not understood."

FLASHING LIGHT SIGNALS

6-30.   Sending messages by flashing lights, using the Morse code, is one form of visual communication from ship-to-ship and from ship-to-shore. The flashing or blinker light has several advantages. It may be used when a radio is not available or when security prevents the use of radio. Brief messages may be sent with considerable speed. A portable flashing light is useful on small craft where size and construction prevent the installation of elaborate equipment. There are certain disadvantages in the transmittal of messages by blinker light. This method is not very good for sending long messages because it is comparatively slow. Range is limited even under ideal conditions. Atmospheric and light factors may prevent its use.

INTERNATIONAL MORSE CODE

6-31.   Morse code (Figure 6-3) is a system of signaling using a series of long and short light flashes or sounds. It is still the only way flashing lights can be used, and in that form is used on almost all Army vessels. All deck personnel should memorize Morse code.

A ·—	N —·	1 ·————
B —···	O ———	2 ··———
C —·—·	P ·——·	3 ···——
D —··	Q ——·—	4 ····—
E ·	R ·—·	5 ·····
F ··—·	S ···	6 —····
G ——·	T —	7 ——···
H ····	U ··—	8 ———··
I ··	V ···—	9 ————·
J ·———	W ·——	0 —————
K —·—	X —··—	
L ·—··	Y —·——	
M ——	Z ——··	

Figure 6-3. International Morse Code

THE BLINKER LIGHT

6-32.   Blinker lights vary in size, shape, and power source, yet they all work the same way. One type of blinker, the Aldis gun (Figure 6-4) consists of a tube closed at one end with a light inside. The light is turned on and off by a trigger-operated switch. The portable type blinker can get its electric power from either batteries or the ship's electrical system. The blinker lamp is fitted with a control knob that is used to dim or brighten the light.

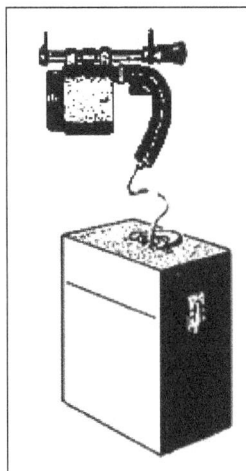

Figure 6-4. Aldis Gun

PROCEDURE SIGNS (PROSIGNS)

6-33. Prosigns are a form of visual communication shorthand. They give in brief form certain orders, requests, instructions, and other information that often comes up in visual communications. Figure 6-5 shows some of the more common ones that may be used. Those prosigns that are overscored are sent as one signal without a break between the letters.

A bar over the letters composing a signal denotes that the letters are to be made as one symbol.	
A̅A̅ A̅A̅ A̅A̅ etc.	Call for unknown station or general call.
E̅E̅E̅E̅E̅ etc.	Erase signal.
A̅A̅A̅	Full stop or decimal point.
T̅T̅T̅T̅ etc.	Answering signal.
T	Word or group received.
A̅R̅	Ending signal or end of transmission or signal.
A̅S̅	Waiting signal or period.

Figure 6-5. Procedure Signs

SENDING BY FLASHING LIGHT

6-34. The flashing light system uses the international Morse code. This is the dot-and-dash system. Most people do not use the term "dots and dashes." Instead they will say "dits and dahs;" the "dah" rhyming with "baa." It is much easier to learn the code by calling out the short flashes of light as "dits" and the longer light flashes as "dahs." When you hear the code called out in this manner, it makes for an easy rhythm.

Note: The dots are one unit long and the dashes are about three units long. The space between the dots and dashes of a letter is one unit; between letters, three units; between words, seven units.

HINTS FOR BETTER SIGNALING

6-35. The following are a few hints you can use when signaling:
- Never send blinker faster than you can receive.
- Have your message written out before you start to send.
- During daylight hours, aim the signal lamp directly at the receiving ship.
- At nighttime, aim the signal lamp at the water, just below the receiving ship's waterline. During times of darkness, you will NEVER aim the signal light onto the other ship's bridge.
- Look to one side when receiving blinker at night. DO NOT look directly at the light.

Note: There are filters available that can be attached to the flashing light during hours of darkness. Then the light can be directed at the receiving station.

6-36.   Write each letter as soon as it is received. If you miss a letter, go to next letter. When the word is completed, you can go back and fill in the space. If you try to figure a letter in the middle of a word, you are apt to miss the next letter or letters being sent.

DISTRESS, URGENT, AND SPECIAL SIGNALS

6-37.   Sailing can be hazardous. Even with today's modern electronic aids to navigation and worldwide radio and satellite communications systems, ships still sink. If you get into trouble, your greatest aid will be the distress signals you send out over the radio. International agreements and U.S. laws have set up special frequencies that will be used for nothing but distress signals. There are also special message formats and key words to use if you are in trouble.

FREQUENCIES

6-38.   Depending upon the type of radio you have, there are different frequencies to use for distress signals. Army TC tactical radios cannot be tuned to any of the frequencies. These frequencies are to be used FOR EMERGENCY CALLS ONLY. Once you have established contact, you will be told to change to another frequency to arrange any necessary rescue. The most widely used emergency frequencies are 2182 kHz and 156.8 MHz on the marine radio (channel 16). For vessels equipped with the longer range CW sets, 500 kilocycles has been established as the emergency frequency. Most vessels and shore stations with the capability will keep watch on the 500 KC frequency, so they can relay or respond immediately if an emergency arises. The 500 KC frequency can also be used for safety and urgent advisory messages. An international silent period on this frequency has been established to enable any vessel in distress to have a clear channel for requesting help. The silent period is observed twice every hour, from 15 to 18 minutes after the hour and from 45 to 48 minutes after the hour. Do not make any calls during these periods unless distress, urgent, or safety matters are involved. If you are transmitting a routine call on 500 KC and notice that you are running into the silent period, transmit "AS" which means "Wait." When the silent period has ended, you may again resume transmission. In addition to being very careful not to transmit on 500 KC during the silent period, you should listen carefully on that frequency for distress messages. If you hear any station making illegal transmissions during the silent period, the nature of the transmission and the call sign of that station should be noted in your radio log.

DISTRESS PROCEDURES

6-39.   The distress signal MAYDAY indicates that a ship or aircraft is threatened by grave or imminent danger and requests immediate assistance. The distress call has absolute priority over all other transmissions--and need not be addressed to any particular station. If you hear a distress call, immediately cease transmissions that might interfere with the distress traffic and continue to listen on the frequency over which the call was heard. Distress transmissions are normally made on the distress frequencies 156.8 MHz (channel 16) or 2182 kHz. They may be handled over other frequencies if the need arises. A distress call consists of the following:

- Distress signal MAYDAY, spoken three times.
- The words THIS IS.
- Call sign of the distressed unit, spoken three times.

Example
"MAYDAY, MAYDAY, MAYDAY
—THIS IS—LCM 2348, LCM 2348, LCM 2348."

6-40.   Normally the distress message will immediately follow the call. Be prepared to copy all information heard. A distress message consists of the following:

- Distress signal MAYDAY.
- Distress unit's call sign. Particulars of position, nature of distress, type of assistance desired, unit's description, persons on board, and any information that might aid the rescue.

> ### Example
>
> "MAYDAY—LCM TOO THUH-REE FO-WER ATE—MY POSITION IS TOO MILES WUN TOO SIX DEGREES TRUE FROM WINDY POINT—I LOST MY POWER AND THE SEAS ARE GETTING ROUGH--REQUEST COAST GUARD ASSISTANCE—I AM AN ARMY LANDING CRAFT—THUH-REE PERSONS ON BOARD—OVER."

DISTRESS MESSAGE REPETITION

6-41.   The distress message, preceded by the distress call, should be repeated until you receive an answer. The repetitions should be preceded by the alarm signal whenever possible. If you receive no answer to a distress call on a distress frequency, the message may be retransmitted on any frequency available on which attention might be attracted.

ATTRACTING ATTENTION TO A DISTRESS CALL

6-42.   If you receive a distress call and are unable to make contact with the distressed unit, take all possible action to attract the attention of stations in a position to give assistance. Also pass along as much information concerning the call as possible.

RECEIPT OF DISTRESS MESSAGES

6-43.   When a distressed unit is in your vicinity, answer the message immediately. However, if the unit is some distance from you, pause a few moments to allow ships or stations nearer the scene to answer. In areas where communications with one or more shore stations are practicable, ships should wait a short period of time to allow them to answer. Receive distress messages in the following manner:

- The call sign of the unit in distress, spoken three times.
- The proword THIS IS.
- The call sign of the unit acknowledging receipt, spoken three times. The words RECEIVED MAYDAY.
- Request essential information needed to effect assistance. Obtain less important information in later transmission.
- Inform unit to stand by.
- The proword OVER.

6-44.   If you receive distress traffic, you should do the following by the most rapid means:

- Forward distress information to the harbormaster or higher HQ.
- Set a continuous radio watch on frequencies of the distress unit.
- Maintain communications with the distressed signal.
- Maintain distress radio log.
- Keep higher HQ informed of new developments in the case.
- Place additional men (if available) on watch (if necessary).
- Obtain radio direction finder bearing of distressed unit if equipment and conditions permit.

VISUAL INTERNATIONAL DISTRESS SIGNALS

6-45.   A vessel or seaplane that is in distress can use the following distress signals to tell other vessels or people ashore that help is needed. The vessel in distress may use more than one distress signal. The distress signals are from the international regulations for preventing collisions at sea. The following signals are recognized by all maritime nations:

- Firing rockets or shells, which throw red stars once every minute.
- Constant sounding of any fog signal apparatus.
- Creating flames on the vessel (burning a tar barrel, oil barrel, and so on).

- Firing a gun or other explosive, once every minute.
- A piece of orange-colored canvas with either a black square and circle or other appropriate symbol for identification from the air.
- Firing a rocket parachute flare with a red light.
- A dye marker of any color.
- Hoisting the international signal flags which indicate "November Charlie." November: White and blue checkered flag. Charlie: horizontal striped (blue, white, red, white, blue) flag.
- Hoisting a square flag with a ball, or something that looks like a ball, above or below it.
- Standing on deck with arms outstretched to each side slowly and repeatedly raising and lowering arms.
- Igniting an orange smoke signal.

PRIORITY OF MESSAGES

6-46.   There are several types of messages which are peculiar to marine communications and which you must be thoroughly familiar. By international agreement, the order of precedence for these messages is as follows:

- Distress call (including the auto-alarm distress signal), distress messages, and distress traffic (SOS or MAYDAY).
- Urgent signals and messages (PAN).
- Safety signals and messages.
- Communications relative to radio direction-finding bearings.
- Communications about the navigation and safe movement of aircraft.
- Communications about the navigation, movements, and needs of ships, including weather observation messages for an official weather service.
- Government communications for which priority rights have been claimed.
- Service communications relating to the working of the radio communication service or to communications previously transmitted.
- All other communications.

6-47.   You should know this precedence, but you should remember that your primary responsibility is the proper handling of military traffic. If it does not interfere with the proper completion of your military mission, you may assist commercial stations (but only with the permission of the vessel master). Except in cases of emergencies, operators aboard Army vessels are not authorized to transmit commercial or personal messages.   Only official messages may be handled and they should be transmitted only through Government facilities. Under certain conditions, it may be necessary for you to send a message through a commercial station, especially if distress, urgent, or safety matters are involved. Commercial stations make no charge for handling such messages.

URGENT MESSAGES

6-48.   The urgency signal PAN (pronounced PAHN) indicates that the station calling has an urgent message concerning the safety of a vessel, aircraft, or person on board or within sight. Send the signal and message on a distress frequency (156.8 MHz [channel 16] or 2182 kHz) or any other frequency that may be needed to get the required help. The urgency signal has priority over all other communications except distress traffic. The message preceded by the urgency signal is usually addressed to a specific station. However, it may be addressed to ALL STATIONS. If you hear the signal, listen on that frequency for at least 3 minutes. If nothing is heard following the urgency signal, you may resume normal communications. Do not interfere with urgent traffic. Normal work may continue on frequencies other than that on which the urgency signal was heard provided the message was not addressed to ALL STATIONS. The urgent message should contain all details concerning the particular case and be in plain language form. If you receive an urgent message, deliver it by the most rapid means to your next higher HQ or harbormaster.

CANCELLATION OF URGENT TRAFFIC

6-49.   When the urgency signal has been sent before transmitting a message to ALL STATIONS, which calls for action by the stations receiving the message, the station responsible for its transmission will cancel it when action is no longer necessary. This message of cancellation shall likewise be addressed to ALL STATIONS.

SAFETY MESSAGES

6-50.   The safety signal consists of the word SECURITE (pronounced SAY-CURE-E-TAY). It means that the station is about to send a message concerning the safety of navigation or it is giving important weather warnings. When you hear this message, inform your vessel master at once. The safety signal and call will be sent on the distress frequency or one of the frequencies that may be used in case of distress.

SAFETY CALL AND MESSAGE

6-51.   The safety message will normally be sent on a working frequency, but an announcement to this effect will be made at the end of the silent period on the distress frequency.

Example: (Preliminary call on distress frequency)

"SECURITE SECURITE SECURITE—HELLO ALL STATIONS—THIS IS (Voice call sign twice) COAST GUARD MARINE INFORMATION BROADCAST (or) HURRICANE ADVISORY/STORM WARNING, and so forth, LISTEN (2670 kHz and/or Channel 22A) OUT."

6-52.   When you hear the safety signal, listen to the safety message until you are satisfied that the message is of no concern to you. Do not make any transmission that will likely interfere with the message.

WEATHER BROADCASTS

6-53.   Radio stations specializing in weather broadcasts are operated by the United States Weather Bureau. These stations have a limited range so the weather forecasts they send out are usually local weather. Your Marine band radio has a special weather channel for these weather broadcasts. As you move along the coast, you will find that the stations fade and are replaced with different ones. When overseas, you will have to rely upon local arrangements for weather forecasts, usually obtainable through your company operations section from the Battalion S2.

PUBLICATIONS

6-54.   The following publications are needed for shipboard communications.
- International Code of Signals (Publication 102). This publication lists all the internationally recognized signals, codes, distress signals, and rules to be employed by vessels at sea to communicate a variety of information relating to safety, distress, medical, and operational information. Each signal has a unique and complete meaning. This publication also contains the internationally recognized message formats and complete instructions for all the forms of communication. Publication 102 is published in several different languages to make it easier to communicate with the crew of foreign vessels.
- Radio Navigational Aids (Publication 117). This publication is a selected list of worldwide radio stations which perform services to the mariner. Though this publication is essentially a list of radio stations providing vital maritime communication and navigation services, it also contains information which explains the capabilities and limitations of the various systems.

EMERGENCY RADIOTELEPHONE PROCEDURES

6-55. If you are in distress, that is, if grave and imminent danger threatens you, transmit your emergency on the International distress frequencies: 2182 kHz and 156.8 MHz (channel 16). If you are merely having difficulty (for example, engine trouble, steering failure, and so forth) and need help, the Coast Guard can be reached by calling on either of the two distress frequencies. The distress call sent by voice radio consists of the following:

- The distress signal MAYDAY spoken three times.
- The words "THIS IS" or the letters "DE," (spoken as DELTA ECHO in case of language difficulties) and your vessel's call signal and name.

6-56. If you are not in immediate danger, you will be shifted to a common working frequency for further communications. This keeps the distress channel open for other emergencies. After you have made contact, speak slowly and clearly to avoid confusion and delays. Give the following additional information:

- Your vessel's position in latitude/longitude or true bearing and distance in nautical miles from a widely known geographical point. Avoid using local names that are known only in the immediate vicinity, as they can be confusing (for example, "buoy 19" or "the rocks").
- The nature of the distress or difficulty.
- The kind of help needed (for example, medical, air evacuation, damage control, and so forth).
- The number of persons aboard and the condition of any injured.
- The present seaworthiness of your vessel.
- A full description of your vessel including length, type, cabin, masts, power, color of hull, superstructure and trim.

6-57. The voice radio alarm signal, if available, should be transmitted for about 1 minute before the distress call. The voice radio alarm signal consists of two audio tones of different pitches transmitted alternately. The radiotelegraph alarm signal is 12 dashes. Each dash is 4 seconds in duration with 1 second of silence between dashes. As in voice radio, the alarm signal comes before the SOS. The purpose of these two alarms is to attract the attention of persons on watch and should be used to announce that a distress call or message is about to follow. Some Army and most commercial vessels are fitted with auto-alarms. The auto-alarm is placed in operation when the radio operator is not on watch while the ship is at sea. If the alarm is sounded, bells will ring on the bridge, in the radio room, and in the radio operator's cabin. This alarm responds to the alarm signals sent by a vessel in distress.

Chapter 7

Marine Emergencies

INTRODUCTION

7-1.   Fire, sinking, or injuries are constant dangers faced by crewmembers aboard ships at sea. The organization, training, and teamwork of the crew usually determine the difference between a marine emergency and a marine disaster. The emergency training that is given to the crew is the direct responsibility of the ship's master. This responsibility is the same for the coxswain on the LCM-8 as it is for the master of a category A-2 vessel. This chapter discusses the "how" and "what to do" during a shipboard emergency. Learn now—not during the emergency. Teamwork is essential for survival.

STATION BILL

7-2.   The starting point for shipboard survival and survival training is the station bill. The station bill is a muster list that is required by federal regulations. It lists the emergency duty station and duty position for each crewmember assigned aboard ship and also the signals for fire and abandon ship. The station bill is prepared and signed by the ship's master. Each time a new master is assigned to the ship, one of his/her first responsibilities is to prepare a new station bill. When a new crewmember is assigned aboard ship, the crewmember will be assigned to a specific line and station bill number. When transferred, the crewmember's name is removed from the station bill. The ship's master is the only one who can sign the station bill. It is also his/her responsibility to keep it current. Copies of the station bill are posted in conspicuous places in the ship, such as the crew's quarters, crew's mess, and bridge.

FILLING IN A STATION BILL

7-3.   The following information should be included on a station bill (see also FM 4-01.502):
- Vessel's name or number.
- Date station bill was filled out.
- "U.S. Army" or "Name of company".
- Master's signature.
- A numerical listing for each man authorized aboard the vessel. The Master is listed as A, the Chief Mate is number one.
- Crew rating and crewmember's name. The crew rating is listed according to precedence in rating and department. If carried, the sequence for departments is deck, engine, radio, stewards, and medical.
- Location and specific emergency duty to be performed by crewmember.
- Specific lifeboat assigned to crewmember.
- Specific location and task to be performed by crewmember.

STATION BILL CARD

7-4.   The crewmember will also be issued an individual station bill card. This is usually posted next to the crewmember's bunk. The card will list the crewmember's station bill number, name and rating, fire and emergency station, lifeboat number, abandon ship, and boat station.

EMERGENCY DUTIES

7-5.   The emergency duties assigned to a particular crewman should, whenever possible, be similar to the normal work activity of that person. For instance, steward department personnel should be assigned to assist passengers; deck department personnel should be assigned to run out hoses and lifeboats; and the engineering department should be assigned to run out hoses in the machinery space with which they are most familiar.

EMERGENCY SQUAD

7-6.   An emergency squad is a group of crewmembers selected by the master for their special training to deal with emergencies. The chief mate (assisted by the boatswain) is normally in command of the emergency squad. The rest of the squad should be made up of crewmen trained in the use of fire, emergency, and rescue equipment. Candidates for the emergency squad would be crewmembers who are highly knowledgeable in emergency procedures. A mustering location for the emergency squad should be included in the station bill. The mustering location could be on either wing of the bridge, at a designated position on the main deck, or wherever the master feels would be best. However, the chosen location should be one that the members of the squads can reach promptly—for example, in less than 2 minutes.

EMERGENCY SQUAD TRAINING

7-7.   An emergency squad is a team. A team is a group of people brought together to accomplish a common goal. The word team brings to mind the words: coordination, cooperation, and training. Training is absolutely essential, since without it there can be little coordination or cooperation. Training consists basically of two parts and must be taught in the following order:
- A teaching-learning process in which the necessary knowledge is communicated to the trainee.
- Practice and demonstration of the necessary skills, using the proper equipment. As an example, fire drills are practice and demonstration sessions. They must come after crewmen have learned what to do; otherwise, they can serve no purpose except to reinforce bad habits.

7-8.   Under an able and understanding leader, proper training will gradually produce coordination and cooperation among members of the emergency squad. After several practice sessions they will be operating as a team. The master is responsible for all ship's functions, including those he/she assigns to subordinates. Although the master assigns the training of the emergency squad (and the rest of the crew) to his/her chief mate, he/she should review and approve the plans for proposed lessons and drills. These sessions are made more meaningful when the master personally observes them and then discusses them with those in charge. The members of the emergency squad should attend periodic instructional sessions dealing with the variety of emergencies that could occur aboard ship. At each session, a problem could be presented, solutions discussed (until a satisfactory one is found), and the necessary tools and equipment should be handled for familiarity. Then the regularly scheduled fire drills would be demonstrations of efficiency rather than training sessions.

CREW FIRE FIGHTING TRAINING

7-9.   The emergency squad may be called upon to deal with many emergencies, such as collision, man overboard, and a lost or damaged rudder. When the fire signal is sounded, all hands are involved. The station bill lists an assigned task and station for each member of the crew. Therefore, all crewmembers should receive some training in fire fighting.

ABANDON SHIP PROCEDURES

7-10.   During all shipboard drills and emergency operations, crewmembers must wear their life jackets. It is one of the most important pieces of equipment for survival in the water. It will hold you in the upright floating position without you having to swim. Another safety point during a drill or the real thing is to always wear a hat or some type of headgear to protect you from the elements.

DONNING THE LIFE JACKET

7-11.   If you have time, put on extra clothing. Include an outer layer of wind and waterproof clothing fitted if possible with head-cover and gloves. Then put on the life jacket.

Note: Practice putting on and securing your life jacket until you are able to don and secure it within 2 minutes.

ENTERING WATER FROM A HEIGHT WEARING A LIFE JACKET

7-12.   Make sure that your jacket is well secured. If it is not well secured, you could hurt your head when you jump. Then get down to a height of less than 30 feet if you can. Below 15 feet is ideal. If you jump from higher than 30 feet, you can hurt yourself (this depends on the height from which you jump and the angle at which your body hits the water). If worn, remove false teeth, eyeglasses, or contact lenses. Also remove any sharp objects from your pockets. Get in the jump position (see Figure 7-1) and do the following:

- Stand on the gunwale and check the water for debris.
- Check to see if the life jacket is tied and all the straps are secured.
- Hold your nose and cover your mouth with your left hand.
- Cross over your left hand with your right hand and hold the life jacket collar securely.
- Hold your elbows into your side as much as possible.
- Keep head and eyes straight ahead. Do not look down.
- Take one step out using either foot.
- Bring your trailing leg up behind your leading leg so that they cross at the ankles. This will protect you if you should land on any floating debris.

Figure 7-1. Jumping in Water

- Get away from the ship once you are in the water. Swim as slowly as possible toward the survival craft. DO NOT swim or thrash about any more than you need to because of the following:
  - You will lose your body heat.
  - You will lose your strength. You will need all your strength to pull yourself up and into the survival craft.
- Use your life jacket to support you in the face-up position.

DROWNPROOFING

7-13.   Drownproofing, also called water survival, is based on the natural buoyancy of the human body when the lungs are filled with air. It is intended to keep anyone alive in the water indefinitely, even a nonswimmer who is fully clothed. Drownproofing saves energy for the potential drowning victim. It is much easier to do the steps on drownproofing for long periods of time than to stay afloat by swimming. Each crewmember should know drownproofing since it is an excellent way to stay afloat without a life preserver. This method can best be described in the following five steps.

- **Step 1. Resting Position.** The swimmer takes a deep breath and then sinks below the surface. The face is kept down with the back of the head even with the water surface (Figure 7-2). In this position, he/she will sink no deeper.

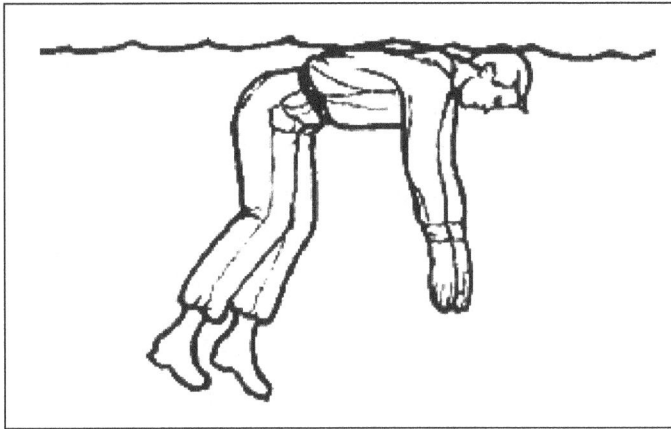

Figure 7-2. Resting Position

- **Step 2. Preparing to Exhale.** When ready for another breath (in about 6 to 10 seconds), maintaining the body and head position as shown in Figure 7-3, the swimmer slowly lifts the arms to about shoulder height. The legs slowly separate into a scissors-type kick.

Figure 7-3. Preparing to Exhale

- **Step 3. Exhalation.** The head of the swimmer is raised just high enough for the mouth to be out of the water (Figure 7-4). The swimmer now exhales through the nose, the mouth, or both. To give the swimmer bearing, the eyes should be open.

Figure 7-4. Exhalation

- **Step 4. Inhalation.** As the head becomes vertical, the swimmer presses his/her arms downward and brings his/her legs together (Figure 7-5). The air is then inhaled through the mouth. The action of arms and legs should be done slowly.

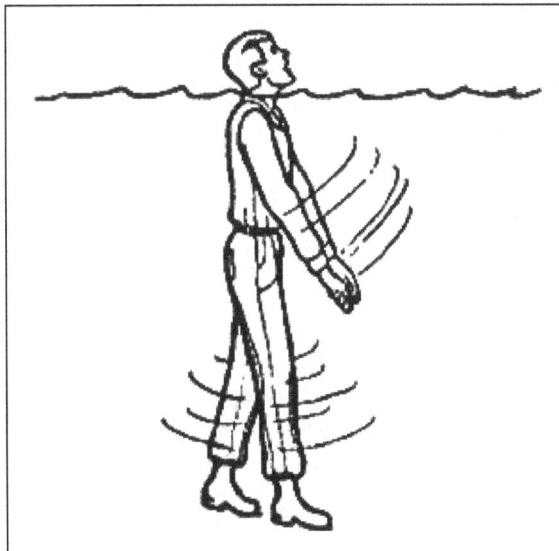

Figure 7-5. Inhalation

● **Step 5. Return to Rest Position.** The swimmer relaxes his/her arms, and at the same time his/her legs move back to a dangling position. The swimmer's face goes back into the water and he/she "rests" once again (Figure 7-6). The cycle is then repeated. While most persons can master drownproofing easily, skill is involved in breathing close to the water and instruction is necessary.

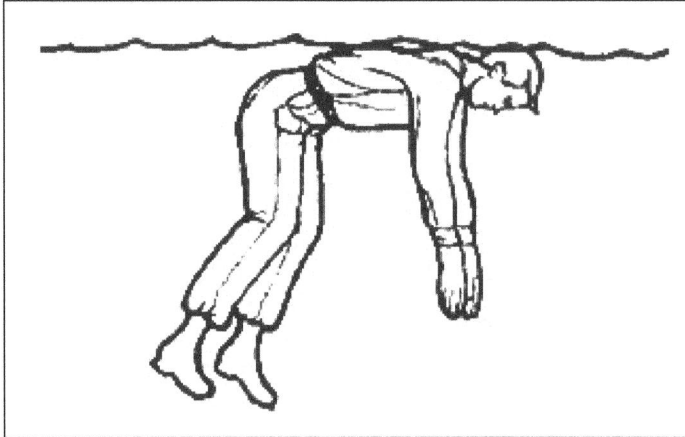

Figure 7-6. Return to Rest Position

TRAVEL STROKE

7-14.   This stroke is used in a water survival situation when you are required to swim, while conserving as much energy as possible. Here is how it is done:
● Enter the water.
● Take a deep breath.
● Put your face in the water, arms at your side, feet together, and body horizontal.
● Prepare to breathe; move your hands up alongside of your body to a position in front of your forehead and palms down. At the same time, spread your legs in scissors fashion in preparation for a kick.
● Kick and exhale. Bring your feet together quickly and exhale through your nose and mouth. Raise your head slowly out of the water.
● Stroke and inhale. Stroke a heart-shaped stroke with your hands, then bring your hands back in front of your chest; at the same time, inhale through your mouth.
● Put your head back in the water and spread your legs for another kick.
● Kick and level. Extend your hands out in front and at the same time kick, bringing your feet together.
● Stroke and glide. With your elbows locked, sweep your hands to the side.
● Continue the glide until your feet start to drop or another breath is required.
● Repeat the process.

SWIMMING THROUGH A THICK OIL FIRE

7-15.   The most important thing to remember for your survival if you are forced to swim through a thick oil fire is to keep calm. The proper procedure for swimming through a thick oil fire is described in the following steps:

```
┌─────────────────────────────────────────────────────────────┐
│                        WARNING                                │
│                                                               │
│   NEVER wear a life jacket while swimming in a thick oil fire.│
└─────────────────────────────────────────────────────────────┘
```

- **Step 1.** Enter the water on the windward side of the vessel (windward is the direction from which the wind blows), feet first with one hand over the nose and mouth and the other hand covering the eyes.
- **Step 2.** Level out under the surface of the water and start swimming.
- **Step 3.** When you must breathe, surface in a straight up-and-down position with your hands extended above your head (Figure 7-7).

Figure 7-7. Break the Surface

- **Step 4.** Exhale about 75 percent of the air in your lungs before breaking the surface.
- **Step 5.** As soon as your hands break the surface, start beating away the burning oil with a circular thrashing motion.
- **Step 6.** Fully inhale before submerging (Figure 7-8).

Figure 7-8. Fully Inhale

● **Step 7.** Continue swimming in this manner until you are clear of the burning oil (Figure 7-9).

Figure 7-9. Swimming Through Thick Oil Fire

SWIMMING THROUGH A THIN OIL FIRE

7-16. The most important thing to remember for your survival if you are forced to swim through a thin oil fire is to keep calm. The proper procedure is shown in Figure 7-10 and described in the following steps:

WARNING

Wear your life jacket at all times while swimming through a gasoline or thin oil fire. Keep your head above water at all times.

● **Step 1.** Enter the water from the windward side of the vessel, feet first with one hand over the nose and mouth and the other hand covering the eyes.
● **Step 2.** Bring your hands up in front of your face, elbows extended with the palms halfway out of the water.
● **Step 3.** Push the water out in front and continue the stroke until your arms are straight out from the body.
● **Step 4.** Continue swimming until you are out of danger.

Figure 7-10. Swimming Through Thin Oil Fire

COLD WATER VICTIMS

DROWNING DATA

7-17.   After reading the following, you and your crew, as would-be rescuers, should be more willing to attempt to revive a person who is supposedly "drowned." Due to recent medical research, it has been discovered that the bodies of people "drowned" in cold water (below 7 degrees F, 14 degrees C) may go into a diving reflex. In this condition, the nervous system cuts off the flow of blood to all parts of the body except the brain and lungs. The heart slows so much that it cannot be heard without special instruments. The result is that a person can exist in seemingly a "dead" state for up to an hour, depending on their age and the temperature of the water. The basic trigger that starts the diving reflex is cold water touching the face, specifically the area around the eyes and forehead. However, the diving reflex does not always work. Studies show that the person's age combined with the temperature of the water are the main factors in deciding whether the reflex will start, and, if so, how long it will be effective. Its effectiveness is measured by how long it works before permanent brain damage begins. The reflex is extremely active in youngsters. In infants and small children, it can be started by a water temperature of 65 degrees F (20 degrees C) and can, in theory, last for as long as an hour. As a person gets older, the water must be colder to start it and it is effective for a shorter time. The diving reflex may be one of those natural systems, which protects small children from their own inexperience. A person's body weight also comes into play when the reflex is connected with hypothermia.

7-18.   Hypothermia simply means that the body temperature is below normal. However, when you take your temperature and it is 97.5 F instead of 98.6 F, it does not necessarily mean that you are a hypothermia victim. Hypothermia usually refers to the lowering of inside body temperature because of coldness outside the body such as cold water or a cold wind. Your arms and legs will become numb and you will lose the use of them if your body temperature gets down to about 93 F. When it reaches 80 to 86 F, you may lose consciousness; should it drop to about 79 to 77 F, it becomes fatal. A person's weight is a factor when figuring how long it will take for all this to happen. Generally, the bigger a person is, the longer it takes for his/her body to lose heat because he/she has better insulation. Other factors that affect this heat loss are age, clothing, and physical activity. It is the result of cold rather than the effect of drowning that begins the diving reflex. Hypothermia victims, even those not in cold water, often get some assistance from the reflex. In reaction to the cold, the vital body functions slow down to an almost immeasurable level and thereby save body heat as well as oxygen. Again, this lengthens the time before serious brain damage begins. This extension can make the difference between whether or not a "drowned" person, or hypothermia victim, can be successfully rescued. Tests during World War II revealed that a thin person in a flight suit and life jacket could survive up to 72 minutes in 40 degrees F (5 degrees C) water. However, he/she would be unconscious and apparently dead some time before that. While this knowledge of the diving reflex may be consoling to a person drowning and going down for the final time, it is primarily important to his/her rescuers. Should your vessel be the first one on the scene of a cold water drowning, the things you and your crew do can determine whether the victim lives or dies. Since you can never assume that medical assistance will be on the scene when a drowning victim is pulled from the water, his/her life may depend on you. The diving reflex stops as soon as the victim is taken out of the water. That means that you may have less than 4 minutes to get his/her blood flowing. Some DOs and DO NOTs to remember when reviving cold water drowning victims:

DO

- Start CPR immediately. This is a form of mouth-to-mouth resuscitation and external heart massage. Only a **qualified** person should attempt CPR.

- Keep the victim warm with a light blanket or jacket, and so on, but do not waste necessary time on this.

- Keep giving CPR until medical assistants take over or until the victim revives.

DO NOT

- Give the victim any alcoholic drink.

- Try to re-warm the victim with anything more than a light blanket or jacket. Uncontrolled re-warming can cause severe injury.

FIRST AID

7-19.   Treatment will depend on the condition of the survivor and the facilities available. In more serious cases where the victim is semiconscious or unconscious, contact should be made immediately with a ship or shore medical facility for detailed information on the care and handling of the victim. Administer the following first aid while waiting for medical instructions:

- After removing the victim from the cold water, gently transfer him/her to a warm environment. Rough handling of the victim can cause further harm.

- Remove his/her clothes only if it can be done with a minimum of movement of the victim's body. Do not massage him/her.

- Lay the victim in a face up and slightly head down position, unless vomiting occurs. This is important because a hypothermia victim has low blood pressure, and the head down position ensures an adequate supply of blood to the brain.

- If available, administer warm, humidified oxygen by means of a face mask. The oxygen will not only assist victims if they are having difficulty breathing or have a low respiratory rate, but it will also provide core re-warming. Mouth-to-mouth resuscitation is always advisable if the victim is having problems breathing and no other form of assistance is available.

7-20.   In some cases, you should re-warm the victim actively; in other cases, you should not re-warm him/her at all. Before deciding what to do, you need to know something about the following two different types of hypothermia.

CHRONIC OR SLOW ONSET HYPOTHERMIA

7-21.   This type comes from being exposed (from a few hours to several days) to cold weather-. Most chronic hypothermia cases develop in air temperatures of 30 to 50 degrees F. The victim usually overestimates how long he/she can withstand the cold and fails to recognize the danger of being wet at such temperatures. A victim can get wet from sweat, rain, or from the splash and spray of water from working on the deck of a vessel. Because chronic hypothermia takes some time to develop, the victim may undergo dangerous fluid and biochemical changes. For these reasons, you do not want to re-warm the victim. As with a cold water drowning victim, victims of chronic hypothermia should be taken to a hospital as quickly as possible. REMEMBER, DO NOT RE-WARM A CHRONIC HYPOTHERMIA VICTIM!

ACUTE OR RAPID ONSET HYPOTHERMIA

7-22.   This type of hypothermia is different and is the result of immersion in cold water. Since water can withdraw heat from the body 25 or more times faster than air, we can estimate that in water temperatures of 72 degrees F and lower, the body cannot generate enough heat to offset heat loss to the water. Depending on water temperatures and body condition, acute hypothermia may begin to develop in as little as 10 to 15 minutes. Because of the rapid onset, acute hypothermia victims do not generally have time to develop dangerous chemical and fluid imbalances. Therefore, without delay, begin to RE-WARM ACUTE HYPOTHERMIA VICTIMS IMMEDIATELY. Even conscious hypothermia victims have died following apparently successful rescues when attempts at re-warming were delayed or were inadequate. Any of the following warming methods are recommended, preferably in the order given.

- Place the survivor in a hot shower or bath at 105 degrees to 110 degrees F or a temperature in which an observer can comfortably leave his/her arm. Keep the arms and legs out of the bath. If you are warming a victim in a shower, keep his/her limbs out of the spray to delay the return of blood circulation to the extremities since heating the limbs causes cold blood to flow from them to the body core. This further cools the core. Rather, direct the spray on the center of the back or the chest.

- Apply hot, wet towels or blankets at 115 degrees F to the victim's head, neck, groin, chest, and abdomen. Again, do not attempt to warm his/her arms and legs.

- Apply your own body warmth by direct body-to-body contact with the victim. A blanket should then be wrapped around you and the victim to conserve the heat you are supplying. Unless he/she is in a warm environment, just wrapping a hypothermia victim in a blanket without a heat source is ineffective. This is because he/she is not generating sufficient heat to re-warm himself and the blanket insulates him/her from the warm environment.

COLD WATER SURVIVAL

7-23. The following will teach you how to improve your chances of survival in cold water. As mentioned before, body heat loss is a gradual process and the diving reflex provides some protection. The loss of body heat is probably the greatest hazard to the survival of a person in cold water. Knowing what steps to take to help your body delay the damaging effects of cold stress will help you stay alive in the event of cold water exposure. Try protecting your head, neck, groin, and the sides of your chest. These are areas of rapid heat loss in cold water. Locate and wear personal flotation equipment such as a life jacket. If you are not wearing it when you enter the water, put it on as soon as possible after entering the water. This is probably the single most important item of survival equipment. Survival in cold water is tough enough without having to contend with staying afloat. Learn how the flotation device is worn and used before an accident occurs. Try to enter the water in a lifeboat or raft. This will avoid getting your insulation wet and loss of valuable body heat to the water. Abandoning a ship by means of a lifeboat or raft will greatly increase the chance for survival. This is better than jumping overboard and attempting to be rescued. Wear several layers of clothing. If you are fortunate enough to stay dry and enter the water in a lifeboat or raft, the trapped air within your layers of clothing will provide excellent insulation. However, if you become wet in abandoning your ship, the layers of clothing, although wet, will slow down the rate of body heat loss. If conditions prevent you from abandoning ship in a lifeboat and you must enter the water directly, try to cut down the shock of a sudden cold plunge in the water. Rather than jumping into the cold water, try to lower yourself gradually. A sudden plunge into cold water can cause rapid death as a result of the severe shock to your nervous system. It may also cause an uncontrollable rise in breathing rate resulting in an intake of water into the lungs. If jumping is necessary, try to hold your breath, pinch your nose, and avoid swallowing water during the plunge. The body position you assume in the water is very important in conserving your body heat. Tests show that the best body position is one where you hold your knees up to your chest in a "doubled up" fashion with your arms tight against the side of your chest (Figure 7-11). This position reduces the exposure to the coldwater of your groin and chest sides, both areas of high heat loss. Try to keep your head and neck out of the water.

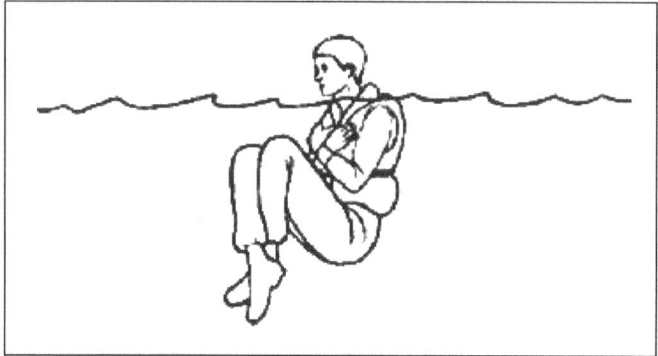

Figure 7-11. Double-Up

7-24.   Another heat conserving position is to huddle closely to one or two others afloat, making as much body contact as possible (Figure 7-12). You must be wearing a life jacket to be able to hold these positions in the water. You should also wear a life jacket in the lifeboat or life raft.

Figure 7-12. Buddy-Up

7-25.   Try to board a lifeboat, raft, or other floating platforms or objects, as soon as possible, in order to shorten immersion time. Remember that you lose body heat about 25 times faster in water than you do in air. Since the effectiveness of your insulation has been seriously reduced by water soaking, you must now try to shield yourself from wind to avoid a wind-chill effect (convective cooling). If you manage to climb aboard a lifeboat, shielding can be accomplished with the aid of a canvas cover, a tarpaulin, or an unused garment. Huddling close to the other occupants of the lifeboat or raft will also conserve body heat. Keep a positive attitude about your survival and rescue. This will improve your chances of extending your survival time until you are rescued.

MAN OVERBOARD PROCEDURES

7-26.   Immediately on seeing a crewmember fall over the side, shout an alarm! Call out the words "Man overboard!" to personnel on the bridge. Be sure to include where on the vessel the person fell overboard. For example:

- On the right side of the vessel call out: "Man overboard, starboard side!"
- On the left side of the vessel call out: "Man overboard, port side!"
- At the front of the vessel call out: "Man overboard at the bow!"
- At the rear of the vessel call out: "Man overboard at the stern!"

7-27.   Immediately after these vocal alarms are given, three things must happen at the same time:

- Get the stern away from the victim.
- Mark the spot.
- Post a lookout.

7-28.   These things do not happen 1-2-3; they are done at the same time. That is why teamwork is a necessity. The first action of the person in charge of the vessel is to get the stern away from the victim. On a small craft it may be necessary to cut the throttle immediately and swing the stern away from the person in the water to avoid hitting him/her with the screws. When the bridge watch hears the man overboard signal, the helmsman must be told immediately to put the rudder hard over to swing the stern away from the victim. If the victim falls overboard on the starboard side, the helmsman would turn the helm "hard right rudder." If the victim falls over on the port side, then naturally the helm will be put to "hard left rudder."

MARK THE SPOT

7-29.   There are two procedures for marking the spot. One is used during hours of daylight the other is for hours of darkness. To mark the spot during daylight—

- Throw a life preserver or life ring immediately.
- Drop a smoke float.
- Get anything that floats into the water near the person that he/she can hang onto.

7-30.   To mark the spot during darkness:

- Immediately throw a life preserver or buoy ring with water lights.
- Keep the vessel's searchlight trained on the victim.

POST A LOOKOUT

7-31.   Keep the victim in sight. It is easy to lose sight of the victim's position, especially in rough weather or at night. The person who saw the victim fall overboard usually makes the best lookout. It is also a good idea for the lookout to be posted on the forward part of the vessel for easier viewing.

RAISE THE OSCAR FLAG

7-32.   The Oscar Flag is raised to let other vessels in the area know that you have a man overboard.

PICK UP THE VICTIM

7-33.   If circumstances permit (such as if you are not limited by narrow channels or landfalls, and so forth), the Williamson turn, used by large vessels, has proven to be the preferred maneuver for picking up victims (Figure 7-13). To make the turn you must do the following:

- Put the rudder over to the side from which the victim fell. This action swings the stern away from that person.
- Hold the rudder hard over until the vessel begins to turn.
- Then steady her on a course about 60 degrees off the original course.

- When the vessel heads on the new course, turn the rudder hard over to the opposite side until she is on a course 180 degrees from the original course.
- Maintain the original speed until the vessel is steady on the reverse course.

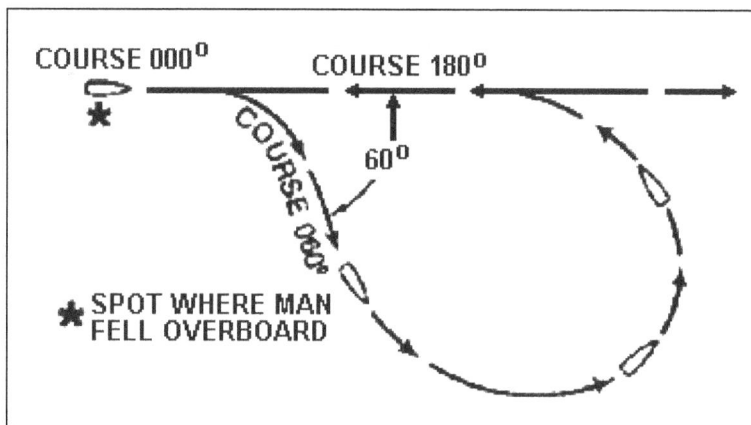

Figure 7-13. Williamson Turn

OTHER MANEUVERS

7-34. On large vessels equipped with lifeboats, a lifeboat is lowered and the lifeboat crew maneuvers into position to rescue the victim. A small vessel, especially one with two screws, often is so maneuverable that it is simpler, safer, and quicker to maneuver the vessel back to the person in the water and throw them a line than to make the Williamson turn. For example, an LCM or LCU could make the pickup at the ramp. The vessel must turn around until its course has been reversed. At slow ahead, when it has been determined that the propellers will not endanger the person in the water, the vessel can be maneuvered toward them. The vessel must approach slowly on the windward side of the victim. If the vessel is placed so that it shields the victim from the waves and the wind, the water around the victim will be calm. However, caution must be exercised to prevent the vessel from coming too close to the victim (Figure 7-14). Lines with life rings must be prepared so they can be thrown as soon as possible. The only time maneuvers of this type may be used at night are when weather conditions make launching a boat impossible.

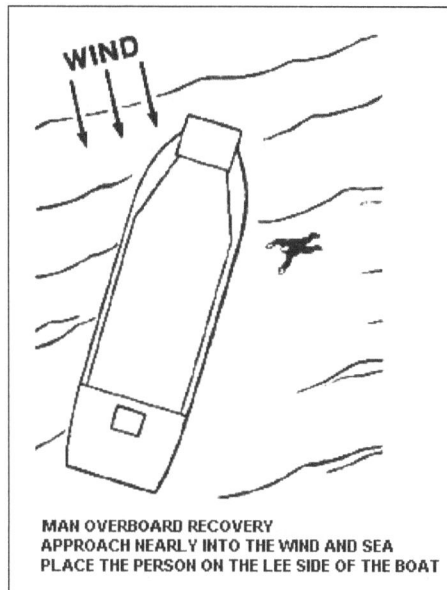

Figure 7-14. Bringing Ship into Wind

7-35.   When having man overboard drills, it is a good idea to identify all crewmembers that are good swimmers, and designate them for special emergency duty. When an overboard victim is unconscious, a good swimmer with life preservers and lifelines can jump in and help with the rescue. The first thing a man overboard should try to do is to get clear of the vessel, especially the stern, so that he/she does not get sucked under or hit by the screw. The overboard victim should stay in the same general area where he/she fell, especially at night and in foul weather. Staying in the same general area will make it easier for the ship's lookouts to spot him/her since they will generally know where to look. Victims of a fall overboard can help the ship's lookouts by:

- Making them more visible.
- Making them heard.

7-36.   A victim can help to make himself more visible by waving his/her arms, a handkerchief, his/her T-shirt (if the water is not too cold) or any brightly colored object he/she might have been holding or wearing when he/she went overboard. If he/she was wearing a life preserver when he/she fell, and he/she does not need it to stay afloat, he/she can float the preserver alongside himself in the water. However, if the water is rough or if he/she is a poor swimmer, he/she should NEVER remove his/her preserver. A victim can make themselves heard by the following:

- Shouting (if in hearing distance).
- Splashing the water (which can also improve his/her visibility).
- Sounding the whistle (if he/she was wearing a life preserver when he/she fell overboard).

7-37.   Up to now, we have covered what to do if an overboard accident happens. Let us go over the different kinds of man overboard accidents, what causes them, and the things we can do to prevent them. Most falls happen while a person is moving, standing, or leaning over the edge of a vessel. However, falls may occur from a wide range of causes which include, but are not limited to the following:

- Limited visibility caused by darkness, fog, or bad weather.
- A sharp turn or acceleration.
- A wave or wake striking the vessel.
- Sitting on a gunwale, the stern, or the bow.
- A slippery surface.

7-38. The following are some things you can do to prevent an overboard accident:

- Do not allow horseplay.
- Make sure everyone uses the handrails.
- Have everyone watch where he/she walks to avoid tripping.
- Make sure lifelines are rigged for crewmembers to use when working near the edge of the vessel.
- Do not allow anyone to sit on gunwales.

LIFE RAFTS

7-39. The inflatable life raft is as important a lifesaving device as the lifeboat. Shipboard drills with the inflatable life raft are not conducted because the raft container is sealed until ready for automatic or manual launching. Therefore, it is important to learn about the current design of rafts and keep informed of future designs.

SIZE

7-40. Inflatable life rafts must be either Navy standard or Coast Guard approved. Life rafts have a range of sizes. Ships that do not make international voyages might have rafts that can hold 4 to 26 people. Ships that make international voyages might have rafts that can hold 6 to 25 people. The capacity (number of persons it will hold) of the life raft is marked on the container and the raft. The manufacturer's name is also shown on the container. An inflatable life raft (complete with case and equipment) does not weigh more than 400 pounds.

STOWAGE

7-41. Life rafts are kept in a cradle on an open deck (Figure 7-15). This is done so they can float free if the ship sinks before you can manually launch the raft.

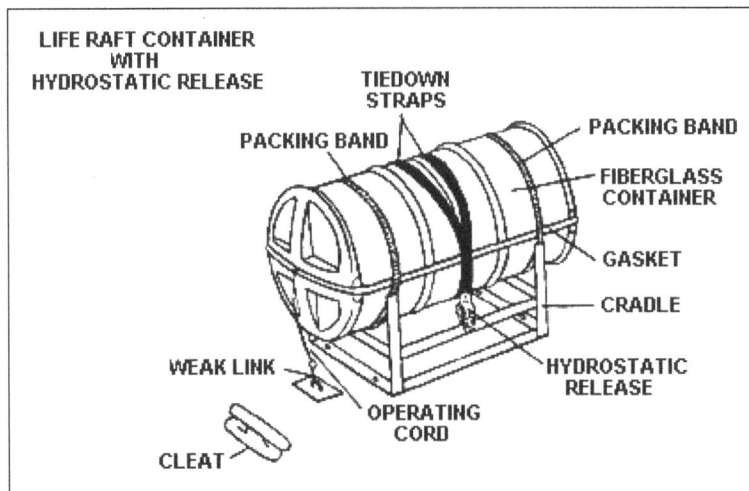

Figure 7-15. A Cradled Life Raft

7-42. The life raft container is strong, weather tight, and tamperproof. The raft container has small holes on the bottom for condensation drainage and air circulation. The container must be stowed with the words "THIS SIDE UP" on top to be sure the holes are on the bottom. Most containers are made of fiberglass. The raft container is usually held together with packing bands, which break when the raft is inflated. A watertight gasket seals the two halves of the container together. The container rests in a cradle. The cradle is permanently secured to the ship's deck. The container may be secured to the cradle with tie down straps.

A tie down strap has a securing device called a hydrostatic release. A cleat provided near the cradle is used for tying the operating cord when launching manually.

DESIGN

7-43.   Buoyancy tubes are located on the outer edge of the raft. They are made of thick nylon-reinforced rubber. The buoyancy tubes make the raft float. They are divided into at least two compartments. The raft is made to support its rated number of persons even if half the compartments in the buoyancy tubes are deflated.

Note: Inflatable life rafts may be designed to be round, oval, octagonal (eight-sided) or boat-shaped. Specific design may vary among manufacturers. A typical oval inflated life raft is shown in Figure 7-16.

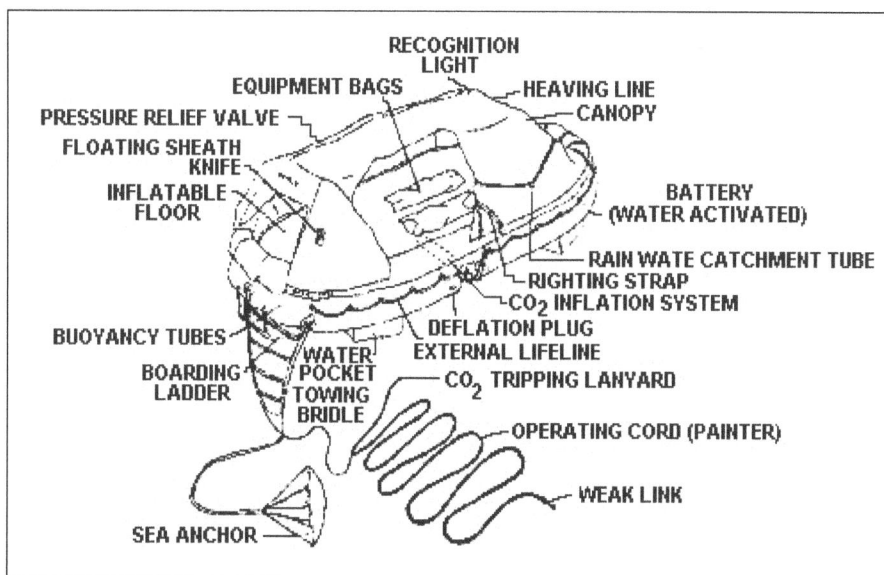

Figure 7-16. A Typical Oval Inflated Life Raft

7-44.   Carbon dioxide (CO2) is usually used to inflate the raft. The CO2 cylinder (container) is on the bottom of the raft. It is activated by a sharp tug on the 100-foot long operating cord. The tug pulls the CO2 tripping lanyard out of the CO2 cylinder to enter the buoyancy tubes. The CO2 can escape through leaks in the tubes. The gas is odorless, tasteless, and colorless, so you must watch for leaks. If you breathe air with a large amount of CO2, you can suffocate, so always leave the curtains open if you know the tubes are leaking. Fix the leak as soon as possible. Pressure relief valves are installed in most rafts. These valves are fitted in the tubes, so excess (extra) gas can automatically escape. It is normal for gas to escape right after the raft is inflated. You can tell it is escaping if there is a hissing sound coming from the valve. The sound will probably stop after a few minutes. During the day, the rise in temperature might cause the gas to expand enough to activate the valves. At night, when the temperature drops, you may have to pump up the tubes with the inflation pumps because the air in the tubes might contract. Sometimes, pressure relief valves do not work correctly. If gas continues to escape from the pressure relief valve, you can fix it with a safety valve plug from the repair kit. Then pump the tube back up. Deflation plugs are provided to deflate the raft after rescue. The floor of the raft is also inflatable. In cold climates, the floor should be inflated with the inflating pump. This will insulate the occupants from the cold seawater. The floor should be left deflated in warm climates. This will allow the cooler seawater to cool the inside of the raft. If necessary, some inflatable floors can be removed and used for an extra emergency float. A boarding ladder and towing bridle are fitted at each end of the raft. The two are usually combined. In addition to boarding and towing the raft, the raft can be hoisted aboard a ship by hooking onto one or both towing bridles. Lifelines are provided inside and outside the raft for survivors to steady themselves. Two lights are installed on the canopy. These lights are automatically activated when the raft inflates. They are powered by either dry cells or water-activated batteries. The lights can operate for at least 12 hours. The external recognition light can be seen from 2 miles away. The other light is inside the canopy. Unscrewing the bulbs during the day will prolong battery life. The canopy has two layers to insulate the inside from extreme temperatures. It erects (pops up) automatically as the arch tubes inflate. The canopy has tubes to collect rainwater. The canopy is colored Indian orange or some other bright color, which will stand out on a white-capped sea. Water pockets are located under the floor. They have holes in them to allow seawater to fill them up when the raft is launched. Water pockets have two purposes: to slow the drifting of the raft and to make the raft more stable (less likely to capsize). The early designs of water pockets were simple, but did not always work well. In heavy seas or high winds, an empty or unevenly loaded raft with three or four small water pockets could easily capsize. Some inflatable life raft manufacturers have improved the basic stabilization design. The Givens raft has a large stability chamber instead of the small water pockets. As the angle of the sea changes, the stability chamber adjusts the raft's center of gravity to compensate for the wave action. When the Givens raft reaches the crest of a wave, the raft bottom should not lose contact with the water, and should not be caught by the wind and capsize. The raft is not easily capsized in high winds with its minimum of 4,800 pounds of water ballast (on the four- to six-person raft). The stability chamber can be de-ballasted (emptied) so the raft can be towed.

Manual Launch

7-45. You are required to know how to manually launch a life raft. Do the following steps to successfully perform this task:

- **Step 1.** Pull open the hook at the hydrostatic release to release the tie-down straps.
- **Step 2.** Secure the operating cord (painter/lanyard) to the cleat. Make sure that the operating cord is free of tangles.
- **Step 3.** DO NOT REMOVE the bands around the container. They will automatically break open when the operating cord is pulled.
- **Step 4.** With two or more crewmembers, throw the life raft in its container overboard (Figure 7-17).

Figure 7-17. Throwing in Life Raft

- **Step 5.** With the life raft and container in the water, pull on the operating cord (Figure 7-18). The bands on the container will break and the life raft will automatically inflate.

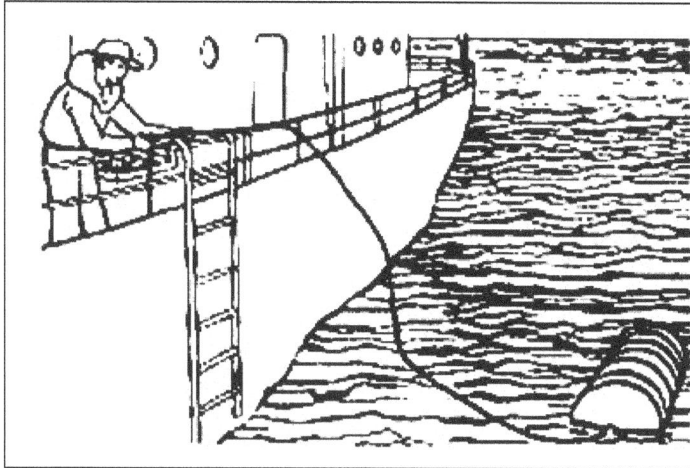

Figure 7-18. Yanking on the Operating Cord

- **Step 6.** Leave the operating cord attached to the cleat aboard the ship.
- **Step 7.** Board the life raft as soon as possible (Figure 7-19).

Figure 7-19. Boarding Life Raft

- **Step 8.** Remove the knife from the pocket on the canopy.
- **Step 9.** Cut the operating cord to free the life raft from the sinking ship.
- **Step 10.** Read the survival manuals that are found inside the raft. These will give you complete instructions on what to do while you are in the life raft.

AUTOMATIC LAUNCH

7-46.   After the ship sinks to a depth of 10 to 15 feet, the hydrostatic release will automatically release and free the life raft container. The container will rise to the surface (Figure 7-20).

Figure 7-20. Container Rises to Surface

7-47.   The pull of the sinking ship will cause the container bands to part and trigger the inflation of the life raft. The life raft will be completely inflated and ready for boarding within 30 seconds. The buoyancy of the life raft will cause the operating cord to part (Figure 7-21).

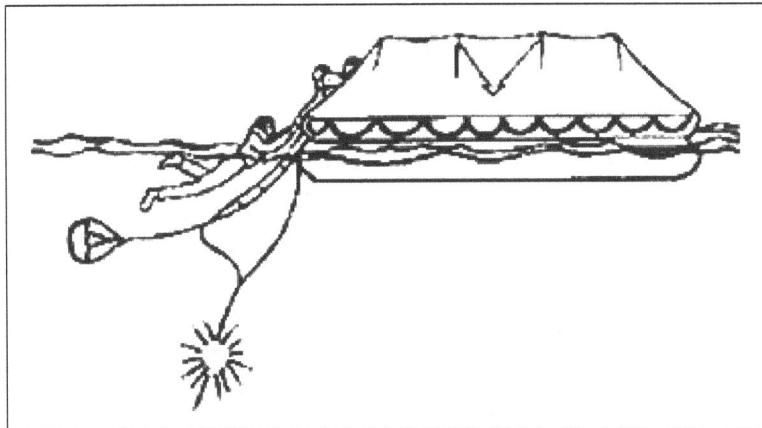

Figure 7-21. Operating Cord Parting

BOARDING

7-48.   The life raft may be boarded by any one of these procedures:
- By climbing down a ladder.
- By jumping into the canopy (Figure 7-22).

Figure 7-22. Jumping into Canopy Opening

- By entering from the ship or from the sea (Figure 7-23).

Figure 7-23. Boarding a Life Raft From the Sea

7-49.   If you can, stay dry when getting off a vessel. Sometimes this is possible by climbing down a ladder, net, or line until you are within 4 feet of the raft. Then jump into the open canopy entrance and land on the floor of the raft with the balls of your feet. (If you land with your heels first, you could fall backwards into the water.) Stretch out your arms and land with your chest against the inflated canopy arch. DO NOT jump onto the roof of the raft! People inside could be injured.

7-50.   When boarding from the sea, place your feet on the boarding ladder. Reach inside the raft and grab the internal lifelines (if there are no external handholds provided). Pull yourself into the raft headfirst. Do not grab hold of the canopy to pull yourself because it tears easily. Two people can help an injured person board an inflatable life raft as shown in Figure 7-24 by doing the following:

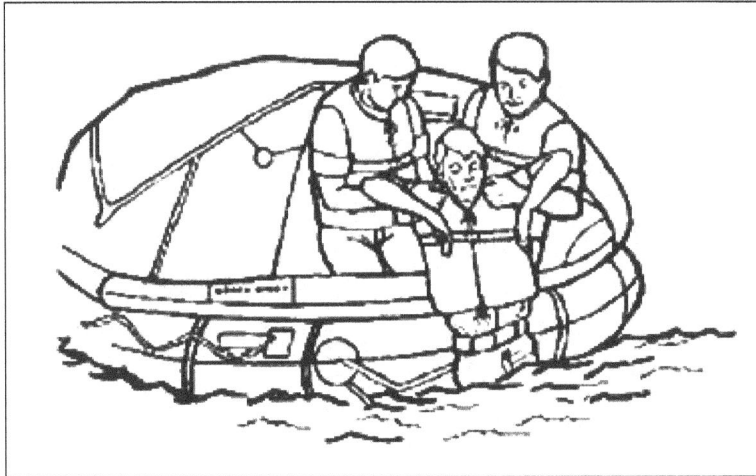

Figure 7-24. Bringing Aboard an Injured Crewman

- Place their outboard knees on the top of the buoyancy tube.
- Turn the injured person with his/her back toward them.
- Grab the injured person's life jacket with their inboard hands.
- With their outboard hands, grab the injured person's upper arms.
- Push the injured person slightly down into the water and, using his/her buoyancy to help them, spring him/her up and over into the life raft, back first.

7-51.   The two rescuers fall to either side on the raft's floor. This allows the injured person to fall between them.

LIFE RAFT SURVIVAL EQUIPMENT

7-52.   Inflatable life rafts are provided with equipment necessary for handling the life raft, surviving at sea, and alerting rescuers. The following list is for inflatable life rafts on ocean service ships. Ships on the lakes, bays, sounds, and rivers have considerably less equipment.

- **Heaving Line.** A buoyant heaving line, 100 feet long, that has a buoyant small floating ring at one end. The other end is attached to the raft near the after entrance.
- **Instruction/Survival Manual.** A booklet printed on water-resistant material hanging in a clear envelope from one of the canopy arch tubes. The manual describes how to use the raft's equipment. It also contains internationally recognized distress signals and survival information.
- **Instruction Card.** A plastic card hangs from the inside canopy. The card shows immediate steps to be taken by survivors upon entering the raft.
- **Jackknife.** One jackknife is provided on rafts holding up to 12 persons. Two are required on larger rafts. The knife has a can opener. One of the knives is in a pocket near the forward entrance. It can be used to cut the painter. If the raft is provided with a floating sheath knife, it can replace the jackknife.
- **Paddles.** Two 4-foot long paddles are included.
- **Inflation/Dewatering Pump.** A pump is provided so survivors can keep the raft inflated. The pump can also be used to pump water out of the raft by switching the hose.

- **Sea Anchors.** Two sea anchors are provided. One sea anchor attaches to the outside of the raft and streams automatically when the raft is inflated. The other one acts as a spare. Each sea anchor has 50 feet of nylon line attached.
- **Bailers.** Two flexible bailers are provided on rafts carrying 13 or more people. One bailer is carried on smaller rafts.
- **Sponges.** Two cellulose sponges are provided.
- **First Aid Kit.** A kit containing first aid supplies is provided.
- **Flashlight.** A flashlight with three spare batteries and two spare bulbs is provided. The flashlight is Coast Guard approved. It is waterproof and has a blinker button for signaling.
- **Signal Mirror and Whistle.** A mirror and whistle for signaling rescue units are provided.
- **Red Rocket Parachute Flares.** Two red rocket parachute flares are provided. They are approved for 3 years of service.
- **Hand-Held Red Flares.** Six hand-held red flares are provided. They are approved for 3 years of service.
- **Provisions.** One pound of hard bread or its approved nutritional equivalent is provided for each person. The food is packed in sealed cans.
- **Water.** One and one-half quarts of water are provided for each person. The water comes in sealed cans. The cans are approved for 5 years after packing. Some rafts may contain a desalting kit for each person. The contents of the kit can be mixed with saltwater to produce 1 pint of drinking water.
- **Can Openers**. Three can openers are provided.
- **Drinking Cup.** A flexible drinking cup marked in ounces is provided.
- **Fishing Tackle Kit.** A kit containing fishing tackle is provided.
- **Anti-seasickness Tablets.** Six anti-seasickness tablets are provided for each person.
- **Repair Kit.** A repair kit for repairing the buoyancy tubes is provided. The repair kit contains a roughing tool, five rubber tube patches (2-inch diameter), and cement. The cement is flammable. There must be no smoking while making repairs. The patches are used for patching small holes. Use patches only if the area around the hole can be kept dry while you are repairing the hole. Roughen the surface of the area that needs patching. Apply cement to both the patch and the area around the hole. Be sure the patch is 1 inch larger than the hole. Allow both to dry. Apply a second coat of cement to both. When both are tacky, press the patch on the hole. Do not completely inflate the raft until the patch has had 24 hours to dry.

LIFE RAFT PATCHING

7-53.   Six sealing clamps are also included in the kit for plugging large holes and any hole which cannot be kept dry enough to use cement. Use the following procedures:
- Loop the cord on the clamp around your wrist to prevent losing the clamp.
- Dip the clamp into the water. This makes the clamp slippery, so it can be inserted easily.
- Push the bottom plate through the hole (see Figure 7-25).

Note: If the hole is too small, carefully enlarge it so the clamp can be forced in.

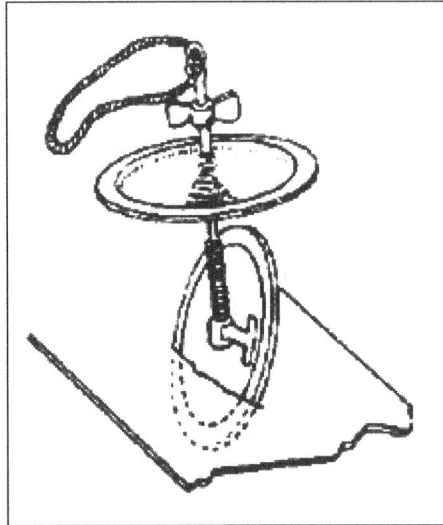

Figure 7-25. Pushing Bottom Plate Through Hole

7-54.   Pull the bottom clamp against the inner surface of the tube, and slide the top clamp over it (Figure 7-26).

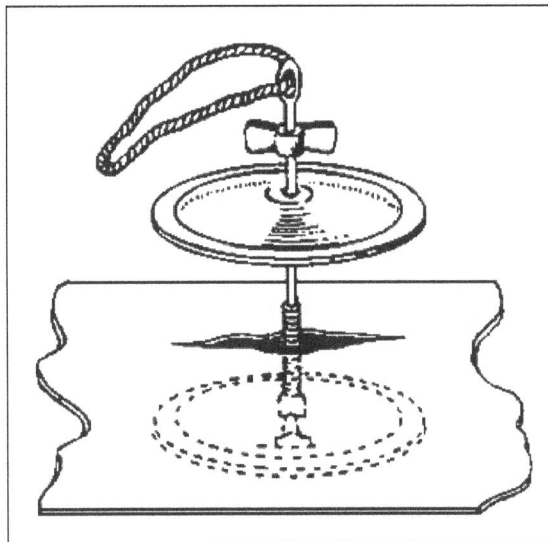

Figure 7-26. Sliding Top Clamp Over

7-55.   Adjust the clamp to completely cover the hole. Hold it in place and screw down the wing nut until it is tight (Figure 7-27). Break off the wire holding the cord.

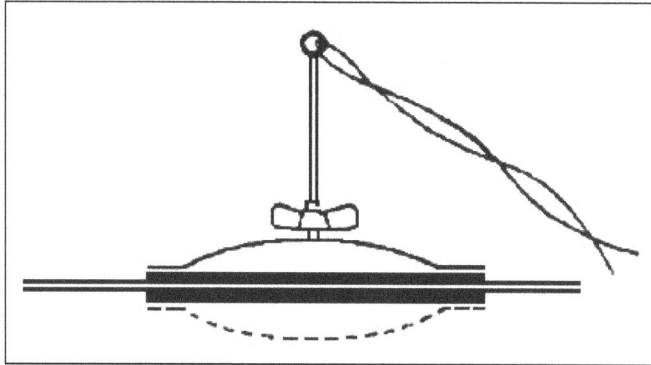

Figure 7-27. Tightening Wing Bottom Clamp Nut

SIGNALING

7-56.   The importance of a good lookout cannot be overstated. Remember, when in a life raft, you are so small and the sea is so big that it is very easy for a search ship or plane to overlook you. An alert lookout will make the difference in survival. Once you have sighted a rescue ship or aircraft, use the following to attract their attention:

* **Signaling mirrors.** Read the instructions for the particular kind of signaling mirror in your survival equipment. Do not wait until you see a rescue craft to use the signaling mirror. When the sun is shining, flash the mirror all around the horizon (Figure 7-28). An aircraft can spot the flash long before you would see the aircraft. The signaling mirror may save your life. Use it as long as the sun is shining.

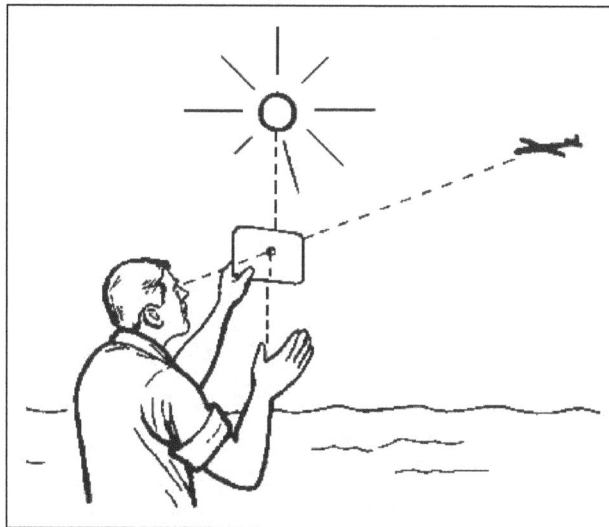

Figure 7-28. Signaling Mirror

* **Whistles.** In calm weather, your voice can be heard only a few hundred yards away. If you keep screaming, you will become hoarse and lose your voice. A whistle, on the other hand, can be heard up to 4 miles away in favorable weather conditions. It can come in handy when you are floating in the water trying to attract the attention of nearby rescuers. A whistle can be used over and over again. It can be used in fog, at night, or during the day.

- **Pyrotechnics.** These are signals such as rockets, flares, and smoke. Instructions for operating various brands of pyrotechnics are written by the manufacturers. Once you are settled in your survival craft, read the instructions on each type of pyrotechnic so you will know how to use them when a ship or aircraft is spotted. Keep the pyrotechnics close by for immediate use, so you can signal when necessary. Heed the following when using pyrotechnics:
  - Be sure to fire the signals downwind on the lee side of the survival craft. When firing, hold them at a slight angle over the water. Pyrotechnics have burning particles that might fall, which may burn you or damage the raft.
  - Only use smoke signals during the daytime. Smoke does not glow in the dark. Only use pyrotechnics when you can see a ship or plane. Do not waste smoke signals.
  - Rockets should be used when a vessel is spotted far away on the horizon. A rocket will get the signal higher, where it can be seen from a greater distance.
  - An aircraft directly overhead would be more likely to spot a hand flare than a flare covered with a parachute.
- **Emergency Position Indicating Radio Beacons (EPIRB).** Your ship may also have at least one EPIRB. There are different makes of EPIRBs. They all have the following things in common:
  - EPIRBs float. They are stowed on the outside of the ship, so they will float free if the ship sinks.
  - They are small (approximately 6 inches thick and 1 to 3 feet long).
  - They transmit signals automatically on two international distress frequencies for military and civilian aircraft. These frequencies are 121.5 and 243 MHz.
  - Most EPIRBs work on one-way automatic operation only and cannot be used for two-way communication.
  - They transmit a continuous two-tone (hi/lo) signal.
  - EPIRBs are easy to use.

7-57. If your EPIRB is floated, tie it to the survival craft, so it will not drift away (Figure 7-29).

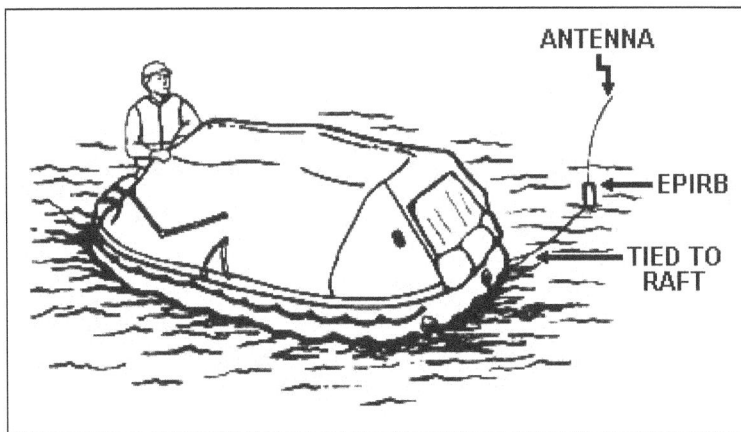

Figure 7-29. EPIRB Secured to a Life Raft

RIGHTING AN OVERTURNED LIFE RAFT

7-58. If a capsized raft can be righted (turned right side up) before the inverted (upside down) canopy fills with seawater, one person can easily right it using the following procedure:
- Swim to the side marked "RIGHT HERE." If it is not marked, go to the side where the CO2 cylinder is located. Reach up and grab the righting strap. Start pulling yourself up onto the raft.

It may help to kick your feet out as if swimming (Figure 7-30). If this does not work, try putting your feet or knees into the external lifelines to help you pull yourself up on the raft. Some rafts may right while you are climbing onto them. Others are more difficult to right.

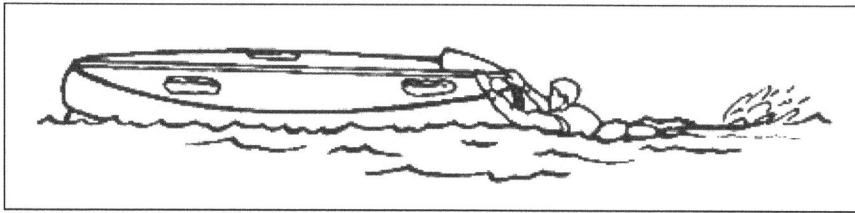

Figure 7-30. Getting Aboard an Overturned Life Raft

Note: A righting strap is fitted on the underside of the raft to right the raft if it capsizes or inflates upside down. The righting strap runs the full width of the oval or round raft.

- Stand on the very edge, where the CO_2 cylinder is located. Lean back with all your weight while pulling on the righting strap (Figure 7-31).

Figure 7-31. Standing on Edge

7-59. If the canopy is clear of water, the raft will begin to follow you. If the raft is large, it will land on your head unless you bend your knees and spring backwards just as the raft begins to free fall (Figure 7-32). This should allow your head to clear the raft.

Figure 7-32. Knees Bent

7-60.   Do not panic if the raft does land on top of you. Because the bottom of the raft is soft and flexible, you can create an air pocket by pushing your arms or head against the floor. This will give you a chance to catch a breath of air. Use your arms and swim face up to get out from underneath the raft. If you try to swim out face down, the raft may hang up on the back of your life preserver. If this happens, it will be difficult for you to get out from underneath the raft. If one person cannot right a capsized raft, the canopy probably has filled with seawater that cannot escape. Try two persons pulling on the righting strap. If this does not work, then get several persons in the water on the opposite side of the raft (Figure 7-33). These persons should work the water out of the canopy by pushing up on the canopy while two people pull on the righting strap.

Figure 7-33. Several People Righting an Overturned Life Raft

7-61.   If the inverted canopy fills with seawater, the raft may be more difficult to right. Generally, round rafts have the righting strap parallel to the canopy openings. This allows the water to flow freely out of the raft while the raft is being righted. If the raft is oval with the righting straps at right angles to the canopy openings, water tends to stay trapped in the canopy. It may take several persons to right this type of raft.

Note: Figure 7-34 shows the overhead views of round and oval rafts.

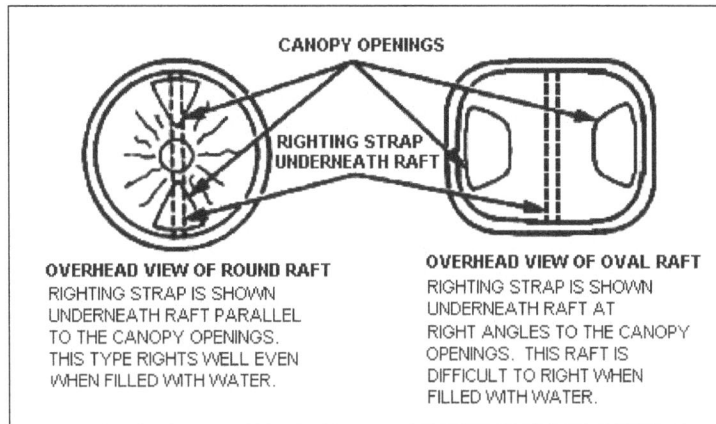

Figure 7-34. Overhead Views of Round and Oval Rafts

7-62.   A single person may be able to right a waterlogged raft. He/She can try by pulling and walking the righting strap through his/her hands until the opposite side is pulled over. This takes a lot of strength and may be very hard to do. It might be done without climbing aboard the raft.

SURVIVAL ABOARD A LIFE RAFT

7-63.   Life rafts are important in a marine emergency. The life raft is the primary means of escape in a shipboard emergency. Survival aboard a life raft starts with the proper launching and inflating of the life raft. Survival can also include how to correctly board the raft, avoiding hypothermia, how to right an upside down life-raft, know how to properly use safety equipment, anchoring the raft, plugging leaks, dealing with seasickness, establishing a chain of command, and rescue.

PRESERVE BODY FLUIDS--AVOID SEASICKNESS

7-64.   Riding in a life raft is very uncomfortable. Your raft will be in constant motion even on a calm sea. A raft wiggles every time someone moves inside or the water moves underneath. You will be confined in a cramped and stuffy space. Even the most experienced seafarers tend to get seasick in a raft. Seasickness must be avoided if at all possible. It is a very miserable illness and can affect your will to survive. If on hand, take a seasickness pill (if you can) before you abandon ship. If unable to take, take a seasickness pill found in the raft's supply kit as soon as all of your shipmates have been helped into the raft. The pills will keep you from vomiting. Vomiting empties your stomach of valuable fluids. You must preserve those body fluids. If you lose them, they will be difficult or even impossible to replace as long as you are in the raft. Remember how cramped your survival conditions may be. If one person vomits, others will probably do the same.

URINATE SOON AFTER BOARDING

7-65.   If you did not urinate within a few hours before boarding the survival craft, you should do so within 2 hours. The traumatic effects of a disaster at sea may make urination difficult. You could damage your bladder if you do not pass urine. There are two methods that might help you urinate:
- Have someone pour seawater slowly back and forth from cup to cup in front of you.
- Hang over the side with the water waist high. The cool water should help.

7-66.   After several days with little drinking water, do not be alarmed if your urine appears dark and thick. Such a reaction to dehydration is normal.

SIT ON A LIFE JACKET FOR PROTECTION

7-67.   In moderate seas, when there is no danger of the raft capsizing, you should take off and sit on your life jacket. The rubber raft constantly moving under you tends to wear your skin until soreness occurs. Your life jacket will provide a cushion that will prevent such soreness.

COVER UP

7-68.   The dangers from exposure to cold are obvious, BUT do not forget the sun, wind, rain, and sea. The life raft comes with a built-in canopy to protect you. Do not cook yourself in the sun! Serious burns and loss of valuable body fluids could result from sunburn. Wear light clothing or stay under the cover. If possible, keep breezes blowing through the survival craft. Sometimes you can change the position of the sea anchor to increase ventilation (movement of air).

AVOID SUNBURN

7-69.   Reduce need for water by avoiding any extra exertion. If you exert yourself, you will sweat and use a lot of fluids. Keep the outside of the raft wet. Wet your clothing during the day with seawater.

CAUTION
Take seasickness pills as soon as you can!

DRINK WATER

7-70.   The normal, healthy body (at rest) can stay alive for over 40 days with no food and as little as 11 ounces (one ration can) of fresh water each day. As little as 2 or 3 ounces of drinking water each day can keep a person healthy for up to 10 days. Without fresh water, a person often becomes delirious in about 4 days and might die in 8 to 12 days.

No Water for First 24 Hours

7-71.   Do not issue water during the first 24 hours unless you have an unlimited supply. The body is already full of water. If you drink more, it will probably be wasted in the form of urine. After 24 hours, your body will be drier and will absorb the water you drink, just like a dry sponge will hold water, but a wet sponge will not. If a survivor is injured, you may give him/her water during the first 24 hours. The survivor will need it to replace the fluid he/she lost through his/her bleeding or burns. Only give water if he/she is conscious. After 24 hours, you may issue a full ration (1/3 of a 1-quart can) of water for each person. The ration should be divided into three equal parts. Drink one part at sunrise, one at midday, and one at sunset.

Rainwater

7-72.   You may collect more water by catching rainwater. Some parts of the inflatable life raft canopy are designed to catch water. Rainwater catchment tubes will take the water into storage bags on the inside of the raft. The storage bags are in the raft's equipment container. Salt spray may dry on the canopy. The salt might be washed in with the first few ounces of rainwater. It might be very difficult to collect uncontaminated rainwater when the seas are rough and waves are constantly being blown onto the canopy. The lookout should alert everyone when it rains. Use and fill all available containers with rainwater (such as equipment accessories bag, ration packs, and empty tin cans). After all of the containers have been filled, everyone should drink as much of the rainwater as they can.

Condensation

7-73.   Water might condense on the inside canopy of the inflatable life raft. Use one of the cellulose sponges that are provided in the raft equipment to soak up the water. Squeeze the water out of the sponge to drink or store. Be sure to keep a sponge clean for this purpose.

Snow and Ice

7-74.   In the Arctic Sea, you can collect "old saltwater ice." It is bluish in color with smooth, rounded corners. It is usually pure enough to eat or drink. Do not make the mistake of eating "salt ice." "Salt ice" is gray and milky. It should not be eaten. Remember, ice and snow will tend to chill your stomach and reduce your body temperature. If you are on the verge of hypothermia, you should not eat ice or snow. Allow it to melt and get as warm as possible. Warm it in your mouth before swallowing.

Never Drink Seawater or Urine

7-75.   Rain, ice, and condensation are good sources of water. Do not mix saltwater, urine, or animal fluid with fresh water to stretch your water supply. Drinking seawater will only worsen your thirst and increase water loss by drawing body fluids from the kidneys and intestines. The salt will go to the brain and cause delirium and convulsions. Drinking seawater and urine during a survival situation could cause madness and death.

OBTAIN FOOD

WARNING

DO NOT eat during the first 24 hours. After 24 hours, you may eat 4 ounces each day. In a life raft, the food will last 5 days. You will have extra rations (food and water) if the boat or raft is not carrying its full number of passengers. DO NOT eat food if you do not have water. Your body needs water for digesting food. Eating without drinking fresh water could cause death.

7-76.   The sea has many different forms of life. If you have enough fresh water, you will probably not starve to death. Remember that water is a MUST! Because fish and birds are rich in salt and protein, more water is needed to digest them. Do not eat food from the sea unless you have two or three times more water than your daily ration. DO NOT panic if you do not have enough water to drink with your seafood or if you cannot catch any seafood right away. You probably abandoned ship with excess body fat. Your system will begin to use the fat if you do not eat. One pound of body fat will probably keep your system working at about the same rate as two meals. The rate at which body fat and protein are changed to heat and energy depends upon air temperature, your activity, and your mental state. You can live longer on your stored energy if you keep your mind and body relaxed. It also helps if you do not overwork yourself or expose your body to very hot or cold temperatures.

Fish

7-77.   Most fish that are found in the open sea can be eaten. If they are found closer to shore they might be poisonous. The puffer, porcupine, and parrot fish are poisonous. They are fish that blow themselves out or have spines or bristles. The flesh of fish caught in the open sea is good to eat whether cooked or raw. The heart, liver, and blood of fish are good to eat. Intestinal walls are edible, but the contents may be dangerous unless they are cooked. The stomachs of large fish may contain small fish partly digested, which are good to eat. Fish eyes also contain a lot of water. You can catch fish by using the fishing kit provided with your equipment. Complete instructions are inside the kit. If you have lost your fishing kit, you could use the following methods to catch fish:

● By tying your knife to a paddle, oar, or boat hook, you may be able to spear large fish near the surface. Slash with your knife in schools of small fish.

● Fishhooks can be made from wood split from the lifeboat. This wood is notched and held together with thread from the equipment or unraveled from cloth (Figure 7-35).

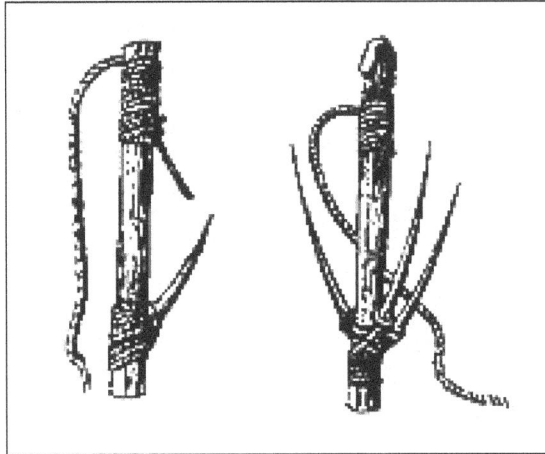

Figure 7-35. Fishhooks Made From Wood

● A jackknife can be made into a large fishhook. Wedge the blade open with a piece of wood and tie as shown in Figure 7-36.

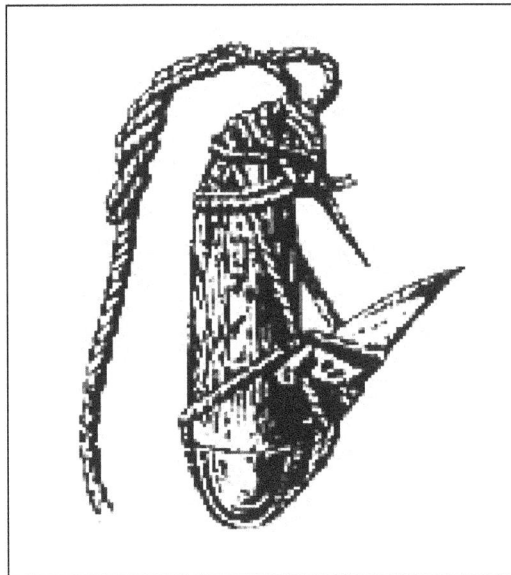

Figure 7-36. Fishhook Made From a Jackknife

7-78.   Flying fish are probably the most available food. Many survivors have lived on them alone. Some may glide into or against your craft. At night, flying fish (and most other fish) are attracted by light. Shine your light on the side of your white craft or cloth and the flying fish will often glide toward the light and into the boat. Often, a bright moon shining on a white object will draw them. If and when you catch more fish than you can eat, in order to drink, squeeze or chew out the juice of the flesh. Fish juice tastes much like the juice of raw oysters or clams. To squeeze it out, cut a piece of fish without bones or skin. Cut it into fine (tiny) pieces. Wrap it in a cloth with long ends. Have two people twist the ends as tight as possible. The juice will drip out. To chew it out, chew a small piece of fish in your mouth. Suck out the juice and swallow it. Spit out the remaining flesh. Cut fish into thin, narrow strips and hang them out in the sun to dry. If it is completely dried and kept dry, it will often stay good for several days. It may even taste better dried.

IMPORTANT!

Fish that are not cleaned may spoil in half a day. Clean and immediately eat or dry your fish.

Turtles

7-79.   All of the meat, blood, and juice of a turtle are good. The best meat is found against the shell, under the backbone. Cut through the ribs to get to this meat. A hot sun brings a clear oil out of turtle fat in which you can dip your food.

CAUTION

A turtle can still bite and scratch even after you have cut off its head.

Seaweed

7-80.   Raw seaweeds are tough and salty. They are difficult to digest. Eat them only if you have plenty of fresh water. Small edible crabs, shrimp, and fish often live in the seaweed. Lift them out of the water slowly and carefully. Shake them over the survival craft. Get rid of the jellyfish and eat the remaining morsels.

Birds

7-81.   All sea birds are nourishing and can be eaten. The blood and liver are also good to drink and eat. Try to catch birds that will sometimes land on you or on or in the survival craft. Catch birds by dragging a baited fishhook behind the craft. Pull on the line after they have swallowed the hook. The hook catches the bird like a fish. Catch every bird you can. Use the feathers as fishing lures and the meat and guts for fish bait. Birds can also locate fish for you. When feeding, they usually follow schools of fish. This will give you an opportunity to get right up to the birds to catch them. Also, do not forget to catch the fish they are feeding on.

SEARCH AIR RESCUE

7-82.   Upon receiving a signal from any source that a ship or aircraft is in distress, it is the responsibility of all vessels in the area to go to the site and give help to the ship, aircraft, or persons in distress. This signal can range from a ship that is sinking or on fire, a downed aircraft, man overboard, or serious illness or injury aboard ship.

TYPES OF DISTRESS SIGNALS

7-83.   A ship at sea can be alerted to an emergency by the following:
- Radio or radiotelephone.
- Visual international distress signals.

- Aircraft.

HOW AIRCRAFT DIRECT SHIPS TO DISTRESS SCENE

7-84. These procedures are used by an aircraft to direct a ship toward another ship or aircraft in distress:

- The airplane circles over the ship at least once. Then the aircraft crosses the bow of the ship as close as possible. At a low altitude, the pilot opens and closes the throttle or changes the propeller pitch (Figure 7-37).

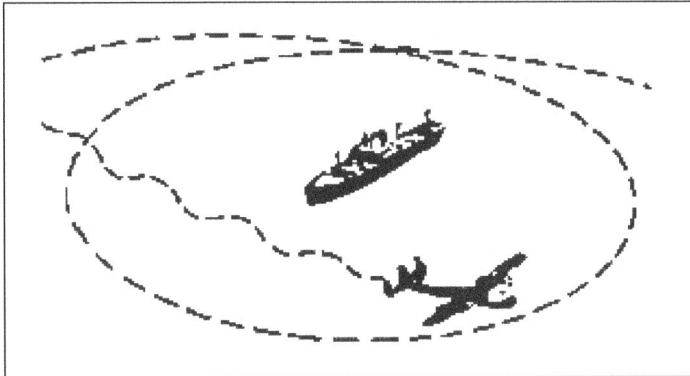

Figure 7-37. Aircraft Signal

- Then the airplane will head in the direction that the ship is to follow in order to find the ship or aircraft in distress (Figure 7-38).

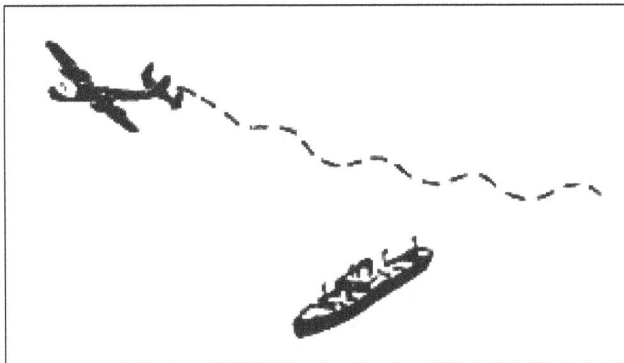

Figure 7-38. Aircraft Dismissal Signal

7-85. If help is no longer required, the aircraft will return and fly, opening and closing the throttle or changing the propeller pitch, across the wake of the ship at a low altitude and as close astern as possible.

SURFACE SHIPS ACTION

7-86. Once your ship has been alerted to the distress situation, you will acknowledge the receipt of the message. You will also provide a continuous radio guard on 2182 kHz and/or channel 16 on the radiotelephone, and, if required, retransmit the distress message to the ships in the area. The next step is to determine your exact position and the position of the vessel or aircraft in distress. If it is possible, you should communicate the following to the ship in distress:

- Your identity.

- Your position.
- Your speed and ETA.
- Your true bearing from the vessel.

7-87. Then the entire crew of your ship is to start preparing for the rescue as follows:
- Get heaving lines, ladders, and scramble nets rigged on both sides of the vessel.
- Prepare to receive survivors who may need medical assistance.
- Put lines over ship's side to assist any lifeboats or rafts that may secure alongside.

ASSISTING AN AIRCRAFT THAT MAY DITCH

7-88. Make a smoke signal if possible to show the pilot direction of surface wind. At night, show deck lights and shine the signal lamp straight up in the air. Do not shine it on the aircraft. You may blind the pilot. Try to make radiotelephone contact with the aircraft and give the following information:
- Wind direction and force.
- Direction, height, and length between the swells.
- Any other information the pilot may require.

7-89. Proceed alongside the aircraft as quickly as possible. The aircraft may break up as soon as it hits the water.

Note: Military aircraft are usually fitted with "ejection seats." Many times the crew will use their ejection seats rather than ditch with the aircraft.

7-90. When picking up survivors from a military aircraft, get the following information as soon as possible and, if necessary, pass the information to other rescue ships by radiotelephone:
- What was the time and date of the casualty?
- Did you bail out or was the aircraft ditched?
- If you bailed out, at what altitude?
- How many others did you see leave the aircraft by parachute?
- How many ditched with the aircraft?
- How many did you see leave the aircraft after ditching?
- How many survivors did you see in the water?
- What flotation gear did they have?
- What was the total number of persons aboard the aircraft before the accident?

PREPARATION FOR MEDICAL EVACUATION OF PERSONNEL FROM YOUR SHIP

7-91. If there is a serious injury aboard your ship, a helicopter may be used to remove the injured crewmembers. This may be a Coast Guard, Army, Navy, or Marine helicopter performing the rescue mission. The ship's master and crew should prepare for removing the crewmember while waiting for the rescue helicopter. The following is a complete helicopter evacuation check list. When requesting helicopter assistance—
- Give accurate position, time, speed, course, weather conditions, wind direction and velocity, voice and CW frequencies.
- If not already provided, give complete medical information, including whether or not the patient can walk.
- If you are beyond helicopter range, advise your diversion intentions so that a rendezvous point may be arranged.
- If there are any changes, advise immediately. Should the patient die prior to arrival of the helicopter, be sure to advise. Remember that members of the flight crew are risking their lives attempting to help you.

7-92.   Make the following preparations prior to arrival of the helicopter:

- Provide continuous radio guard on 2182 kHz or specified VOICE frequency, if possible. The helicopter cannot operate CW.
- Select and clear the hoist area, preferably aft, with a minimum 50-foot radius. This must include securing loose gear, awnings, and antenna wires. Trice up running rigging and booms. If the hoist if aft, lower flagstaff.
- If hoist is at night, light up pickup area as well as possible. BE SURE YOU DO NOT SHINE ANY LIGHTS ON THE HELICOPTER. If there are obstructions in the vicinity, put a light on them so the pilot will be aware of their positions.
- Point searchlights vertically to aid in locating the ship, and secure them when helicopter is on scene.
- Advise location of pickup area BEFORE the helicopter arrives, so that the pilot may make his/her approach aft, amidships, or forward, as required.
- Arrange a set of hand signals among the crew who will assist. There will be a high noise level under the helicopter, making voice communication almost impossible.

HOIST OPERATIONS

7-93.   Hoisting operations are used to rescue or evacuate personnel from a number of dangerous situations. The following are some guidelines to follow when using hoisting operations:

- If possible, move the patient to a position as close to the hoist area as his/her condition permits-- TIME IS IMPORTANT.
- It may be necessary to move the patient by litter. Be prepared to do this as quickly as possible. Be sure patient is strapped in, face up, with life jacket, if his/her condition permits.
- Be sure patient is tagged to indicate what medication, if any, was administered, and when.
- Have patient's medical record and necessary papers in envelope or package ready for transfer WITH him/her.
- Change course so the ship rides as easily as possible with the wind on the bow, preferably on the port bow. Once established, maintain course and speed.
- Reduce speed, if necessary, to ease ship's motion; but maintain steerageway.
- If you do not have radio contact with rescue aircraft, when you are in all respects ready for the hoist, signal the aircraft with a "come on" by hand, or at night by flashlight.
- To avoid static shock, let basket or stretcher touch the deck before handling.
- If the aircraft drops the trail line, guide the basket or stretcher to deck with line. Keep line clear at all times.
- Place patient in basket, sitting with hands clear of sides, or in the litter as described above. Signal hoist operator when ready for hoist. Patient signals by nodding head if he/she is able. Deck personnel give thumbs up.
- If necessary to take litter away from hoist point, unhook hoist cable and keep free for aircraft to haul in. DO NOT SECURE CABLE TO VESSEL OR ATTEMPT TO MOVE STRETCHER WITHOUT UNHOOKING.
- When patient is strapped in stretcher, signal aircraft to lower cable, hook up, and signal hoist operator when ready to hoist. Steady stretcher from turning or swinging.
- If trail line is attached to basket or stretcher, use to steady. Keep feet clear of line.

HELICOPTER HOIST PROCEDURES

7-94.   The wind developed by the helicopter rotor system can be over 70 knots. It is important to have all loose gear, on deck, securely tied down or stowed below decks. The rotor system could be destroyed if any loose objects are blown into the rotor during the hoist. It is important to plan ahead because your voice cannot be heard over the noise made by the helicopter engine. Work out problems that may occur before the helicopter hovers overhead. Do not forget to wear your life jacket! A helicopter might be used to rescue survivors or evacuate injured mariners by rescue basket, rescue sling, and stokes litter (Figure 7-39).

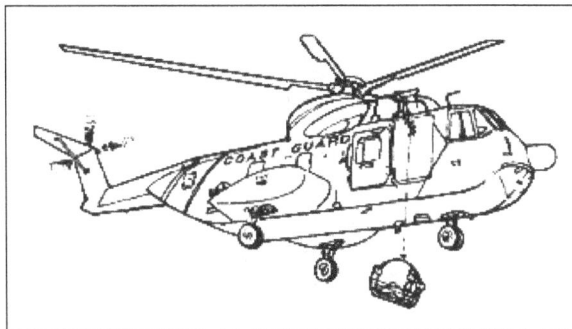

Figure 7-39. Rescue Basket Hoist

RESCUE BASKET

7-95.   The U.S. Coast Guard usually uses a rescue basket for survivors who can help themselves (Figure7-40). The basket is very easy to use. Just climb into the basket after it touches the deck (to discharge static electricity), sit down, and keep hands and arms inside.

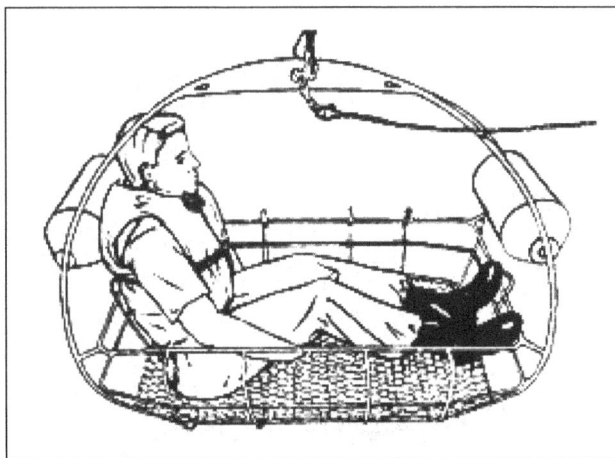

Figure 7-40. Rescue Basket

RESCUE SLING

7-96.   A rescue sling is carried on board helicopters. Rescue helicopters from other countries use the sling more often than the U.S. Coast Guard. The rescue sling is just a padded loop that is placed over the body and underneath the armpits. The arms are held around the sling as shown in Figure 7-41.

Figure 7-41. Rescue Sling

STOKES LITTER

7-97.   This type of litter will usually be used to hoist those who have serious injuries or illnesses or who are unable to walk. To use the litter, it is necessary to get help from other crewmembers. The straps must be disconnected and spread out. The blankets must be removed. The patient should be put in the litter and covered with the blanket. The straps are then snugly fastened with the pad on top of the chest as shown in Figure 7-42.

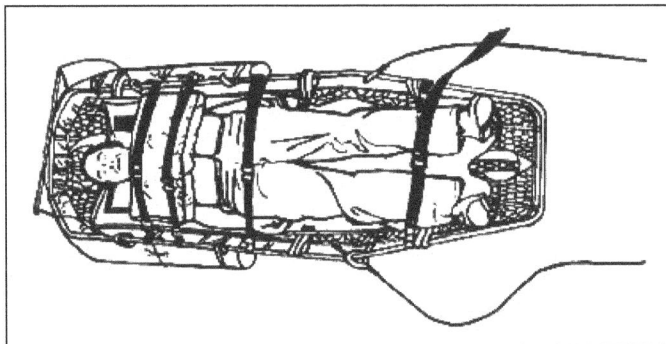

Figure 7-42. Stokes Litter

7-98.   If the litter has to be taken below decks to the patient, it must be unhooked from the cable. This hook must not be attached to any part of the vessel. There is always a possibility that there may be an emergency aboard the helicopter itself. The helicopter may have to move unexpectedly. To decrease this type of danger, the pilot may hover off to one side of the vessel while waiting. If a steadying line is attached to the basket, horse collar, or litter, it must be tended. This will stop the rescue device from swinging too much. It is very important that the rescue device touches the vessel before anyone touches it.

As soon as the object being lowered touches the deck, static electricity (which builds up in the helicopter during flight) will be discharged. Never shine lights on the helicopter. It will blind the pilot.

READY TO HOIST

7-99.   To signal the helicopter pilot that all is ready for hoisting, give him/her a thumbs-up signal, or if you are a patient, nod your head if you are able.

SHIPBOARD CBRNE DEFENSE

7-100.   Much of your military training is dedicated to CBRNE training in a land combat situation and the protective measures to be taken for survival. This paragraph will discuss CBRNE countermeasures to be taken aboard ship for survival. Although a nuclear detonation is devastating, survival is possible, and aboard ship, it is probable. Your survival will depend upon the actions taken before, during, and after the attack.

TYPES OF NUCLEAR BURSTS

7-101.   The energy yield of a nuclear weapon is described in terms of the amount of TNT that would be required to release a similar amount of energy. A nuclear weapon capable of releasing an amount of energy equivalent to the energy released by 20,000 tons of TNT is said to be a 20-kiloton (KT) weapon. A nuclear weapon capable of releasing an amount of energy equivalent to the energy released by 1,000,000 tons of TNT is said to be a 1-megaton (MT) weapon. Weapon yields may range from a fraction of a KT to many MTs. Although a weapon's total yield is not significantly influenced by the environment about the burst point, the relative importance of weapon effects depends greatly on where the detonation takes place. The four types of bursts are high altitude, air, surface, and underwater. Although the four types of bursts are defined below, there is actually no clear line of demarcation between them. Obviously, as the height of burst is decreased, the high altitude burst becomes an air burst, an air burst will become a surface burst, and so forth. The significant military effects associated with each type of burst follow.

- **High Altitude Burst.** This explosion takes place at an altitude in excess of 100,000 feet. It produces air-blast, thermal radiation, an electromagnetic pulse (EMP), initial nuclear radiation, and atmospheric ionization. At altitudes above 100,000 feet, the proportion of energy appearing as blast decreases markedly, while the proportion of radiation energy increases. Due to the low density of the atmosphere, the range of the initial nuclear radiation increases. In contrast to explosions below 50,000 feet, the attendant atmospheric ionization from bursts above 100,000 feet lasts for minutes to hours. The important consequences of high altitude bursts are the damage to weapons systems or satellites operating in the upper atmosphere or in space, and the effects on electromagnetic waves (communications and radar) relying on propagation through or near the region of the burst.
- **Air Burst.** In this type of burst, the fireball does not contact the surface. An air burst produces air-blast, thermal radiation (heat and light), EMP, and initial nuclear radiation (neutron and gamma rays) about the burst point. There will be no significant residual nuclear radiation (gamma and beta radiations from airborne or deposited radioactive material) except when rain or snow falls through the radioactive cloud.
- **Surface Burst.** The fireball touches or intersects the surface. A surface burst produces air-blast, thermal radiation, EMP, initial nuclear radiation around surface zero (SZ), and residual (transit and deposit) nuclear radiations around SZ and downwind from SZ. Transit radiation is produced by airborne radioactive material (base surge/fallout) and deposit radiation is produced by radioactive material (base surge/fallout) collection on exposed surfaces. Surface bursts over water will also produce underwater shock and surface water waves, but these effects will be of less importance. Over land, earth shock will be produced, but will not be an important effect at any significant distance from the burst point.
- **Underwater Burst.** This burst occurs below the water surface. It produces underwater shock and a water plume, which then causes a base surge. Very shallow bursts may also produce air-blast, initial nuclear radiation, fallout, and possibly some thermal radiation. These effects will be

reduced in magnitude from those of a water surface burst and will become rapidly insignificant as the depth of burst is increased. The damage range due to shock is increased as depth of burst is increased. For a given weapon yield, greater hull and machinery damage will be produced by shock from an underwater burst than by air-blast from an air or surface burst. When a high yield weapon is detonated underwater in the deep waters adjacent to a continental shelf, large breaking waves may be generated by the upsurge along the shelf slope. These waves will appear on the shallow water side of the shelf edge. They are characterized by a long period with a sharp, possibly breaking, crest. They dissipate in amplitude as they progress toward the shore. Calculations and simulation experiments with the East Coast U.S. continental shelf indicate that, in the near vicinity of the shelf edge (shallow water side only), these waves may be large enough to damage the largest combatant ships and swamp or capsize smaller ships. This shoaling phenomenon does not appear in deep water. Except in shoaling waters, water waves normally will not be a major hazard.

UNDERWATER SHOCK

7-102. Underwater shock is the shock wave produced in water by an explosion. The shock wave initially travels several times the speed of sound in water, but quickly slows down to sonic speed (about 5,000 feet per second). Underwater shock produces rapid accelerations that may result in equipment and machinery disarrangements, hull rupture, and/or personnel injuries. Both the directly transmitted shock wave and the shock wave reflected from the sea bottom can be damaging. An underwater explosion produces a shock wave similar to an air burst. Four factors determine whether the greater damage will be caused by the direct wave or the reflected wave:

- Distance from burst.
- Depth of burst.
- Depth of water.
- Bottom configuration and structure.

THERMAL RADIATION

7-103. Thermal radiation is the radiant energy (heat and light) emitted by the fireball. Thermal radiation travels at the speed of light and persists as long as the fireball is luminous. The duration of thermal radiation emission depends on weapon yield. It usually lasts less than 1 second for 1-KT yield and about 8 or 9 seconds for a 1-MT yield. Thermal radiation is effectively shielded by anything that will cast a shadow (opaque materials). Thermal radiation can produce combat ineffectiveness (that is, individuals unable to man battle stations) among exposed personnel by skin burns, flash blindness, or retinal burns. Thermal radiation is modified by the height of burst, weapon yield, cloud cover, and terrain features. As height of burst is increased, the area of the earth's surface exposed to thermal radiation increases. This happens because there are fewer shadows from existing structures (such as vegetation, terrain features, and so forth). As weapon yield increases, the range at which thermal radiation can cause skin burns and eye injuries to exposed individuals extends well beyond the range where blast and initial nuclear radiation are of significance. The rate at which thermal radiation is emitted from a high-yield weapon is slower than for a low-yield weapon. Therefore, the high-yield weapon must deliver more thermal energy to do an equivalent degree of damage because a target has more time to dissipate the heat being received.

NUCLEAR RADIATION

7-104. The four basic types of nuclear radiation given off during a nuclear explosion are alpha particles, beta particles, gamma rays, and neutrons.

- **Alpha particles.** These do not travel more than a few centimeters in air without being stopped. They cannot penetrate even a thin sheet of paper.
- **Beta particles.** These may travel several feet in the air, but they cannot penetrate a sheet of aluminum more than a few millimeters in thickness. Beta particles cannot penetrate the normal combat uniform.
- **Gamma rays.** These are a form of electromagnetic radiation, indistinguishable from X-rays.

- **Neutrons.** These are electrically neutral particles. Gamma rays and neutrons can travel comparable distances in the air, up to several hundred meters. Gamma rays and neutrons have the greatest penetrating power of all the forms of nuclear radiation, and their injurious effects on personnel are quite similar.

7-105. Nuclear radiation does not affect most materials in any visible manner. Therefore, the essential value of ships, vehicles, electronic equipment (except transistors), and other equipment is not impaired by radiation. However, radioactive contamination does pose a danger to operating personnel. The term CONTAMINATION is used to mean radioactive material that has been deposited in a location where it is not desired. All radioactive contamination gives off nuclear radiations.

INITIAL NUCLEAR RADIATION

7-106. Initial nuclear radiation is defined as the radiation (essentially neutrons and gamma rays) emitted by the fireball and the cloud during the first minute after detonation. Depending on weapon yield, all significant neutron radiation is emitted in less than 0.1 second, gamma radiation up to 20 or 30 seconds. The 1- minute time limit is set as the maximum time for the nuclear cloud to rise beyond the range in the air at which gamma radiation is a significant hazard. Initial nuclear radiation generally may not produce significant material damage, but will produce combat ineffectiveness.

NUCLEAR RADIATION INJURY

7-107. The radiological hazards described are those which might be of significance to the military effectiveness of marine personnel in combat operations. Injuries to personnel can result from exposure to sufficient quantities of either initial or residual radiation, or a combination of the two. Unlike injuries from other weapon effects, nuclear ionizing radiation injuries may not become evident immediately unless a high dose is received. All nuclear radiation, even in very small doses, has some harmful effect on the body and should be avoided whenever possible. The biological injury to an individual from nuclear radiation depends on many factors. Some of these factors include the following:

- Radiation dose received.
- Partial or whole-body exposure (all radiation doses referred to are due to external whole-body exposures to penetrating radiation).
- Period over which the dose is received.
- Variations in the body's resistance to radiation injury including those due to physical condition, sex, and age.
- Previous radiation exposure.
- Presence or absence of other injuries.
- Periods of recuperation between periods of radiological exposure.

FALLOUT

7-108. Fallout, a major effect of a shallow underground and underwater burst, is the radioactive material that falls from the nuclear cloud and deposits on exposed surfaces. The fallout primarily consists of fission products (gamma and beta emitters) mixed with material vaporized by the fireball and drawn up into the nuclear cloud. Fallout, whether airborne or deposited, is a hazard because it emits gamma radiation that can penetrate ship structures, buildings, and aircraft. It can also cause radiation injury or death to personnel. Deposited fallout also presents a personnel contamination hazard. The area of fallout is determined by the wind structure up to the top of the cloud. In complete calm, the fallout pattern is roughly circular. A constant wind direction leads to an elongation of the pattern. Complicated wind patterns (wind shear) as well as variations in wind pattern in time and space lead to complicated ground patterns. Fallout is difficult to predict accurately except under calm and very stable wind conditions. Reduction in yield or changing the height/depth of burst to a point where the fireball does not intersect the ground will reduce fallout, as will complete containment of an underground burst. Fallout landing on water will sink and will not constitute a hazard to ships passing through the area after fallout cessation. Fallout over a land area will remain on the surface and will be a hazard to personnel living in or passing through the area. In time, the fallout on a land surface will decay to an insignificant level.

PROTECTIVE SHIELDING

7-109. Protective shielding is one method of defense against nuclear radiation. The tremendous penetrating power of gamma rays makes it difficult to provide enough shielding to protect personnel from gamma rays. However, the structure of the ship provides some protection against them. The main materials likely to provide shielding aboard a ship are steel plating, piping, machinery, water, fuel oil, and some types of wood. Shielding materials at storage facilities include concrete and earth. The amount of shielding required to stop gamma rays is measured in half-value layer thickness or "half-thickness," for short. A half-thickness is defined as the amount of material necessary to cut down the amount of radiation to one half of its original value. The half-thickness value for each material is different. For example, a concrete shield about 6 inches thick or an earth shield about 7 1/2 inches thick will cut the gamma radiation in half. Suppose that you are standing at a plate where the gamma radiation is 400 roentgens. If you are behind a half-value layer thickness of some kind at the time, you will receive a dose of 200 roentgens. Now suppose you are standing behind two shields, each of which is a half-value layer. The 400 roentgens of gamma radiation is reduced to 200 roentgens by the first half-thickness and to 100 by the second half-thickness. With each additional half-thickness shield, you reduce the remaining gamma radiation by half. Remember that these thicknesses do not stop gamma radiation altogether; instead, they cut it in half. In a nuclear attack, one-half value layer of steel or concrete might be just enough of a shield to keep you from getting a lethal dose of gamma radiation. The approximate half-thickness of some materials, listed in order of their effectiveness as shields against gamma radiation, are shown in Table 7-1.

Table 7-1. Materials Effectiveness Against Gamma Radiation

	Initial	Residual
Steel	1.5 inches	0.7 inches
Concrete	6.0 inches	2.2 inches
Earth	7.5 inches	3.3 inches
Water	13.0 inches	4.8 inches
Wood	23.0 inches	8.8 inches

PREVENTIVE MEASURES (BEFORE ATTACK)

7-110. Personnel should take preventive measures before an attack. The steps that are listed are not in a required sequence, they only list the things that should be performed. The situation at the time will determine the sequence. Notify all ship's masters and coxswains. They must take immediate charge of the situation aboard their vessel.

- Sound the CBRNE alarm.
- Shut all watertight doors, ports, and ventilation systems.
- Cease all cargo operations.
- Get away from the pier or beach, and put out to sea.
- Get all "soft" items off the decks, such as wood, hawsers, line, canvas, and so on.
- If the vessel is not equipped with a wash-down system, rig the fire hoses for wash-down.
- Commence washing the vessel down.
- Secure all loose gear inside the vessel.
- Have personnel don their protective clothing.
- Get personnel to take cover in interior of the vessel. Get them as far down below the centerline and in between the engines as possible.

Note: If the vessel is 1,000 yards or more from "ground zero," the crew should survive. With the crew below the waterline and in between the engines, the bulkheads, engines, ship's hull, and the water all provide a shield against radiation. Personnel should also take the necessary actions against a nuclear attack. Table 7-2 shows the actions personnel should take during nuclear denotations.

Table 7-2. Recommended Personnel Action Against Nuclear Detonations

		WITH WARNING	NO WARNING
BURST TYPE	TOPSIDE PERSONNEL	BELOW DECK PERSONNEL	TOPSIDE PERSONNEL
Air	A	B	C
Surface	A	B	C
Underwater	B	B	

A – Lie prone and hold on to solid ship structure.
B – Stand with knees flexed and hold on to solid ship structure.
C – Hands-to-face evasion.

PREVENTIVE MEASURES

DURING AN ATTACK

7-111. Personnel should take preventive measures during an attack. The following are some precautions to take:

- DO NOT eat, drink, or smoke.
- Brace yourself and hold on to a secure object.
- Continue the wash-down system for at least 1 hour.
- Keep all cuts or open wounds bandaged.
- At the sign of brightness, shut your eyes and turn your face away.

Note: When you are in an open topside area (where you can quickly drop to the deck) and you see the flash or see the sky light up, close your eyes and immediately raise your hands to cover your face. Meanwhile, drop to the deck as quickly as possible. Use your shoulder, not your hands, to break a fall. Curl up to present a minimum target. You may feel the heat from the detonation. Two to 5 seconds after the flash (depending on the weapon yield) or after the heat sensation is over, remove your hands from your face. Then immediately and firmly, grab a solid ship structure to prevent the air-blast winds from blowing you overboard or against the ship's structure. You may suffer flash blindness for a period up to 30 minutes.

AFTER AN ATTACK

7-112. Personnel should take preventive measures after an attack. The following are some precautions to take:

- Maintain maximum speed. Put the vessel on a course that is crosswind and away from the point of detonation.
- Continue the wash-down for at least 1 hour.
- Do not eat, drink, or smoke.
- Check the interior of vessel for watertight integrity.
- Observe the fallout pattern and continue to leave the ventilation system shut off.

- If radiation instruments are on board ship, monitor interior of vessel, all open food, liquids, and cigarettes.
- Don protective clothing before going on deck.
- Commence vessel wash-down.

RADIOLOGICAL DECONTAMINATION

7-113. This neither neutralizes nor destroys the contamination. Instead, it merely removes the contamination from one particular area and transfers it to an area in which it presents less of a hazard. At sea, dispose of radioactive material directly over the side. Flushing with water, preferably water under high pressure, is the most practicable way of rapidly decontaminating topside surfaces. Aboard ship, a water wash-down system is used to wash down all the exterior surfaces (from high to low and from bow to stern). The wash-down system consists of piping and a series of nozzles that are specially designed to throw a large spray pattern on weather decks and other surfaces. Permanent wash-down systems are being built into ships under construction or conversion. Interim wash-down system kits are provided for ships already in service. If the wash-down system is turned on before the arrival of contamination, the system prevents heavy contamination of the ship by coating the weather surfaces with the flowing stream of water. The flowing stream of water carries away radioactive particles as they fall on the ship and keeps radioactive particles from settling into cracks and crevices. If some areas of the ship become heavily contaminated before the wash-down system is activated, it will probably be necessary to hose down such areas vigorously, using seawater under pressure. Hosing should proceed from higher to lower surfaces, from bow to stern, and, if possible, from the windward side to the lee side. Every possible precaution should be taken to see that contaminated water does not flow back over cleaned areas. Precautions must also be taken to see that contaminated water is not hosed into the interior of the ship through vents, doors, or hatches. The hose should be directed so that the water strikes the surface about 8 feet from the nozzle. The hose stream should sweep horizontally from side to side, moving lower on each sweep. The hosed areas should be overlapped somewhat on each sweep to ensure complete washing. The runoff should be directed into scuppers and deck drains as rapidly as possible to keep the contaminated water moving and to prevent pools of contaminated water from forming. Hosing down will be most effective if it is done before metal or painted surfaces have dried after contaminating material has been deposited. However, contamination that has been deposited despite wash-down will also resist hosing alone. Vigorous scrubbing with deck brushes and detergents, followed by hosing, is required. Ships without wash-down systems will initially decontaminate by hosing down with seawater as soon as the tactical situation permits.

CONTAMINATION MARKERS

7-114. Areas or objects that are contaminated by CBRNE attack must be clearly marked to warn personnel approaching the area of the existence of hazards. Contamination markers should outline dangerous areas and establish boundaries within which safety control must be exercised. Radiation hot spots—that is, areas having radiation intensities significantly greater than the general radiation level of the surrounding areas—should be identified. The standard NATO system for marking areas that are contaminated by CBRNE attack is used. Figure 7-43 shows these standard survey markers. Each marker is in the shape of a right triangle; one side of the triangle is about 11 1/2 inches long, and the other two sides are about 8 inches long. The markers may be made of wood, metal, plastic, or other rigid material.

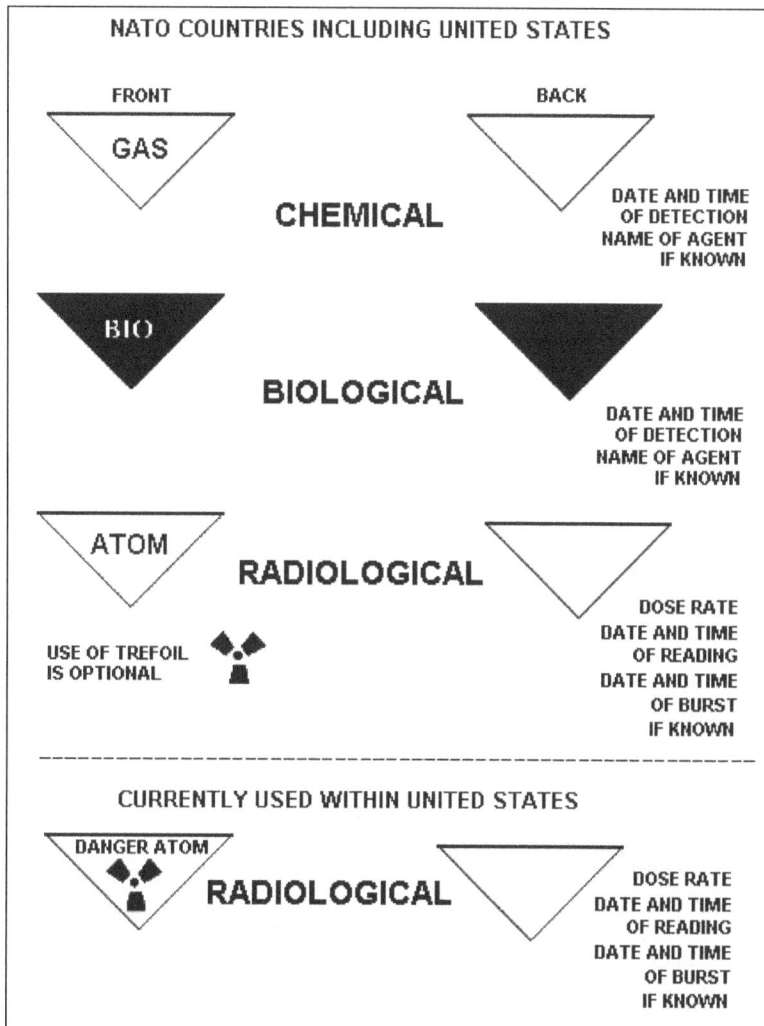

Figure 7-43. Standard NATO CBRNE Markers

This page intentionally left blank.

Chapter 8

Marlinespike Seamanship

INTRODUCTION

8-1.   Marlinespike seamanship is a general term for the handling, knotting, whipping, splicing, and caring for fiber line and wire rope used aboard ship or in other marine operations. The knowledge and practical application of marlinespike seamanship principles and procedures are important to the crewman. A person who is truly dedicated to his/her work takes pride in the handling and caring of fiber line and wire rope to make operations safe and satisfactory. This chapter covers all phases of marlinespike seamanship required by the watercraft operator.

FIBER LINE

8-2.   One characteristic of a competent watercraft operator is his/her ability to work with fiber line. In order to do this, he/she must know the characteristics and properties of fiber line, how to handle and care for the line, and tie basic knots, bends, and hitches.

8-3.   Fiber line is made of either vegetable or synthetic fibers. Vegetable fibers include manila, sisal, hemp, cotton, and flax. Synthetic fibers include nylon, Dacron, polyethylene, and polypropylene. Nylon is the primary synthetic fiber line used in the Army, so this text covers only nylon and none of the other types of synthetic fibers. These materials are described below.

- **Manila.** Manila is a strong fiber that comes from the leaf stems of the abaca plant that is in the banana family. Varying in length from 4 to 15 feet in their natural state, the fibers have the length and quality to give manila rope relatively high elasticity, strength, and resistance to wear and deterioration. Most lines used in the Army are manila.
- **Sisal.** Sisal is made from sisalana, a species of the agave plant. Although sisal is seldom used in the Army, it is covered here because it is a good substitute for manila. Sisal fibers are 2 to 4 feet long, and withstand exposure to seawater very well.
- **Hemp.** Hemp is a tall plant that has useful fibers for making rope and cloth. It was used extensively before manila was introduced. Now hemp's principal use is in fittings such as ratline and marline. Because hemp is absorbent, the fittings are invariably tarred to make them more water resistant. Uses of marline include lashings and whippings.
- **Nylon.** Nylon is made from mineral products. It is waterproof, absorbs shocks, stretches, and resumes its original length. It also resists abrasion, decay, and fungus growth.

CONSTRUCTION AND SIZE

8-4.   Figure 8-1 shows how a fiber line is made by twisting fibers into yarns, yarns into strands, and strands into the finished line. The fibers are twisted from left to right to spin the yarn. The yarn is twisted from right to left to form the strands. The strands are then twisted from left to right to lay or form the line. Three stranded nylon line is constructed in the same way as fiber line.

8-5.   Fiber line is measured, in inches, by its circumference. One exception is "small stuff". Small stuff has fiber line that is 1 ¼ inches or less in circumference. It also has three strands. The number of threads it contains determines its size. Small stuff will range in size from 6 to 21 threads. To determine the number of threads, count the number in one strand and then multiply it by three. Small stuff is used for lashing material and heaving lines. Fiber line between 1 3/4 and 5 inches in circumference is referred to as line. Line over 5 inches in circumference is referred to as hawser. Hawsers are used for mooring and towing.

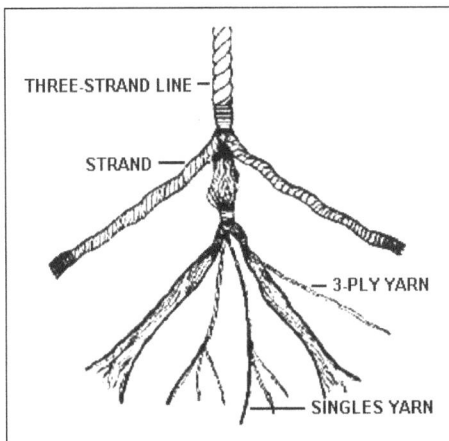

Figure 8-1. Fabrication of Fiber Line

STRENGTH

8-6.   Manila is the standard line against which all other types of fiber line are measured. Implied in the measurement is that all the other lines have the same circumference as that of the manila line against which each is measured. With manila line having a strength of 100 percent, the strengths of the other lines are shown in Table 8-1.

Table 8-1. Line Strengths

Type Line	Strength
Manila	100 percent
Three-strand nylon	250 percent
2-in-1 braided nylon	300 percent
Sisal	80 percent

8-7.   Nylon line is fast replacing natural fiber line for mooring and towing because of its strength and durability. With three-stranded nylon being 250 percent stronger than manila, size for size, it allows the use of smaller and lighter mooring and towing lines. Three-stranded nylon line will stretch 30 to 35 percent under an average load or a load that does not exceed the safety factor for that size line. Three-stranded nylon line will stretch 40 percent without being damaged and draw back to its original length. The nylon line will part at 50 percent.

USEFUL FORMULAS

8-8.   The manufacturer states the size and breaking strength (BS) of its lines. If this information is available, use the manufacturer's figures for determining the strength of line. If this information is not available, then use the rule of thumb to compute the safe working load (SWL) and the BS. These rules of thumb give only approximate results (because of the constants that are used in Table 8-2) but the error will be on the side of safety.

Table 8-2. Line Constants

Type Line	Constant
Sisal	160
Manila	200
Three-strand nylon	500
2-in-1 braided nylon	600

8-9.   With "C" meaning circumference in inches, the formula for SWL in pounds is as follows:

C2 x constant for line = SWL

Examples

3-inch sisal:	3 x 3 x 160 = 9 x 160 = 1,440 pounds SWL
3-inch manila:	3 x 3 x 200 = 9 x 200 = 1,800 pounds SWL
3-inch three-strand nylon:	3 x 3 x 500 = 9 x 500 = 4,500 pounds SWL
3-inch 2-in-1 braided nylon:	3 x 3 x 600 = 9 x 600 = 5,400 pounds SWL

8-10.   In marine operations, a safety factor of 5 is generally used. If you multiply this times the SWL, you will find the BS of the fiber line. This is the amount of weight in pounds required to part the line. If you are given the BS of a line and divide it by the safety factor 5, you will find the SWL.

Note: The safety factor of 5 is valid when using new line or line that is in good condition. As line ages and wears out through use, the safety factor drops. Old line may have a safety factor of 3.

Examples

Example 1:	Find the BS of 3-inch manila line:
Solution:	C2 x constant = SWL: 3 x 3 x 200 = 9 x 200 = 1,800 pounds
	BS = SWL x SF = 1,800 x 5 = 9,000 pounds
Example 2:	Find the SWL for a 6-inch hawser that has a BS of 36,000 pounds:
Solution:	SWL = BS / SF = 36,000 / 5 =7,200 pounds

8-11.   Table 8-3 shows the SWL and BS of the various sizes of lines used on Army watercraft.

CORDAGE

8-12.   In marine usage, cordage is a collective term that includes all cord, twine, line, and string made from twisted vegetable or synthetic fibers. Cord, string, and twine are loosely used to mean small line.
- **Cotton Twine.** This is like the string found in homes. It is used for temporary whippings and should be run through beeswax before use.
- **Sail Twine.** This is made of flax or a better grade of cotton that's used in cotton twine. It is waxed during manufacture. Measured by the number of plies and comes in three to seven plies. Like a yarn, a ply has a certain number of fibers. Sail twine is used for whippings.
- **Marline.** This is tarred hemp. It is made of two yarns with fibers making up the yarns. Marline is used for whippings on lines 3 inches and larger.
- **Flax.** This is braided. It is used for halyards or the lines for flags and pennants. Flax is stronger than cotton and lasts longer.

INSPECTION

8-13.   The outside appearance of the line is not always a good indication of its internal condition. Therefore, it is necessary to inspect the inside as well as the outside. Overloading a line may cause it to break with possible damage to material and injury to personnel.

8-14.   Inspect line carefully at regular intervals to determine its condition. Slightly untwist the strands to open the line so you can examine the inside. Mildewed line has a musty odor and inside fibers have a dark, stained appearance. It is ordinarily easy to identify broken strands of yarn. Dirt and sawdust-like material inside the line means it has been damaged. If the line has a core, it should not break away in small pieces. If it does, the line has been overstrained. If the line appears to be satisfactory in all respects, pull out two fibers and try to break them. Sound fibers should offer considerable resistance to breakage. When any unsatisfactory conditions are found, destroy the line or cut it up in short pieces. Make sure that none of these pieces is long enough to permit its use in hoisting. This not only prevents the use of line for hoisting, but saves the short pieces for miscellaneous use such as lashings, whippings, and seizing.

Table 8-3. Line Strength Table (Safety Factor of 5)

Size in Inches	Manila		Three-Strand Nylon		2-in-1 Braided Nylon	
	SWL	BS	SWL	BS	SWL	BS
1	200	1,000	500	2,500	600	3,000
1 1/2	450	2,250	1,125	5,625	1,350	6,750
2	800	4,000	2,000	10,000	2,400	12,000
2 1/2	1,250	6,250	3,125	15,625	3,750	18,750
3	1,800	9,000	4,500	22,500	5,400	27,000
3 1/2	2,450	12,250	6,125	30,625	7,350	36,750
4	3,200	16,000	8,000	40,000	9,600	48,000
4 1/2	4,050	20,250	10,125	50,625	12,150	60,750
5	5,000	25,000	12,500	62,500	15,000	75,000
5 1/2	6,050	30,250	15,125	75,625	18,150	90,750
6	7,200	36,000	18,000	90,000	21,600	108,000
6 1/2	8,450	42,250	21,125	105,625	25,350	126,750
7	9,800	49,000	24,500	122,500	29,400	147,000
7 1/2	11,250	56,250	28,125	140,625	33,750	168,750
8	12,800	64,000	32,000	160,000	38,400	192,000
8 1/2	14,450	72,250	36,125	180,625	43,350	216,750

UNCOILING NEW LINE

8-15. New line is coiled, bound, and wrapped in burlap as a protective covering. The burlap covering should not be opened until the line is to be used. To open, strip back the burlap wrapping and look inside the coil for the end of the line. It should be at the bottom of the coil. If it is not, turn the coil over so that the end will be at the bottom. Put your hand down through the center and grab the end of the line. Pull the end of the line up through the center of the coil. As the line comes up through the coil, it will unwind in a counterclockwise direction (Figure 8-2).

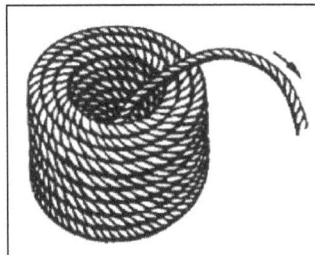

Figure 8-2. Opening View Coil Line

STOWING LINE

8-16.   Hawsers and line must never be stowed when wet. After they are thoroughly dry and cleaned, they should be stowed in a dry, unheated, well-ventilated area or locker. Hawsers should be loosely coiled on wood grating or hung on wood pegs. Hawsers should never be stowed in contact with metal surfaces. Line can be coiled, Flemish down, or faked down. Coiling down a line means laying it up in circles, roughly one on top of the other. Always coil down right-laid line right-handed or clockwise (Figure 8-3). When a line is coiled down, the top end is ready to run off. If you try the bottom end, the line will kink. If for some reason the bottom end must go first, it is necessary to turn over the coil to free it for running.

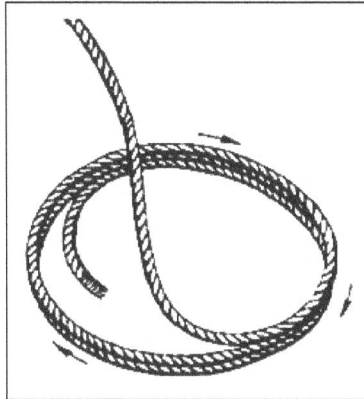

Figure 8-3. Coiling Line

8-17.   To Flemish down a line, start with the bitter end and lay on the deck successive circles of the line. Always Flemish down right-laid line clockwise and left-laid counterclockwise. Figure 8-4 shows the bitter end is in the middle. Short lengths of a line, such as bitter ends of boat painters and guys, usually are Flemish down.

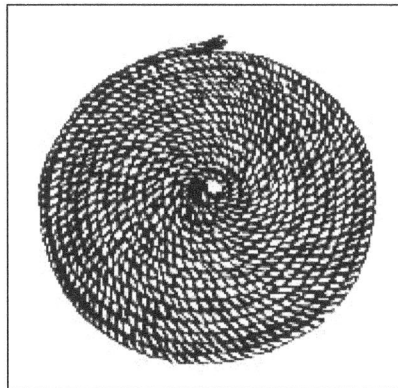

Figure 8-4. Flemish Down a Line

Note: Nylon comes on reels. To uncoil it, place a reel on stands or jacks. Nylon is handled differently from natural fiber line. Coil three-strand nylon clockwise one week and counterclockwise the next week, because continuously coiling three-strand nylon line down one way tends to take the lay out of the strands. With 2-in-1 braided nylon line, simply put it in figure eights.

8-18.   Faking down a line is laying it up the same way as in coiling down, except that it is laid out in long flat bights, one alongside the other, instead of in coils (Figure 8-5). The main advantage of working with line that is faked down is that it runs off easily.

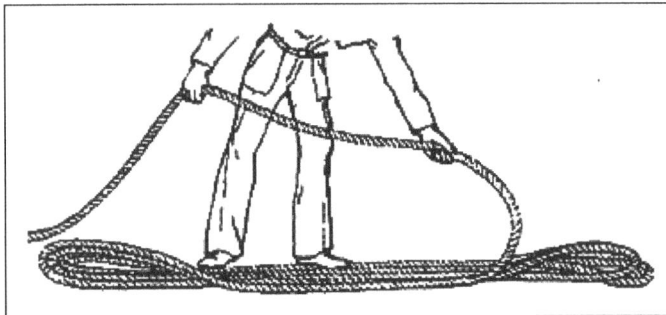

Figure 8-5. Faking Down a Line

WHIPPING A LINE

8-19.   Never cut a line or leave the end of a line dangling loose without a whipping to prevent it from unlaying. A line without a whipping will unlay of its own accord. A frayed line is a painful sight to a good seaman. Whenever a line or hawser has to be cut, whippings should be put on first. Put one whipping on each side of the cut. To prevent fraying, a temporary or plain whipping can be put on with any type of cordage, even with rope yarn. Figure 8-6 shows one of several methods that can be used for putting a temporary whipping on a line.

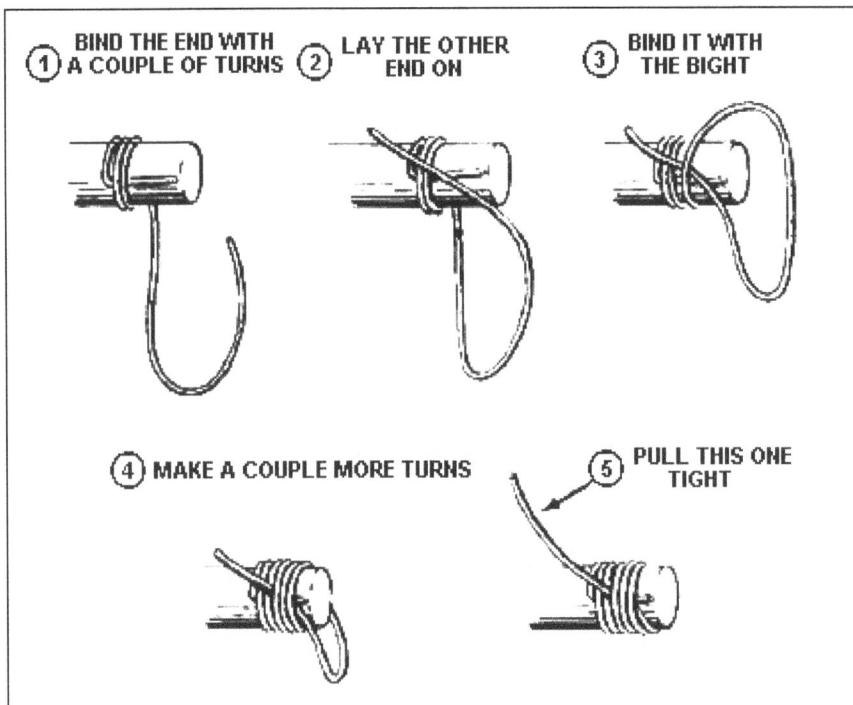

Figure 8-6. Plain or Temporary Whipping

8-20. Do the following to make a temporary whipping (Figure 8-6).
- **Step 1.** Lay the end of the whipping along the line and bind it down with three or four round turns.
- **Step 2.** Then lay the other end on the opposite way.
- **Step 3.** Bind it with a bight of the whipping.
- **Step 4.** Then take a couple more turns.
- **Step 5.** Take the bitter end of the whipping and pull it tight.

8-21. As its name implies, a permanent whipping is put on to stay. One way to put on a permanent whipping is with a needle and a sewing palm. Sewing palms are made for both right- and left-handed people. The width of the permanent whipping should equal the diameter of the line. Two whippings are recommended. The space between the two whippings should be six times the width of the first whipping. Do the following steps to make a permanent whipping (Figure 8-7).

Note: The needle is threaded with sail twine, doubled. Figure 8-7 also shows a single strand for clearness.

- **Step 1.** Push the needle through the middle of a strand so that it comes out between two strands on the other side.
- **Step 2.** Wind the turns toward the bitter end. The number of turns or the width of the whipping will depend on the diameter of the line.
- **Step 3.** Push the needle through the middle of a strand so that it comes out between two strands again.
- **Step 4.** Then go up and down between strands so as to put a cross-seizing between each pair of strands.
- **Step 5.** Pull each cross-seizing taut before taking the next one.
- **Step 6.** Have the thread come out through the middle of a strand the last time you push it through so that after you knot and cut the thread, the strand will hold the end of the twine.

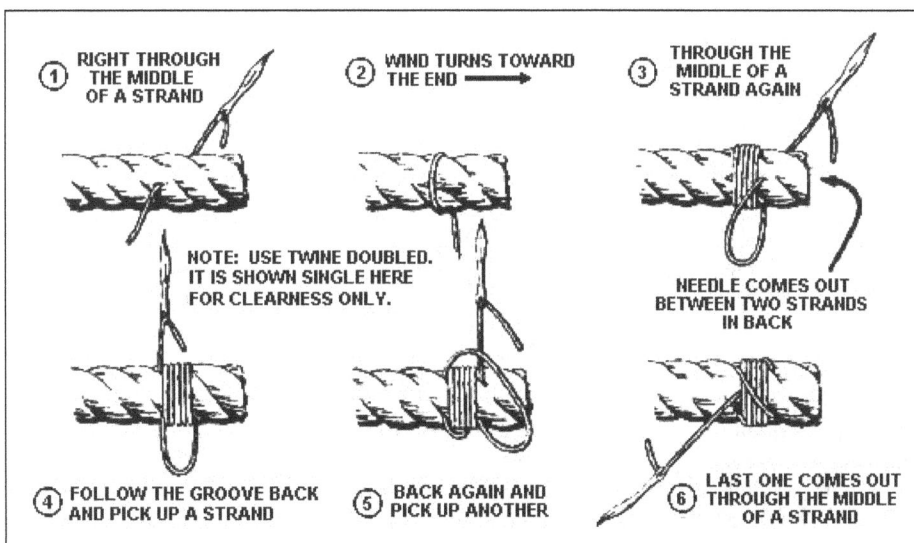

Figure 8-7. Making a Permanent Whipping

KNOTS, BENDS, AND HITCHES

8-22.   A good knot must be easy to tie, hold without slipping, and be easy to untie. The choice of the best knot, bend, or hitch to use depends largely on the job it has to do (Figure 8-8). This TC explains why a given method is used and also gives the efficiency or strength of many of the knots, bends, and hitches. Always follow this rule: never tie a knot on which you are not willing to stake your life. Each of the three terms—knot, bend, and hitch—has a specific definition. In a knot, a line is usually bent or tied to itself, forming an eye or a knob or securing a cord or line around an object, such as a package. In its noun form, a bend ordinarily is used to join the ends of two lines together. In its verb form, bend means the act of joining; bent is the past tense of bend. A hitch differs from a knot and a bend in that it is ordinarily tied to a ring, around a spar or stanchion, or around another line. In other words, it is not merely tied back on itself to form an eye or to bend two lines together. Tying a knot, bend, or hitch in a line weakens it because the fibers are bent sharply, causing the line to lose varying degrees of its efficiency or strength. A general rule to follow is to use a knot, bend, or hitch for temporary work and use a splice for permanent work because it retains more of the line's strength.

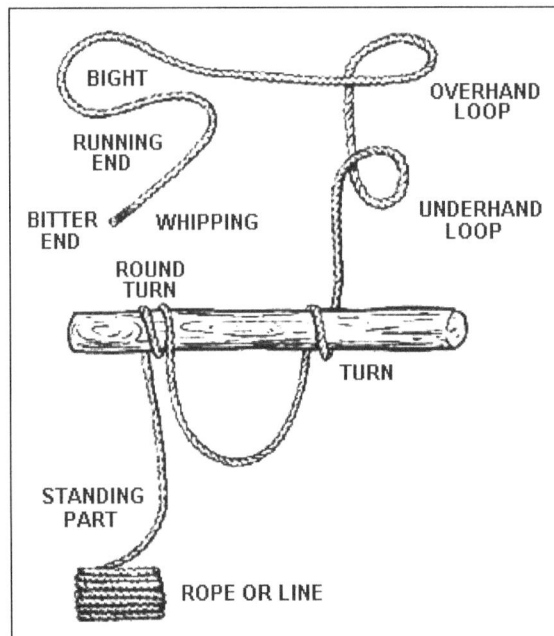

Figure 8-8. Elements of the Knot, Bend, and Hitch

KNOTS

Overhand Knot

8-23.   The overhand knot (Figure 8-9) is the basis for all knots. It is the simplest of all and the most commonly used. It may be used to prevent the end of a line from untwisting, to form a knot at the end of a line, or to be part of another knot. When tied to the end of a line, this knot will prevent it from running through a block, hole, or other knot.

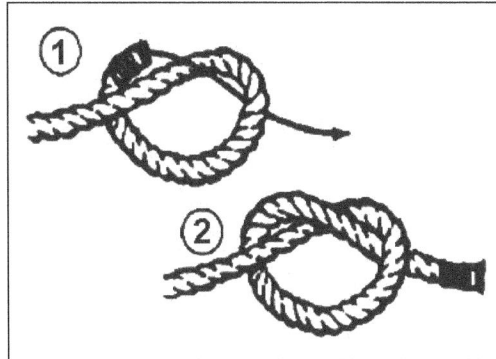

Figure 8-9. Overhand Knot

Figure Eight Knot

8-24.   The figure eight knot (Figure 8-10) is used to form a larger knot at the end of a line than would be formed by an overhand knot. It is used to prevent the end of the line from running through a block. It is an easy knot to tie. To tie this knot, form an overhand loop in the line and pass the running end under the standing part, up the other side, and through the loop. Tighten the knot by pulling on the running end and the standing part.

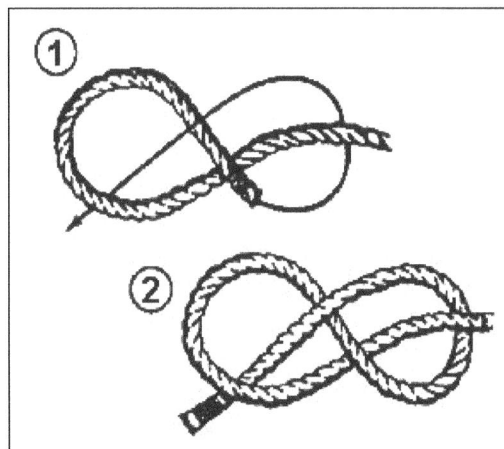

Figure 8-10. Figure Eight Knot

Square Knot

8-25.   Use the square knot (Figure 8-11) to tie two lines of equal size together so that they will not slip. Figure 8-11 shows that for the square knot, the end and standing part of one line come out on the same side of the bight formed by the other line. This knot will not hold if the lines are wet or are of unequal sizes. It will tighten under strain but can be untied by grasping the ends of the two bights and pulling the knot apart. Its strength is 45 percent. To avoid a "granny" or a "fool's knot" which will slip, follow this procedure. Take the end in your right hand and say "over and under." Pass it over and under the part in your left hand as shown in Figure 8-11. With your right hand, take the end that was in your left hand. This time say to yourself "under and over." Pass it under and over the part in your left hand.

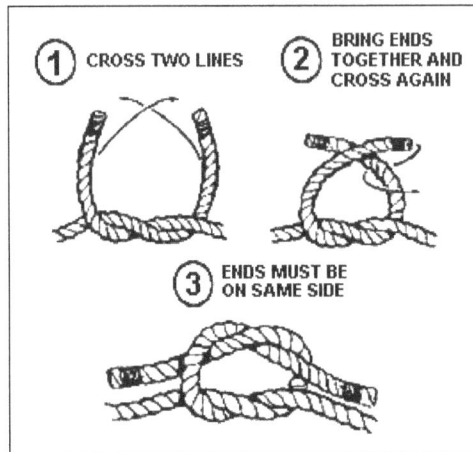

Figure 8-11. Square Knot

SHEET OR BECKET BEND

8-26.   Use a single sheet or becket bend to tie two lines of unequal size together and to tie a line to an eye. Always use a double sheet or becket bend to tie the gantline to a boatswain's chair. The single sheet or becket bend will draw tight, but will loosen when the line is slackened. The single sheet or becket bend is stronger than the square knot, with a strength of 55 percent, and is easily untied in comparison to the square knot. To tie a single sheet or becket bend (Figure 8-12), take a bight in the larger of the two lines. Using the smaller of the two lines, put its end up through the bight. Then put it around the standing part of the larger line first because it will have the strain on it and then around the end of the larger line. Next put the end of the smaller line under its standing part. The strain on the standing part will hold the end. Notice in the double sheet or becket bend that the end of the smaller line goes under its standing part both times.

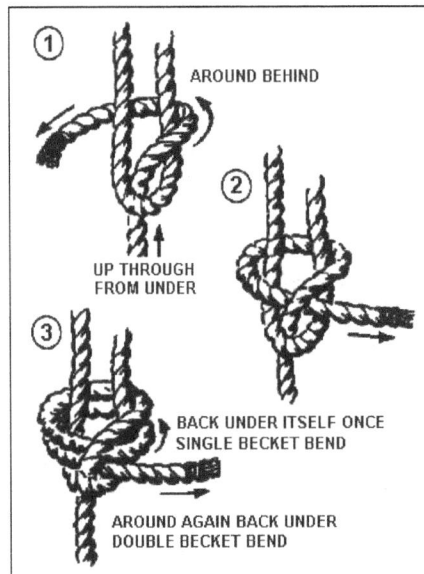

Figure 8-12. Tying the Single and Double Sheet or Becket Bend

BOWLINE

8-27.   Use the bowline to tie a temporary eye in the end of a line. A bowline neither slips nor jams and unties easily. An example of a temporary use is that of tying a heaving line or messenger to a hawser and throwing it to a pier where line handlers can pull the hawser to the pier, using the heaving line or messenger. To tie a bowline (Figure 8-13), hold the standing part with your left hand and the running end with your right. Flip an overhand loop in the standing part, and hold the standing part and loop with the thumb and fingers of your left hand. Using your right hand, pass the running end up through the loop, under the standing part, and down through the loop. Its strength is 60 percent.

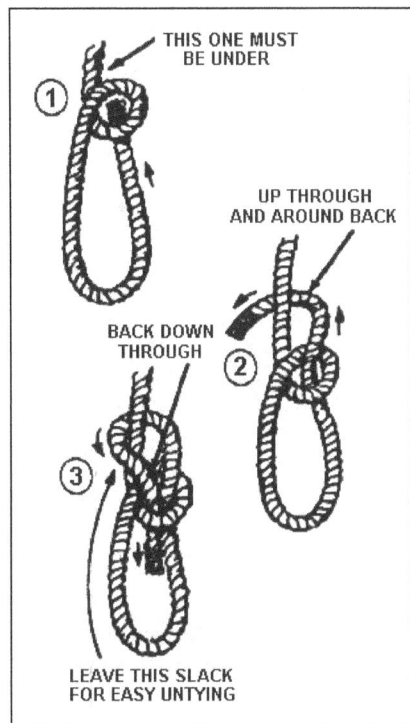

Figure 8-13. Tying a Bowline

Bowline on a Bight

8-28.   A bowline on a bight gives two loops instead of one, neither of which slips. It can be used for the same purpose as a boatswain's chair. It does not leave both hands free, but its twin, non-slipping loops form a comfortable seat. Use the bowline on a bight when—

● Strength (greater than a single bowline) is necessary.
● A loop is needed at some point in a line other than at the end.
● The end of a line is not accessible.

8-29.   The bowline is easily untied and can be tied at the end of a line by doubling the line for a short section. To tie a bowline on a bight (Figure 8-14) double the line, form an overhand loop, and put the end of the bight through the loop. Put your hand through the bight, take hold of the bight under the loop, and pull it through the first bight to tighten the knot.

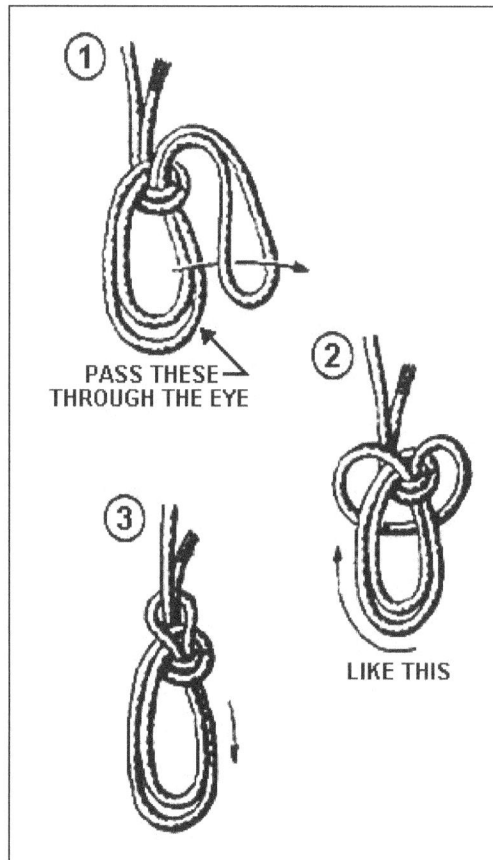

Figure 8-14. Tying a Bowline on a Bight

French Bowline

8-30.   Use a French bowline as a sling for lifting an injured person. For this purpose, one loop is used as a seat and the other loop is put around the body under the arms, then the knot is drawn tight at the chest. Even an unconscious person can ride up safely in a properly secured French bowline, because his/her weight keeps the two loops tight so that he/she will not fall out. It follows, though, that it is necessary to take care not to allow the loop under his/her arms to catch on any projections. Also, use the French bowline when a person is working alone and needs both hands free. The two loops of the knot can be adjusted to the required size. Figure 8-15 shows the step-by-step procedure for tying the French bowline.

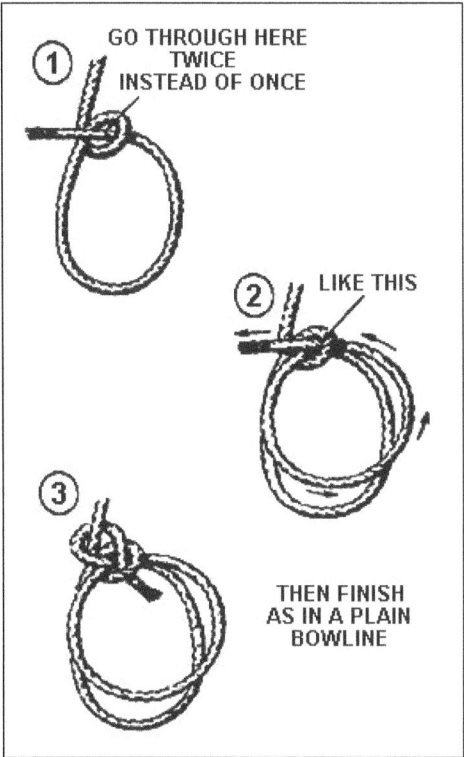

Figure 8-15. Tying a French Bowline

DOUBLE CARRICK BEND

8-31.   A double carrick bend with its ends seized (Figure 8-16) is recommended for tying together two hawsers. Even after a heavy strain, it is easy to untie because it never draws up; its strength is 56 percent. However, a double carrick will draw up if the ends are not seized.

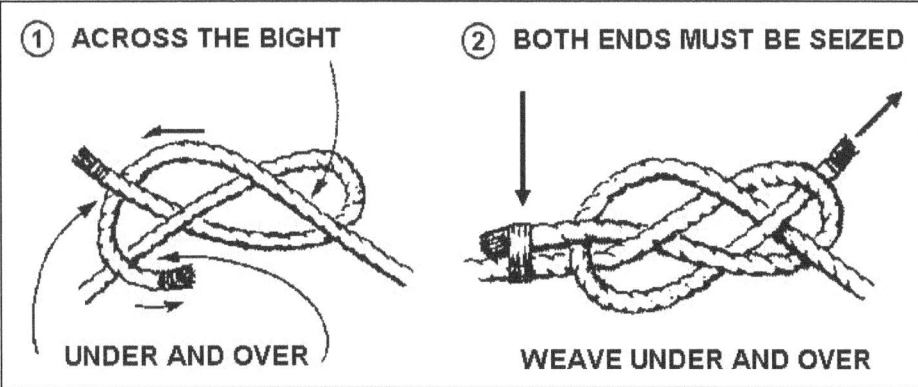

Figure 8-16. Tying the Double Carrick Bend

HITCH

Half Hitch

8-32.   The half hitch is used to back up other knots, and tie with the short end of the line. Never tie two half hitches by themselves. Instead, take two round turns so that the strain will be on the line, not the hitches, and then tie the hitches (Figure 8-17).

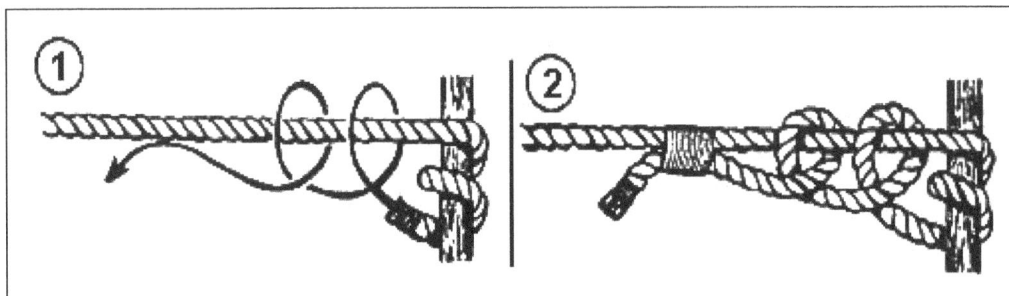

Figure 8-17. Half Hitch

Clove Hitch

8-33.   The best knot for tying a line to a ring, a spar, or anything that is round is a clove hitch (Figure 8-18). It will not jam or pull out. Its strength is 55 to 60 percent.

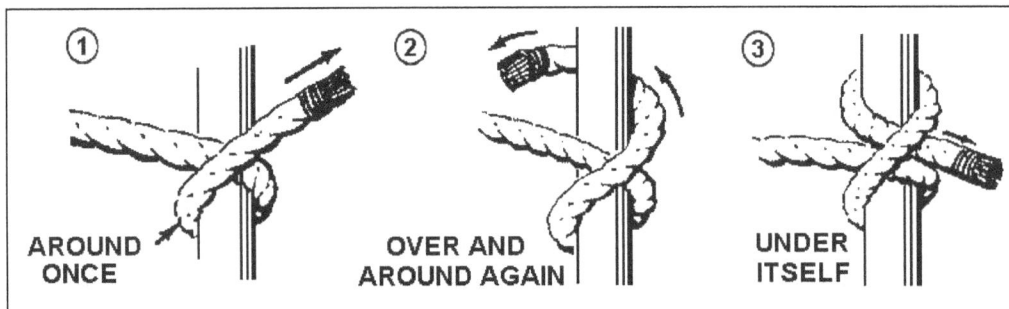

Figure 8-18. Clove Hitch

Stopper Hitch

8-34.   A possible defect of a clove hitch is that it can slide along the round object to which it is tied. To prevent this, use a stopper hitch (Figure 8-19), commonly called a rolling hitch. When tying, make a turn around the line with the stopper (first view). Pull tight and take another turn. This one must cross the first turn and then pass between the first turn and the stopper (second view). This completes the stopper hitch itself, but it must be stopped off in one of two ways.

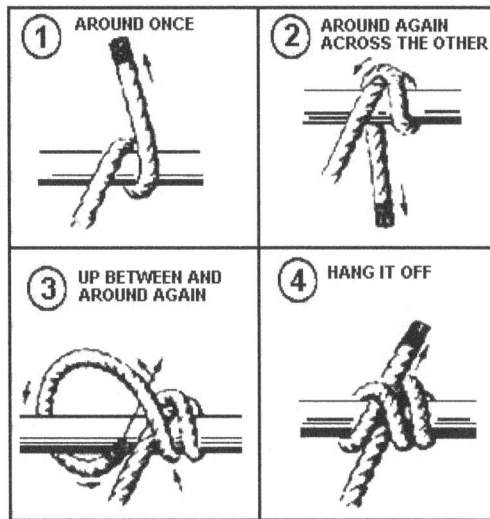

Figure 8-19. Stopper Hitch

8-35.   You may make two or more turns with the lay of the line and then seize the stopper to the line with marline. Another method is to tie a half hitch directly above the rolling hitch (third view), and then take a couple of turns against the lay, and seize the stopper to the line.

Stage Hitch

8-36.   Use a stage hitch (Figure 8-20) for working over the side of a vessel. A stage hitch consists of a plank with a wooden horn attached at a right angle to the plank near each end to keep it away from the side. Note that two parts of the line go under the plank. Therefore, the line supports the plank, as well as the horns. This gives more protection to persons working on the stage.

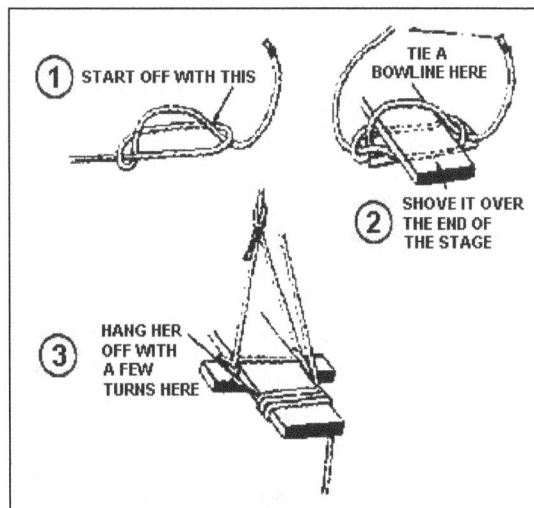

Figure 8-20. Stage Hitch

MONKEY FIST

8-37.   The monkey fist (Figure 8-21) is tied at the end of a heaving line and a weight is put in it so that it can be thrown for a distance with some ease and accuracy. The monkey fist consists of three sets of turns taken at right angles to each other. For clarity, Figure 8-21 shows only three turns in each set; four turns per set are more likely to be used. To tie a monkey fist, start as in view 1, taking a set of turns around your hand. Then slip this set off your hand, hold it as shown in view 2, and pass the running end over your thumb and under and over the first set. Complete this set of turns. Put the last set around the second and through the first as shown in view 3. Note that the first turn of the last set locks the first two sets in place.

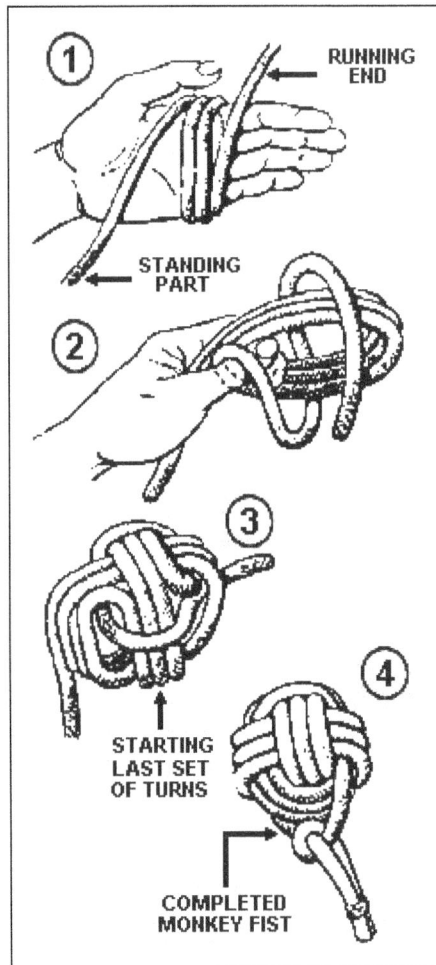

Figure 8-21. Monkey Fist

8-38.   After completing the third set of turns, insert a 5-to-10-ounce weight in the monkey fist. Tighten the turns by working the slack back towards the standing part. In a properly tied monkey fist, the ends come out at opposite corners as shown in view 4. To complete the monkey fist, put a half hitch on the standing part with the running end and seize it to the standing part.

SPLICING THREE-STRAND FIBER LINE

8-39.   Splicing is a method of permanently joining the ends of two lines or of bending a line back on itself to form a permanent loop or an eye. If two lines are going to be spliced, strands on an end of each line are un-laid, and the strands are interwoven with those of the standing part of the line. Small stuff can be spliced without need of a fid. A fid is a tapering length of hickory or some other hard wood used in splicing larger lines. A knife is needed to cut off the ends of the strands. This paragraph explains and shows the back, short, and eye splices.

BACK SPLICE WITH A CROWN KNOT

8-40.   Where the end of a fiber line is to be spliced to prevent unlaying and a slight enlargement of the end is not objectionable, use a back splice. This splice is usually done on small stuff. To make this splice, do the following:

- **Step 1.** Unlay six turns of the line (Figure 8-22).

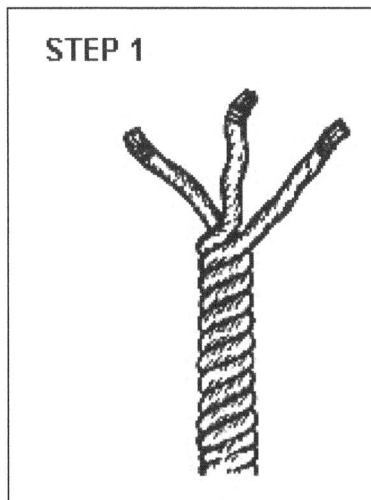

Figure 8-22. Making a Back Splice, Step 1

- **Step 2.** To start the crown knot, form a bight with the left strand and lay the bitter end of the strand between the right and center strand. Then lay the center strand over the running end of the left strand. Take the right strand under the running end of the left strand, over the running end of the center strand, and back through the bight of the left strand. Then take all the slack out of the strands and gently pull the strands tight (Figure 8-23).

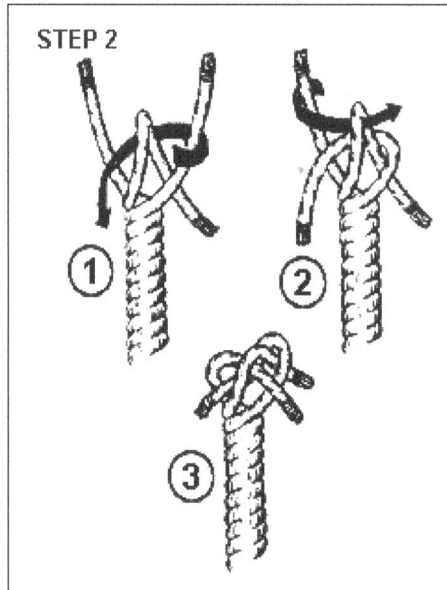

Figure 8-23. Making a Back Splice, Step 2

● **Step 3.** Start the left strand; go over one strand, tuck under the next one, and pull the strand tight (Figure 8-24).

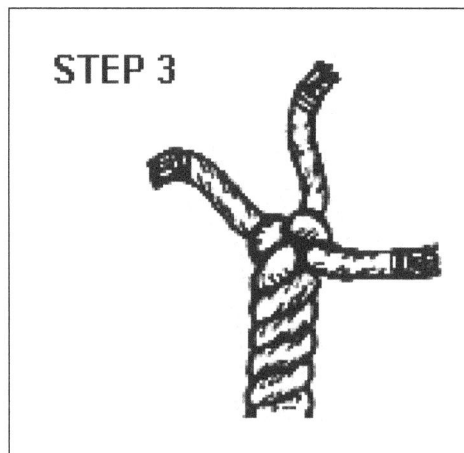

Figure 8-24. Making a Back Splice, Step 3

- **Step 4.** Turn the line and tuck each strand. Three complete tucks are required for each strand (Figure 8-25).

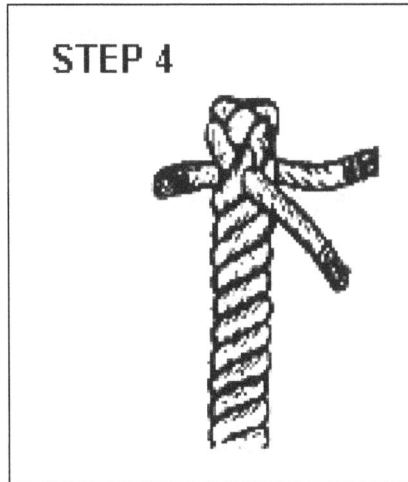

Figure 8-25. Making a Back Splice, Step 4

- **Step 5.** Trim off the ends of the strands. Then lay the splice on the deck, put your foot on it, and roll it back and forth. This will tighten up and smooth out the splice.

SPLICE

SHORT SPLICE

8-41.   The short splice (Figure 8-26) is as strong as the rope of which it is made. However, the short splice will increase the diameter of the line at the splice and can be used only where this increase in diameter will not affect the operation. Use the short splice to repair damaged lines. The damaged parts of the line are cut out and the short splice rejoins the line. Only lines of the same size can be joined together using the short splice. Do the following to make a short splice:

- **Step 1.** Untwist one end of each line five complete turns. Whip or tape each strand. Bring these strands tightly together so that each strand of one line alternates with a strand of the other line. Put a temporary whipping on the lines where they join to keep them from suddenly coming apart. Do this with small lines until you are skilled enough to hold them together while you tuck.
- **Step 2.** Starting with either line, tuck a round of strands in the other line. Then, using the strands of the other line, tuck a round in the first line. These first two rounds of tucks are expressed: "Tuck in one direction. Reverse and tuck in the other direction." When making a round of tucks, regardless of the direction, face where the lines are butted so to always tuck from right to left. Pull each strand as required to tighten the center of the splice.
- **Step 3.** Tuck two more rounds in each direction. After tucking in one direction and reversing and tucking in the other direction, pull the strands as required to strengthen the center of the splice. When finished with three rounds of tucks in each direction, cut off any excess length on the strands. To have a smoother splice, you may cut off one-third of the circumference of each strand before making the second round of tucks and another one-third cut before the third round.
- **Step 4.** When the splice is completed, cut off the excess strands as before. Lay the splice on the deck and roll it with your foot to smooth out and tighten the splice.

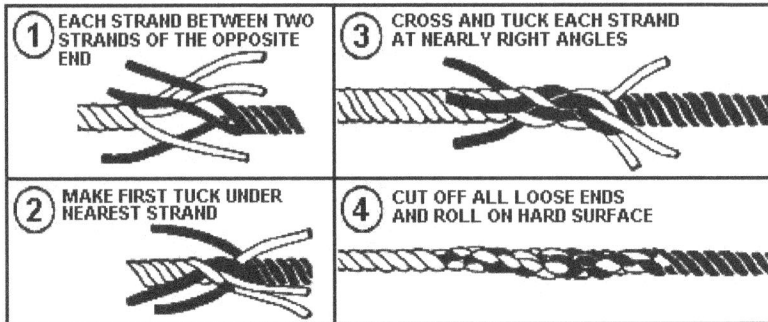

Figure 8-26. Making a Short Splice

EYE SPLICE

8-42.   When a loop is to be permanent, put in the line with an eye splice, which has a strength of 90 to 95 percent. Compare this with the strength of a bowline of 60 percent. Unlay (untwist) the strands four to five turns and splice them into the standing part of the line by tucking the un-laid strands from the ends into the standing part. Whip or tape the ends of the strands. An original round of tucks with two more complete rounds is enough because, if the line parts, it will likely part in the eye rather than in the splice. For this reason, three rounds are as effective as a greater number. Do the following to make an eye splice:

Note: Always whip or tape the ends of the strands before starting; otherwise they will unlay. Seize large lines at the point where unlaying stops to avoid trouble working with them. With up to 21 threads, you can open the strands in the standing part with your fingers. Use the fid for larger lines.

- **Step 1.** Figure 8-27 shows how to make the first two tucks. Separate the strands in the end and hold them up as shown. Place the three un-laid strands against the standing part where they will be tucked, forming an eye the size you need. Always tuck the middle strand facing you first. Put a reverse twist on the standing part so that you can raise the strand under which you will make the first tuck. Pick up the strand that you will tuck, and tuck it under the strand raised. Always tuck from right to left or with the lay of the line.

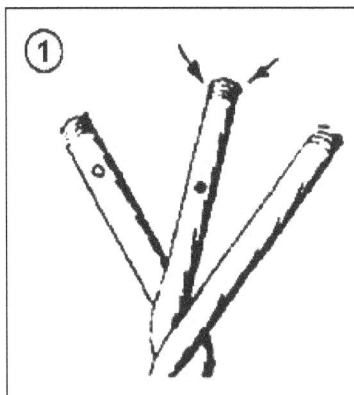

Figure 8-27. Selecting the Middle

● **Step 2.** Be sure to keep the next strand on the side of the line that is facing towards you. Tuck that one next. Put it over the strand under which the first one is tucked, and tuck it under the next one (Figure 8-28).

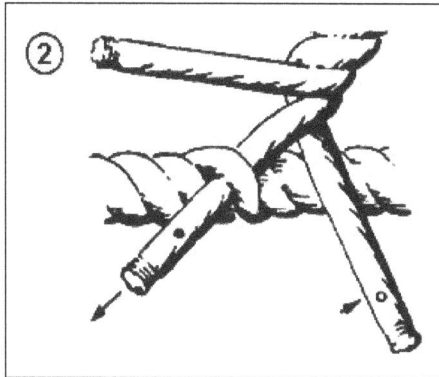

Figure 8-28. First Two Tucks in a Strand Eye Splice

● **Step 3.** Now turn the incomplete eye over as shown. Check the third strand to be sure that it has not unlaid more. If it has, twist it back to where it should be. Take the last strand and put it across the standing part, turn its end back toward you, put it under the strand over which the first tuck was made, and tuck it in a direction toward you. This results in the third tuck going to where the second came out and coming out where the first went in. After this round of tucks, there is a strand in each lay (Figure 8-29).

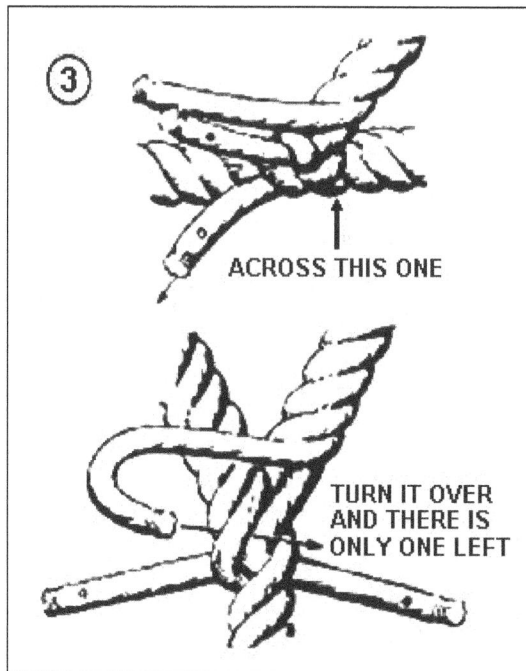

Figure 8-29. Last Tucks of an Eye Splice

8-43.   Pull each of the three strands tucked backward at about a 45-degree angle to the eye to tighten the splice. The first round of tucks is the key to making perfect eye splices. Starting with any strand, simply tuck each one over and under two more times. None of the last two rounds of tucks requires "over and back." However, always tuck from right to left. As required, pull the tucked strands away from the eye and twist the splice and line to tighten them. After finishing the splice, bend the end of each strand back toward the splice and, using a knife, cut it off, up, and away, leaving a one-fourth inch tip.

WIRE ROPE

8-44.   Wire rope is made of steel (the core is likely to be fiber). The grades in descending order of strength are: extra improved plow, improved plow, plow, and mild plow steel. Of these four grades, the Army uses improved plow steel extensively and plow steel to a lesser extent. The manufacturer stamps the grade on the reel. Because you cannot tell the grade of wire rope by looking at it, always treat it as plow steel.

MAKEUP OF WIRE ROPE

8-45.   The basic unit of wire rope is the individual wire. Wires are laid together to form strands. The number of wires in a strand varies according to the purpose for which the rope is intended. Strands are laid around a core to form the wire rope (Figure 8-30). The core may be a wire, hemp, or polypropylene (a synthetic fiber). Use wire rope, with a wire as its core, where high temperatures would damage hemp and polypropylene. New wire rope is made with polypropylene as the core. The core is a foundation to keep the wire rope round, is a shock absorber when the wire rope contracts under strain, and is a reservoir or place where a portion of the lubricant is stored.

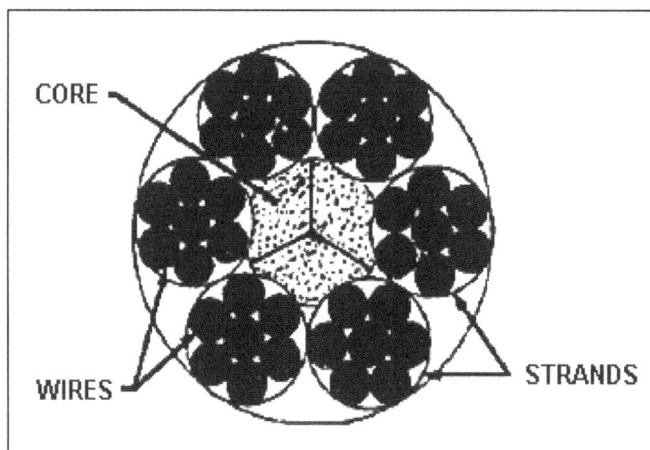

Figure 8-30. Makeup of Wire Rope

CLASSIFICATION

8-46.   Wire rope is classified by the following:
- Number of strands
- Number of wires per strand.
- Strand construction.
- Type of lay.

Strands and Wires

8-47.   Standard wire rope has six strands. The present commercial classifications are not factually descriptive. Rather, they are groupings of wire ropes of similar weight, flexibility, and strength. Therefore, the 6 x 19 classification has 6 strands of wires per strand. The 6 x 37 classification has six strands and 37 wires in each strand. Figure 8-31 shows cross sections of four classifications. The rope is more flexible with smaller and more numerous wires but less resistant to external abrasion. Wire rope made up of a smaller number of larger wires is less flexible and more resistant to abrasion. All else being equal, two ropes of the same size have the same strength even though, for example, one is 6 x 19 and the other is 6 x 37.

Strand Construction

8-48. Wires and strands used in most wire rope are preformed. Preforming is a method of presetting the wires in the strands into the permanent corkscrew form they will have in the completed rope. As a result, pre-formed wire rope does not have the internal stresses found in non-preformed wire rope, does not untwist as readily as nonpreformed wire rope, and is more flexible.

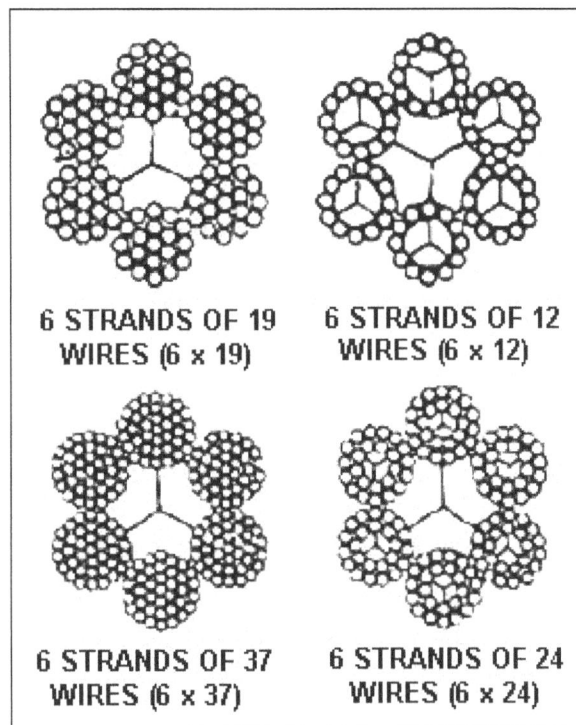

Figure 8-31. Strands and Wires

Types of Lay

8-49. Lay refers to the direction of winding of the wires in the strands and the strands in the rope. Both may be wound in the same direction or they may be wound in opposite directions. In regular lay, the strands and wires are wound in opposite directions. Most common is the right, regular lay in which the strands are wound right and wires left. Use this lay in marine operations. In Lang lay, the strands and wires are wound in the same direction. Use this type of wire rope on the blades of bulldozers and scrapers.

MEASUREMENT

8-50. Whatever its grade, wire rope is usually measured by its diameter. Figure 8-32 shows the correct and incorrect methods of measuring the diameter of wire rope. To measure wire rope correctly, place it in the caliper so that the outermost points of the strands will be touching the jaws of the caliper. Because of friction and tension, the diameter of used wire rope will be 1/64-inch to 1/8-inch less than when new.

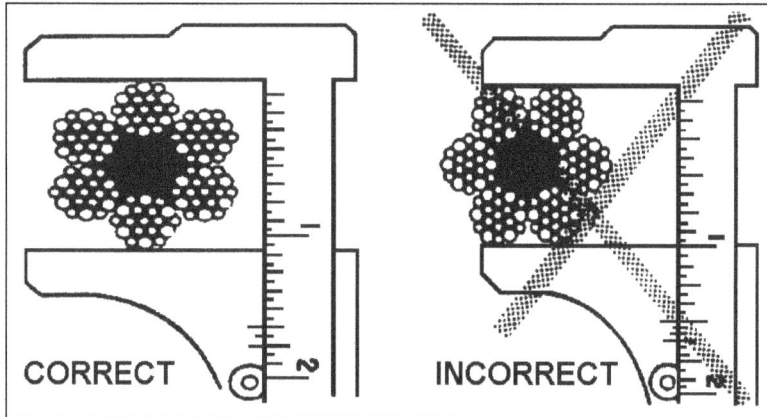

Figure 8-32. Measuring Wire Rope

SAFE WORK LOAD AND BREAKING STRENGTH

8-51. Useful formulas for determining the SWL of several grades of wire rope have constants not to be confused with safety factors. For example, the formula for the SWL in short tons (2,000 pounds) for extra improved plow steel wire rope is diameter (D) squared times 10 or SWL = D^2 x 10. To find the SWL of 1-inch, 6 x 19, extra improved plow steel wire rope:

$$SWL = D^2 \times 10 = 1 \times 1 \times 10 = 10 \text{ STON}$$

8-52. A figure relatively constant in marine operations, especially for new wire rope, is the safety factor. The safety factor (SF) is 5. Use the SF with the SWL to find the breaking strength (BS) or strain.

$$BS = SWL \times 5 = BS = 10 \times 5 = 50 \text{ STON}$$

8-53. The formulas for improved plow steel, plow steel, and mild plow steel (6 x 19 wire rope) are as follows:

Improved plow steel
SWL = D^2 x 8 = STON
BS = SWL x SF = STON

Plow steel
SWL = D^2 x 7 = STON
BS = SWL x SF = STON

Mild plow steel
SWL = D^2 x 6 = STON
BS = SWL x SF = STON

INSPECTION

8-54. Inspect wire ropes frequently and replace frayed, kinked, worn, and corroded ropes. How frequently rope should be inspected depends on the amount of its use. A rope used 1 or 2 hours a week requires less frequent inspection than one used 24 hours a day. The common causes of wire rope failures are the following:

- Using rope of incorrect size, construction, or grade.
- Allowing rope to drag over obstacles.
- Operating over sheaves and drums of inadequate size.
- Over-winding or cross-winding on drums.
- Operating over sheaves and drums out of alignment.
- Permitting rope to jump sheaves.
- Subjecting rope to moisture or acid fumes.
- Permitting rope to untwist.
- Using kinked rope.

8-55.   Carefully inspect weak points and points of greatest stress. Worn or weak spots show up as shiny flat spots on the wires. If the outer wires have been reduced in diameter by one-half, the wire rope is unsafe. Broken wires also show where the greatest stress occurs. If individual wires are broken next to one another, unequal load distribution at this point will make the rope unsafe. These broken wires are called "fishhooks." To determine the extent of damage to the wire rope, slide your finger along one strand of wire for one complete turn, which is equal to the length of one wire rope lay. Count the number of "fishhooks." If you count eight or more "fishhooks," replace the wire rope immediately. Any time you find six to eight "fishhooks" within the measured area, you will consider the wire rope unsafe and should have it replaced.

UNREELING

8-56.   When removing wire rope from a reel or coil, be sure to rotate the reel or coil (Figure 8-33). If the reel is mounted, unwind the wire rope by holding the end and walking away from the reel. If a wire rope is in a small coil, stand the coil on end and roll it along the deck, barge, wharf, or ground. Be sure to remove any loops that may form, although the reason for rotating the reel or coil is to avoid loops.

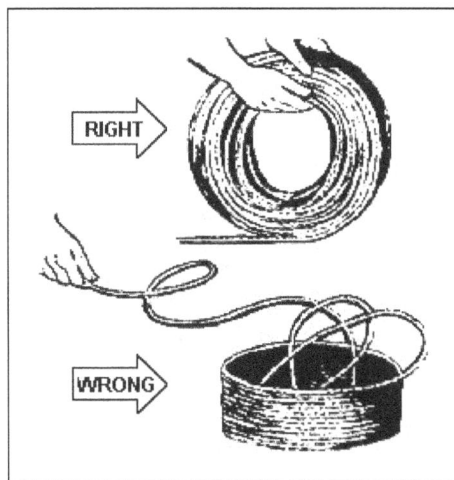

Figure 8-33. Uncoiling Wire Rope

SEIZING

8-57.   Seize all wire rope before cutting. If the ends of the rope are not properly secured, the original balance of tension is disturbed and maximum service cannot be obtained because some strands carry a greater load than others. Use annealed wire for the seizing. Figure 8-34 shows the steps on how to seize wire rope. The turns of the annealed wire rope should be put on close and tight so that it will not be necessary to tighten them when the ends are being twisted together. It is well to twist the ends together at one end of the seizing so that the completed twist can be tapped into the groove between two strands where it is less likely to be knocked off. There are three formulas for determining the number and length of seizing and the space between them. When a calculation results in a fraction, use the next larger whole number. The following formulas are based on a wire rope with three-fourths inch diameter: The number of seizing required equals about three times the diameter of the rope. For example—

$$3 \times 3/4 = 2\ 1/4 \text{ or } 3 \text{ seizing}$$

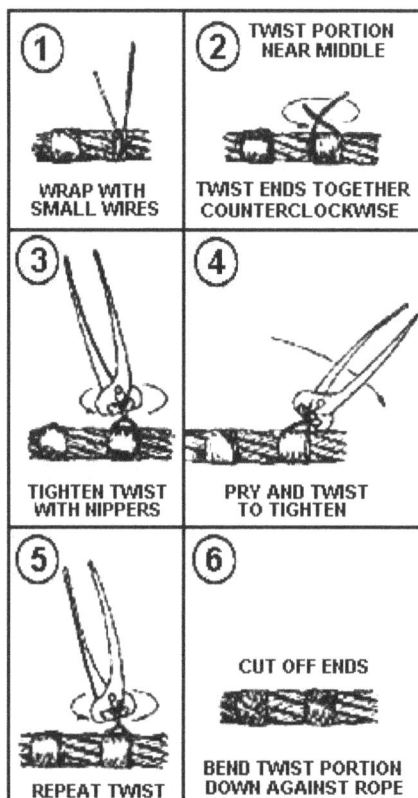

Figure 8-34. Seizing Wire Rope

8-58.   Because the rope will be cut, six seizing are required so that there will be three on each rope end after the cut. The length of a seizing should be equal to the diameter of the rope. For example—

$$1 \times 3/4 = 3/4 \text{ or } 1 \text{ inch}$$

8-59.   The seizing should be spaced apart at a distance equal to twice the diameter. For example—

$$2 \times 3/4 = 1\ 1/2 \text{ or } 2 \text{ inches.}$$

CUTTING

8-60.   Wire rope may be cut with a wire rope cutter, a cold chisel, a hacksaw, bolt clippers, or an oxyacetylene cutting torch. After seizing the wire rope, insert it into the cutter with the blade coming between the two central seizing. Close the locking device. Then, close the valve on the cutter and pump the handle to build up enough pressure to force the blade through the rope. Use the bolt clippers on wire rope of fairly small diameter. However, the oxyacetylene torch can be used on wire of any diameter. Cutting with the hacksaw and cold chisel is slower than cutting with the other tools and equipment.

COILING

8-61.   It may be necessary to take a length of wire rope from a reel and coil it down before using. Small loops or twists will form if the wire rope is coiled in a direction opposite to the lay. To avoid them, coil right lay wire rope clockwise and left lay counterclockwise. When a loop forms in the wire, put a back turn in as shown in Figure 8-35.

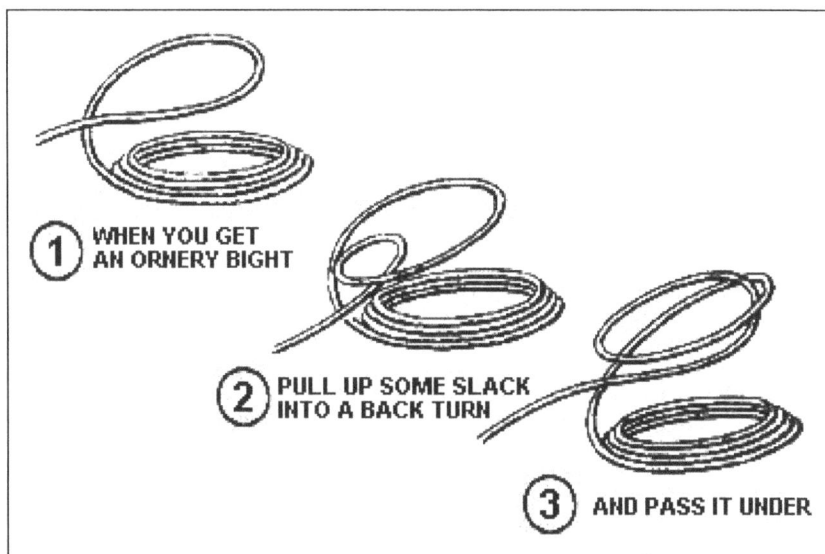

Figure 8-35. Putting a Back Turn in Wire Rope

SIZE OF SHEAVES AND DRUMS

8-62.   Two things happen when a wire rope is bent over a sheave or drum:
- Each wire is bent to conform to the curvature.
- The wires slide against each other longitudinally because the inside arc of the rope against the sheave or drum is shorter than the outside arc.

8-63.   The smaller the diameter of the sheave or drum the greater the bending and sliding. This bending and moving of wires should be kept to a minimum to reduce wear. The minimum recommended sheave and drum diameter is 20 times the diameter of the rope. For example, determine the minimum sheave diameter for 6/8-inch rope:

$$20 \times 5/8 = 12\ 1/2 \text{ inch sheave}$$

8-64.   If a 12 1/2-inch sheave is not on hand, use the next larger size. Never use a smaller size.

LUBRICATION

8-65.   Wire rope is lubricated when it is manufactured. The lubricant generally does not last throughout the life of the rope, which makes re-lubrication necessary. Crater "C" compound is recommended, but oil on hand may be used rather than delaying lubrication. Always lubricate as often as necessary. Heat Crater "C" compound before putting it on the wire rope. When lubricating, use a brush if one is on hand. If not, use a sponge or cloth, but look out for "fishhooks" or broken wires.

REVERSING ENDS

8-66. It is sometimes wise to reverse ends or cut back the ends to get more service from wire rope. Reversing ends is more satisfactory than just cutting ends because frequently the wear and fatigue on a rope are more severe at certain points than at others. Reversing distributes other, stronger parts of the rope to the points getting wear and fatigue. To reverse ends, remove the drum end and put it in the attachment. Then fasten the end taken from the attachment to the drum. Cutting back the ends has a similar effect, but there is not as much change involved. In reversing ends, cut off short lengths of both ends to remove the sections that have sustained the greatest local fatigue.

STORAGE

8-67. Wire rope should be coiled on a spool for storage. Attach a tag to the rope or spool to show its grade, size, and length. Store wire rope in a dry place to reduce corrosion. Do not store it with chemicals or where chemicals have been stored because both chemicals and their fumes might attack the metal. Always clean and lubricate wire rope before storing.

PUTTING AN EYE IN WIRE ROPE

8-68. This paragraph discusses how to put both a temporary eye and a permanent eye in wire rope. A temporary eye can be put in wire rope by using wire rope clips or by using a field expedient known as a "hasty eye" or "Molly Hogan" splice. A Liverpool splice is the accepted method for putting a permanent eye in the end of a wire rope. With the proper equipment, and a bit of practice, a Liverpool splice can be put in wire rope in less than 15 minutes.

TOOLS USED FOR SPLICING

8-69. Except for the knife, Figure 8-36 shows the tools needed for splicing. Use the marlinespike for opening the strands in the standing part of the wire rope and for working the strands to be spliced into the standing part. Use the wire cutters for cutting the strands after the splice is complete. Use the hydraulic wire rope cutter to cut the length of wire rope that will be spliced. Use a thimble to keep the wires from moving and the vise from crushing them when a soft eye is made. An eye splice can be made with or without a thimble. Always use a thimble whenever an eye splice is put in unless special circumstances prohibit it. The thimble protects the wire rope from sharp bends and abrasive action. The efficiency of a well-made splice with a heavy-duty thimble varies from 70 to 90 percent. After splicing the soft eye, remove the thimble. When an eye is to have a thimble as a permanent part, the thimble is the size of the eye desired.

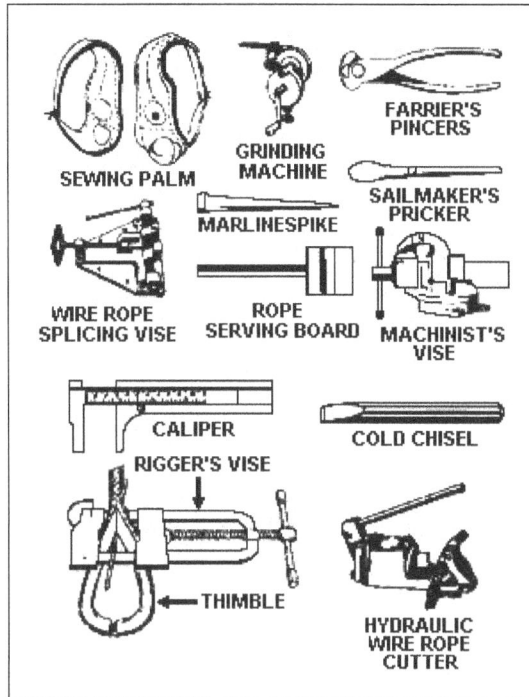

Figure 8-36. Selected Components of Rigger's Cargo Set

TEMPORARY EYE USING WIRE ROPE CLIP

8-70.   A temporary eye may be put in wire by using wire rope clips. Figure 8-37 shows the correct and incorrect ways of using these clips. The U-bolt always goes over the bitter end and the roddle goes on the standing part. Space the clips apart at a distance equal to six times the diameter of the wire. After a rope is under strain, tighten the clips again. On operating ropes, tighten the clips every few hours and inspect the rope carefully. Inspect at points on the rope where there are clips. Pay particular attention to the wire at the clip farthest from the eye, because vibration and whipping are dampened here and fatigue breaks are likely to occur.

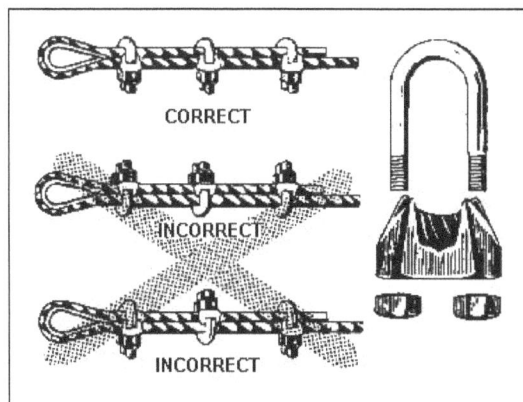

Figure 8-37. Correct and Incorrect Use of Wire Clips

8-71.   To obtain maximum strength from the temporary eye, use the correct size and number of wire clips. Size is stamped on the roddle between the two holes. The correct number of clips to use for various sizes of wire ropes is shown in Table 8-4. Or use the following formula:

3 x diameter of rope + 1 = number of clips (round off)

8-72.   Correct spacing between clips is—

6 x diameter of rope = correct spacing (inches)

Table 8-4. Size and Number of Wire Clips

Size of Rope (Inches)	Number of Clips
1/2	2
5/8	3
3/4	3
7/8	4
1	4
1 1/8	5
1 1/4	5
1 1/2	6

8-73.   The improved type of wire rope clip shown in Figure 8-38 has a few advantages over the older type. Both halves are identical and provide a bearing surface for both parts of the rope. Therefore, it cannot be put on wrong and it does not distort the wire. It also allows a full swing with a wrench.

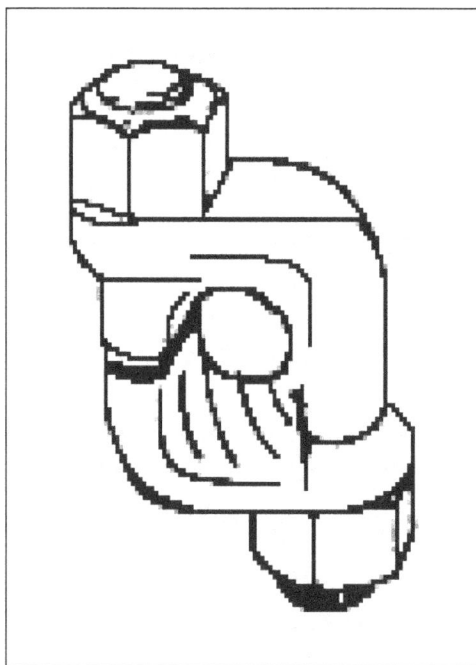

Figure 8-38. Improved Type of Wire Rope Clip

THE HASTY EYE ("MOLLY HOGAN") SPLICE

8-74.   Sometimes it becomes necessary to construct a field expedient, called the hasty eye or "Molly Hogan" splice. This splice can be easily and quickly made, but it is limited to about 70 percent of the strength of the wire rope. Never use this splice to lift heavy loads. Use this splice only when working with preformed wire rope. To make this splice, do the following steps.

- **Step 1.** Using a marlinespike, screwdriver, or if necessary, a nail; separate the wire rope into two three-strand sections. These sections should be unlaid four times the diameter of the desired eye. If you want a 1-foot diameter eye, unlay the sections back 4 feet (Figure 8-39).

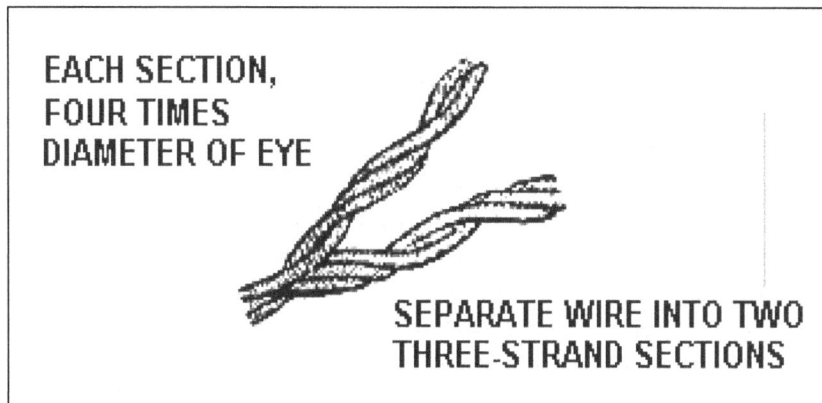

Figure 8-39. Making a Hasty Eye (Molly Hogan) Splice, Step 1

- **Step 2.** Use the two sections to form a loop of the desired diameter for the eye. Then, lay the strands back around each other to form the eye (Figure 8-40).

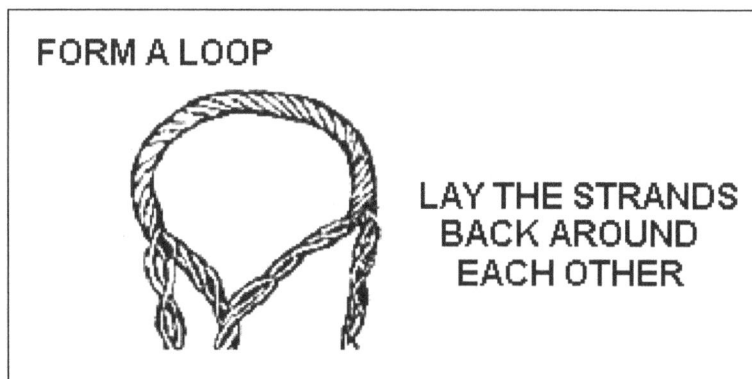

Figure 8-40. Making a Hasty Eye (Molly Hogan) Splice, Step 2

- **Step 3.** After the strands have been laid back around each other and the eye has been formed, seize the wire to complete the splice (Figure 8-41).

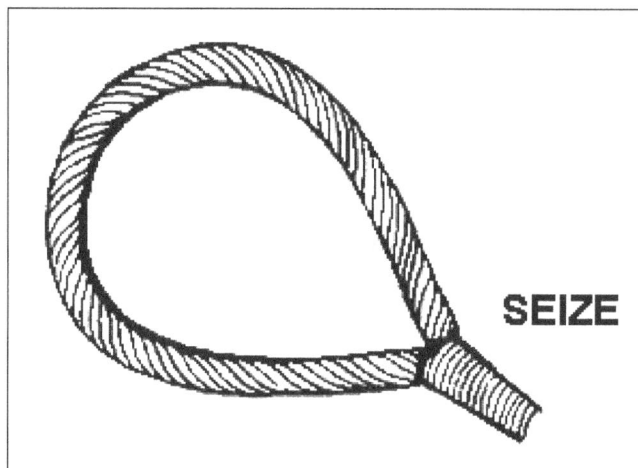

Figure 8-41. Making a Hasty Eye (Molly Hogan) Splice, Step 3

THE LIVERPOOL SPLICE

8-75.   The Liverpool splice is the easiest and most common of the wire splices to make. It is the primary splice used when a permanent eye is required. To find the distance, the strands should be unlaid for an eye splice, multiplying the diameter of the wire by 36 inches. (Example: 5/8-inch wire rope—5/8 x 36/1 = 180/8 = 22 1/2 or 23 inches.) Measure off that distance on the wire rope and put a seizing at that point. Next, cut the end seizing and carefully unlay the strands. Whip the ends of each strand with either sail twine or friction tape. Form the desired size eye and put the eye in the rigger's vise with the un-laid strands to your right as you face the vise. Stretch out the standing part of the wire, clamp and lash it and you are ready to start.

Note: When splicing wire, always insert the marlinespike against the lay of the wire, and make sure not to shove it through the core. The core must be on the left-hand side of the spike.

- **Making the First Tuck of Strands One, Two, and Three.** In the Liverpool splice (Figure 8-42) the first strand goes under three strands, the second strand goes in the same place but only under two strands, and the number three strand goes in the same opening but only under one strand. All of the strands go in at the same point, but come out at different places. Then, run the spike behind the three strands under which the first three are tucked, but above the first three strands as tucked. Holding the marlinespike at a 90-degree angle to the standing part, turn the spike counterclockwise about one fourth of a turn and insert the core through the standing part. This is called "dipping the core." Make sure that the core is inserted under the marlinespike. Pull the core down and run it down into the splice.
- **Tucking Strands Four, Five, and Six.** Remember that the core was between strands three and four and that the strands are numbered clockwise. To tuck strand four, put the marlinespike under the strand to the left of where one, two, and three were tucked through the standing part. Turn the marlinespike counterclockwise around the standing part and tuck the strand. Pull it tight and run it down with the spike. Tuck strand four around the same strand four times. Lock each tuck in place by holding the strand down and running the spike up. Push the marlinespike under the next higher strand on the standing part and tuck strand five around it four times, using the same procedure as with strand four. Then tuck strand six four times. This completes strands four, five, and six.
- **Running the Core Up.** Burying the core in the center of the splice in the standing part is called "running the core up." The entire core does not run up and any excess is cut off. This is done before strands one, two, and three are tucked three more times. Run the spike under the same

three strands under which strand one was passed. With the spike in your left hand and the core in your right hand, move the spike to the left and down, and pull up the core with your right hand to tighten. Then move the spike back to the right. Run up the core into the center of the splice and cut off the excess.

- **Tucking Strands One, Two, and Three.** To avoid kinking the strands on the last tucks, insert the spike and run it up the wire. Follow the spike up with the strand, shove it under the spike, and pull taut. Keeping a strain on the strand, work the spike and strand back around and down together. Hold the strand there and work the spike back up the wire. Follow up with the strand and take the last tuck. Work the strand back down and hold it there. Before pulling out the spike, run it back up until the strands of the standing wire bind the working strand in place (Figure 8-43). Make the second and third tucks with the remaining strands in the same way. The recommended order for finishing the splice is to tuck strands three, two, and one. Each is tucked three times in a row, ending up with a total of four tucks each. Remove the wire from the vise, take a hammer and pound the splice into shape, and cut off the ends of the tucking strands close to the splice.

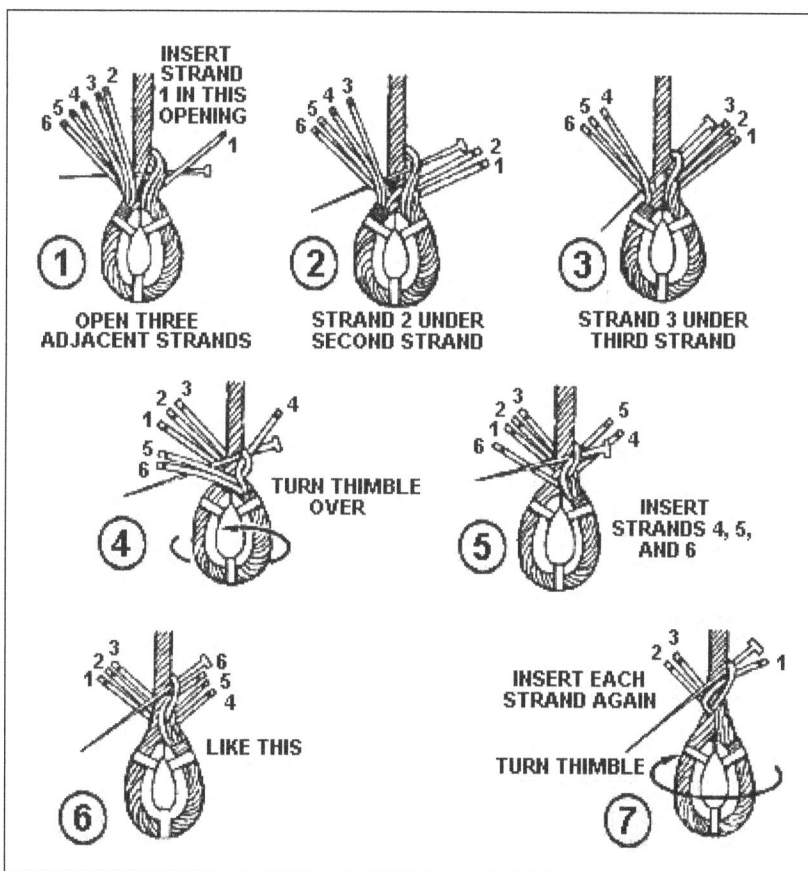

Figure 8-42. Tucking Strands of a Liverpool Splice

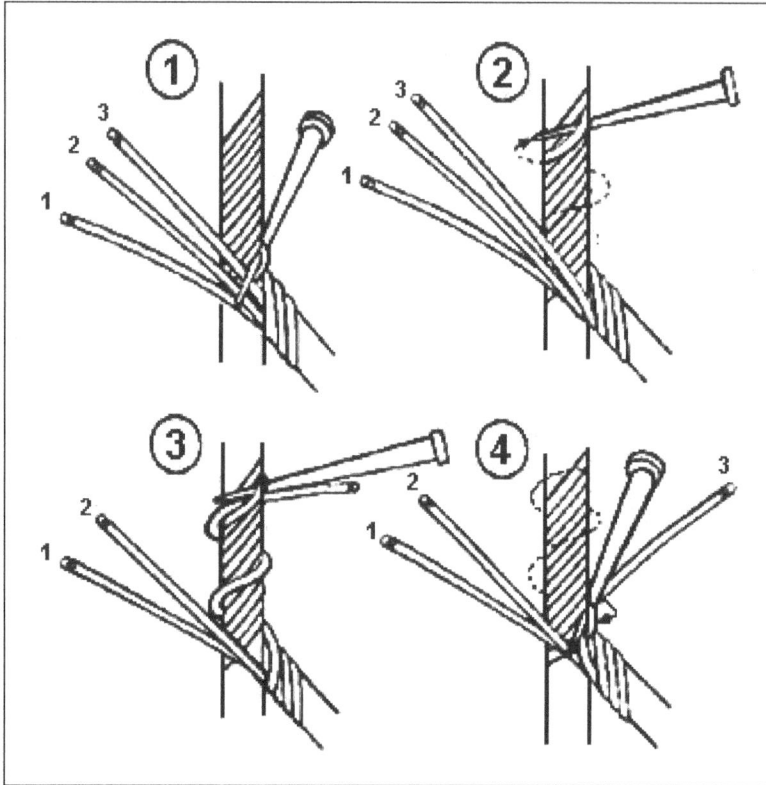

Figure 8-43. How to Avoid a Kink Completing a Splice

SPLICING 2-IN-1 DOUBLE-BRAIDED NYLON LINE (SAMSON 2-IN-1 BRAID-SPLICING PRINCIPLES)

8-76.   Double-braided nylon has a braided core inside a braided cover. It is commonly called 2-in-1 braided nylon line. Special tools and procedures are required to splice this type of line. The following describes the procedures for making the standard eye splice and the end-for-end splice. The Samson Cordage Works developed both splices and the line that is currently being used. The following information is used with permission and through the courtesy of Samson Ocean Systems, Inc., Boston, Massachusetts.

SPECIAL TERMS

8-77.   Refer to the following terms when making the standard eye splice and the end-for-end splice.

* **Tubular fid.** The hollow steel tool used for cover and core insertions (see Figure 8-44).

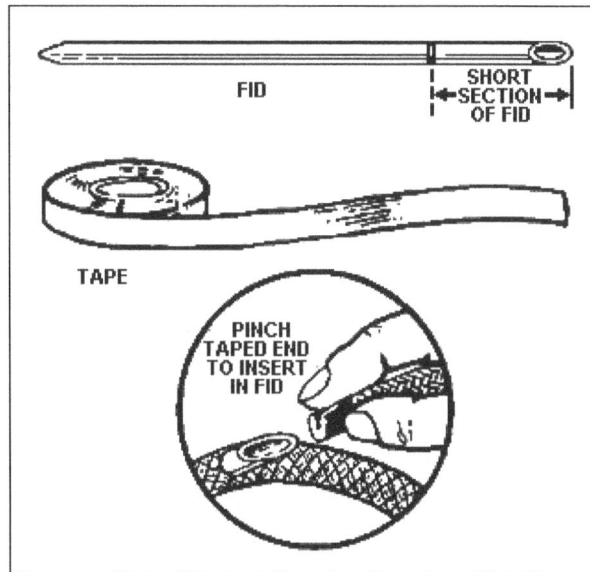

Figure 8-44. Tubular Fid

- **Metal wire fid.** For line over 1-inch diameter (see Figure 8-45).

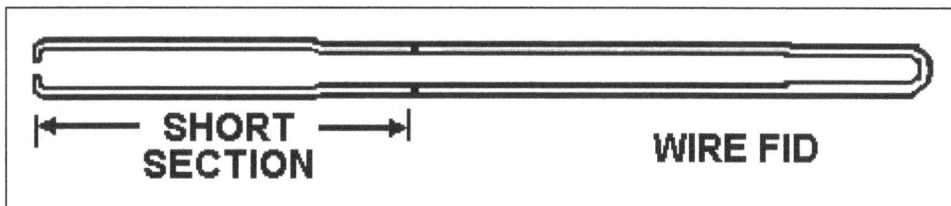

Figure 8-45. Metal Wire Fid

- **One fid length.** The full length of one tubular fid; two full lengths of a wire fid.
- **Short section of fid.** Distance away from open end to the scribe marks on body of fid. Approximately 35 percent of the full fid length.
- **Pusher.** Ice-pick-like tool used to extract core from cover and to aid in sliding fid through rope elements (see Figure 8-46).

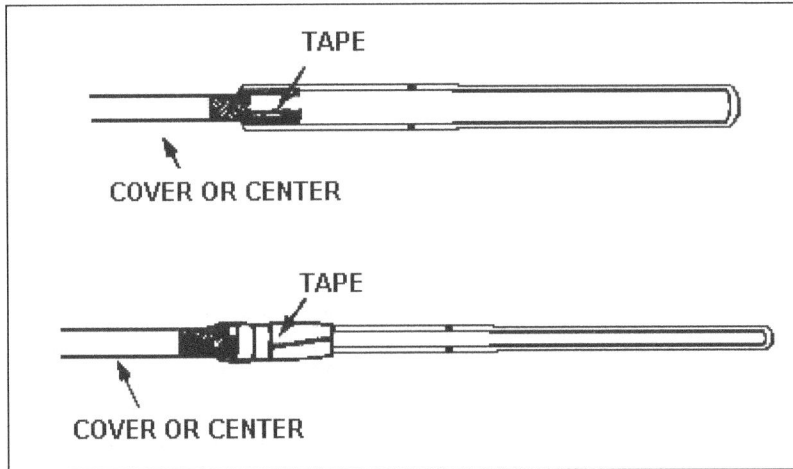

Figure 8-46. Pusher

- **Eye.** The closed loop formed at the end of rope as a result of splicing.
- **Crossover.** The point of intersection of cover and core created during splicing.
- **Milking.** The intermittent squeezing-pulling-sliding movement of the hand used to bring cover over core in forming splice.
- **Smooth Out.** To "milk" the slack out of a particular section during the splicing process.
- **Point X.** The extraction point; place on cover from where the core is initially extracted.
- **Point R.** The reference point; the mark made after measuring one fid length from taped end of cover.
- **Point T.** The point from which the taper count is measured.
- **Point Z.** The point on the cover from which the core tail will emerge; located one-half fid below point X.
- **Strand.** The strand of a braid is a group of one or more plied yarn ends, which make up one stitch or pick. The usual number of cover strands in a Samson braid is 16, 20, 24, or 32.

Note: Since most Samson braid covers have two ends per strand, they are referred to in the splicing book as strand pairs.

- **End.** An end is a plied yarn component of a braid strand. In a cover strand one to four ends are found. In a core strand two to six ends can be found.

Note: On many Samson 2-in-1 braids, it is possible to distinguish between the cover and core as follows: The cover has a light blue tracer strand while the core has no visible tracer strand.

SPECIAL TOOLS AND TECHNIQUES

8-78. The following are the special tools and techniques needed when making the standard eye splice.
- **For Splice with Thimble.** Step 1 in the procedure for the standard eye splice (page 8-38) tells how to determine eye size. Minimum eye and eye sling length with 2-in-1 braid is five fid lengths from extraction Mark X to extraction Mark X, regardless of rope diameter. The size of the eye does not affect the minimum length (see Figure 8-47).

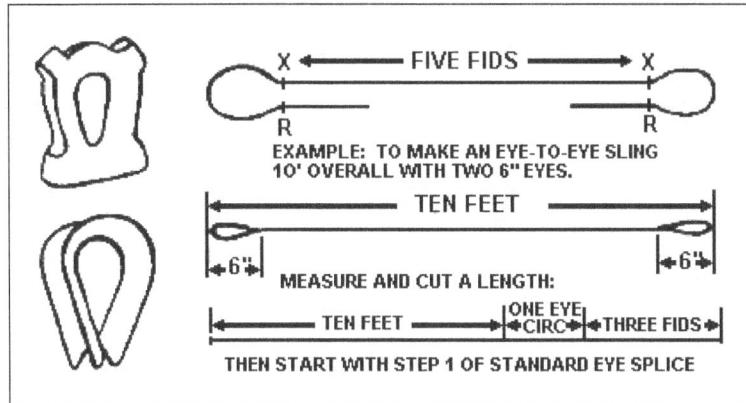

Figure 8-47. Minimum Lengths for Standard Eye Splice

- Exact overall lengths with eye splices are determined by allowing for extra rope to be used in making the splices. For each splice, the length of extra rope is equal to one and one-half fid lengths plus one-half the circumference of the eye. When burying exposed core as in STEP 8 (page 8-41), bury to crossover and insert thimble into eye before milking cover all the way. When using a thimble with ears, as in STEP 5 (page 8-40), insert core through the rings (ears) and slide thimble beyond Mark 3 before inserting cover into core. Proceed to make the splice according to instructions.

Note: Before final burying, slide thimble around to cover side of eye.

- To secure finished eye tightly around thimble, either whip throat or dip the eye in hot water for several minutes. Hot water will shrink eye tightly around thimble. Minimum endless loop (grommet) with 2-in-1 braid is 10 fid lengths between extraction Mark X to extraction Mark X, regardless of rope diameter. Proceed to make the end-for-end splice as shown in Figure 8-48.
- **For Line Less Than 1-Inch Diameter (3-Inch Circumference).** Each size line requires a different size of tubular fid. Use the fid, along with the pusher, to insert the cover into the core and vice versa. Also use the fid as a measuring device. The scribe marks indicate the short section of the fid.
- **For Line Larger Than 1-Inch Diameter (3-Inch Circumference).** Use only a metal wire fid for splicing larger lines (a pusher is not needed). As with the tubular fid, there are different sizes of wire fids for each size of line. Cover and center measurements are made with the wire fid in the same manner as the tubular fids. Tightly tape the end of braided cover or center after extraction (STEP 2, page 8-39). Press prongs of fid into cover or center just behind tape. Tape wire fid to braid by wrapping tape in a tight, smooth, spiral, starting on the braid and wrapping in the direction of the round tip of the fid. Keep tape smooth to ease the fid through braid. The round end of the fid can then be inserted and pushed through without a pusher.

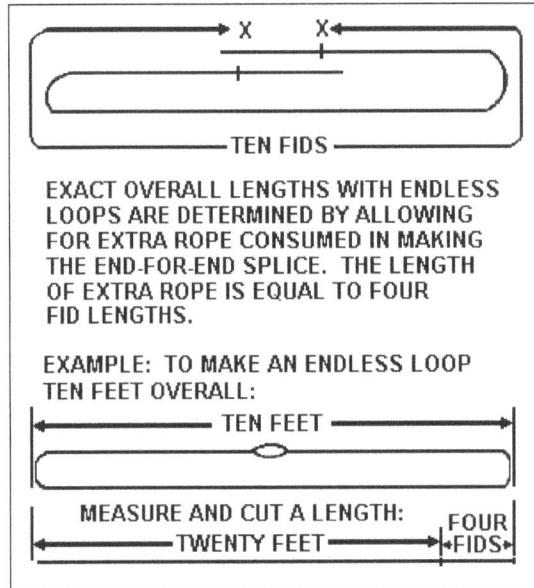

Figure 8-48. Minimum Lengths for End-for-End Splice

STANDARD EYE SPLICE

8-79.   This Samson eye splice is for new line only. It retains about 90 percent of the average new line strength.

- **Step 1. Marking the measurements.** Tape end to be spliced with one thin layer of tape. Then measure one tubular fid length (two wire fid lengths because wire fid is one-half size) from end of line and mark. This is point R (see Figure 8-49). From R, form a loop the size of the eye desired and mark. This is point X (where you extract core from inside the cover). If using a thimble, form the loop around the thimble. Tie a tight slip knot about five fid lengths from point X. THIS MUST BE DONE. If you require the line with the finished splice(s) to be a certain overall length.

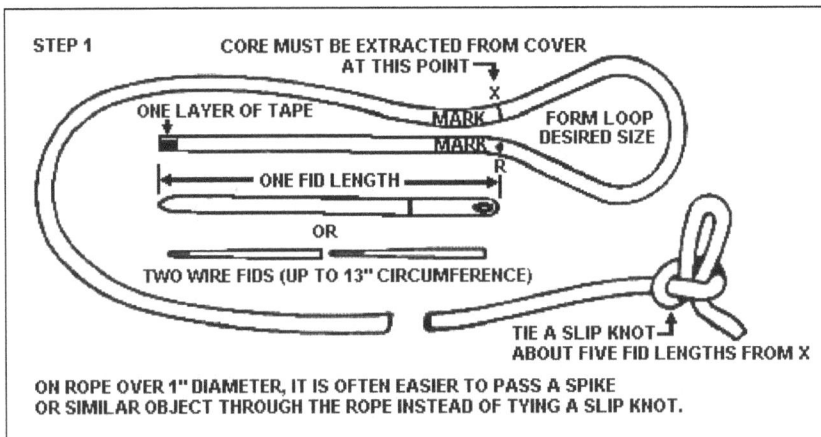

Figure 8-49. Marking the Measurements (Step 1)

- **Step 2. Extract the core.**
  - Bend the line sharply at point X. With the pusher or any sharp tool such as an ice pick, an awl, or a marlinespike, spread the cover strands to expose the core. Pry and then pull the core completely out of the cover from point X to the taped end of the line. Put one layer only of tape on end of the core (see Figure 8-50).

Note: DO NOT pull cover strands away from line when spreading as this will unnecessarily distort the rope.

  - Holding the exposed core, slide cover as far back towards the tightly tied slip knot as you can. Then, firmly smooth the cover back from the slip knot towards taped end. Smooth again until all cover slack is removed. Then, mark the core where it comes out of the cover; this is Mark 1.

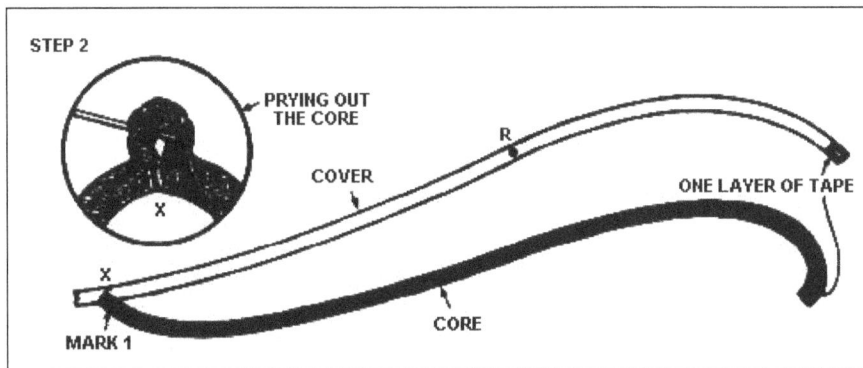

Figure 8-50. Extracting the Core (Step 2)

- **Step 3. Marking the core.** Again, slide cover toward slip knot to expose more core. From Mark 1 following the core towards point X, measure a distance equal to the short section of tubular fid (two short sections with wire fid) and make two heavy marks. This is Mark 2. From Mark 2, measure in the same direction one fid length plus another short section of the fid (with wire fid, double measurements). Make three heavy marks for Mark 3 (see Figure 8-51).

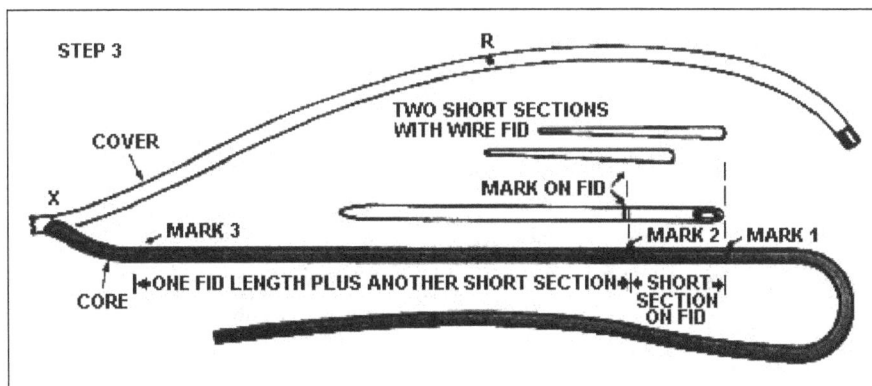

Figure 8-51. Marking the Core (Step 3)

- **Step 4. Marking the cover for tapering.** Note nature of cover braid. It is made up of strands - either one or two (pair). By inspection, you can see half the strands revolve to the right around the rope and half revolve to the left. Beginning at point R and working toward the taped end of the cover, count eight consecutive strands (single or pairs) which revolve to the right (or left). MARK THE EIGHTH STRAND. This is point T (see Figure 8-52 insert). Mark point T completely around cover. Starting at point T and working toward the taped cover end, count and mark every fifth right and left strand (single or paired) until you have progressed down to the end of the taped cover.

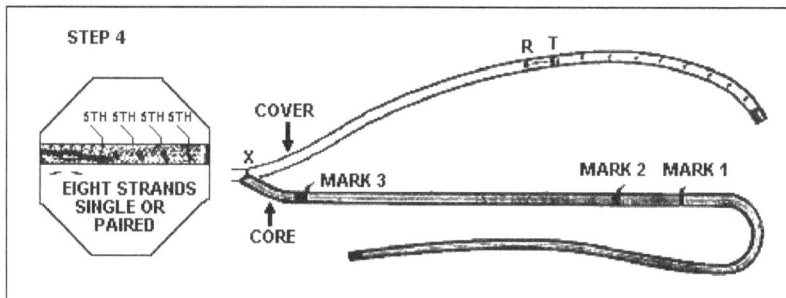

Figure 8-52. Marking the Cover for Tapering

- **Step 5. Putting the cover inside the core**. Insert fid into core at Mark 2. Slide it through and out at Mark 3. Add extra tape to cover end; then jam it tightly into the hollow end of fid (see Figure 8-53 insert). Hold core lightly at Mark 3, place pusher point into taped end, and push fid and cover through from Mark 2 and out at Mark 3. Press prongs of wire fid into cover. Then tape over them. After the fid is on, milk braid over fid while pulling fid through from Mark 2 to Mark 3. Take the fid off the cover. Continue pulling cover tail through the core until point R on the cover emerges from Mark 3. Then remove tape from end of cover.

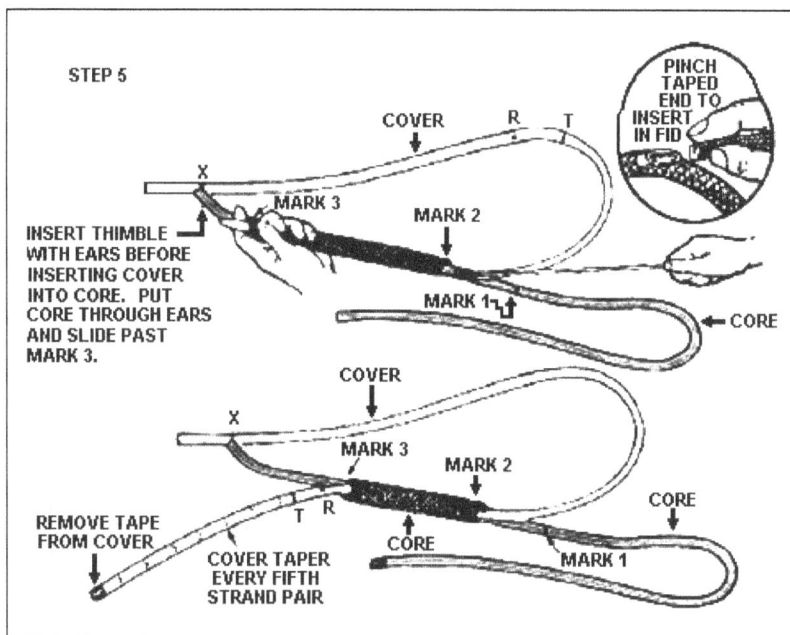

Figure 8-53. Putting the Cover Inside the Core

- **Step 6. Performing the taper.** Make sure tape is removed from cover end. Starting with the last marked pair of cover strands toward the end, cut and pull them completely out (see Figure 8-54 insert). Cut and remove next marked strands and continue with each right and left marked strands until you reach point T. DO NOT cut beyond this point (see Figure 8-54 insert). The result should be a gradual taper ending in a point. Very carefully pull cover back through core until point T emerges from Mark 2 of core.

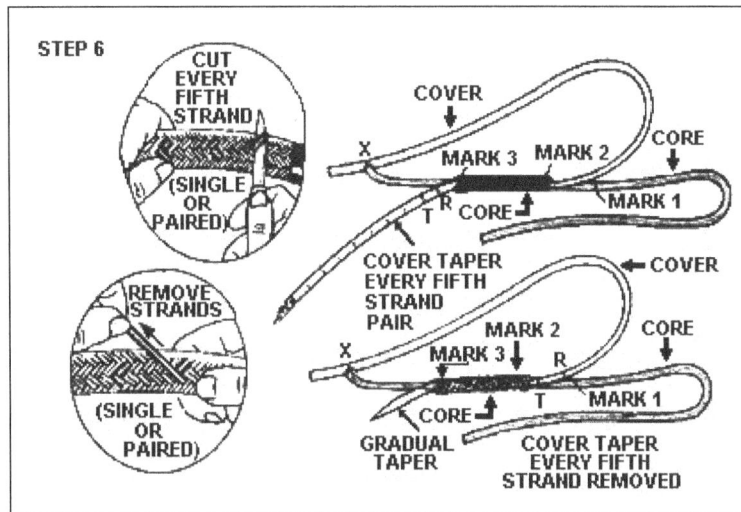

Figure 8-54. Performing the Taper

- **Step 7. Reinserting the core into the cover.** From point X on cover, measure approximately one-half fid length toward slip knot on line and mark this as point Z (see Figure 8-55). You are now ready to put core back into cover from point T to point Z. Insert fid at point T. Jam the taped core end tightly into end of fid. With pusher push fid and core through cover "tunnel," past point X, to and through cover at point Z. When using wire fid, attach fid to taped core. After fid is on, milk braid over fid while pulling through from point T to point Z. When pushing fid past point X to point Z, make sure fid does not catch any internal core strands.

Note: Depending on eye size, fid may not be long enough to reach from point T to point Z in one pass. If not, bring fid out through cover, pull core through and reinsert fid into exact hole it came out. Do this as many times as needed to reach point Z.

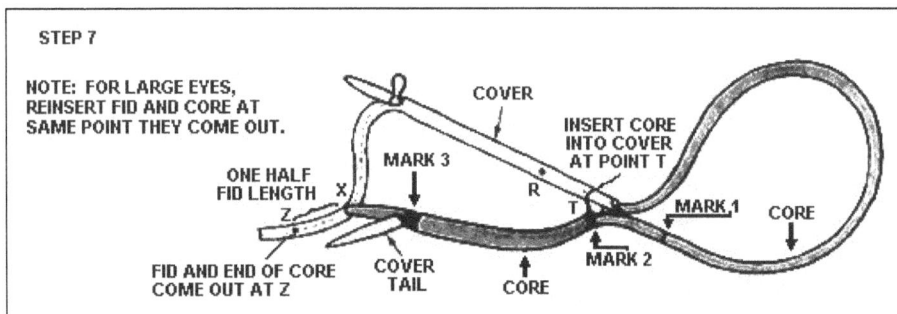

Figure 8-55. Reinserting the Core Into the Cover

- **Step 8. Marking the reduced volume tail core.** Alternately pull on core tail at point Z, and then pull on tapered cover at Mark 3. Tighten the crossover until it is about equal to the diameter of the line (see Figure 8-56). Smooth out cover of eye completely, from crossover at point T toward point X, to get all slack out of eye area. MARK CORE TAIL THROUGH COVER AT POINT X. Pull core tail out until mark on core just made is exposed at point Z. Reduce core volume at this point by cutting and removing one strand at each group, progressing around the circumference of the rope (see Figure 8-56 insert). Measure one-third fid length from start of reduction cuts toward end and mark. Cut off remaining tail at this point. Make cut on a 45-degree angle to prevent a blunt end (see Figure 8-56 insert). With one hand, hold crossover—Mark T. Smooth cover section of eye out firmly and completely from crossover toward X; tapered core tail should disappear into cover at point Z. Smooth out core section from crossover towards Mark 3 and cover taper will disappear into core.

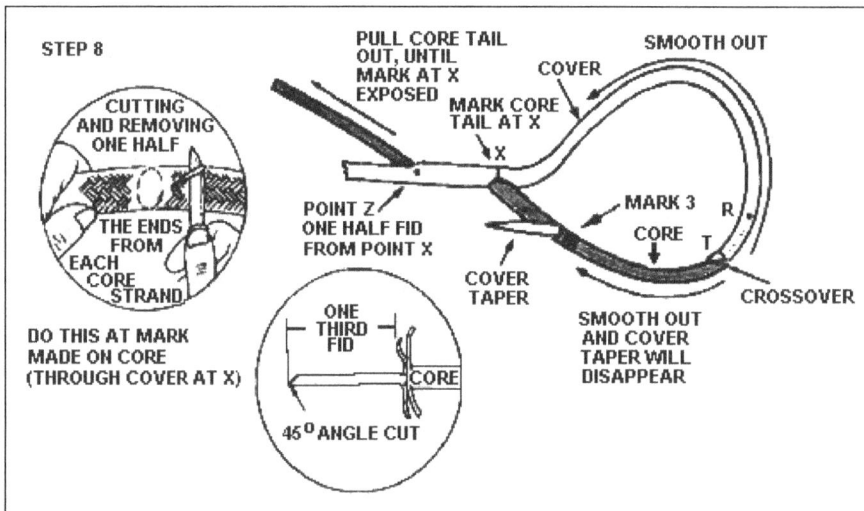

Figure 8-56. Marking the Reduced Volume Tail Core

- **Step 9. Burying the exposed core.** Hold rope at slip knot and with other hand milk cover toward splice, gently at first, then more firmly (see Figure 8-57). Cover will slide over Mark 3, Mark 2, the crossover, and point T and point R. (It may be necessary to occasionally smooth out eye during milking to prevent tapered tail from catching in throat of splice.) If bunching occurs at crossover preventing full burying, smooth cover from point T to point X. Grasp crossover at point T with one hand and then firmly smooth cover slack (female side of eye) with other hand towards throat point X. Repeat as necessary until bunching disappears. Continue milking until all cover slack between knot and throat of eye has been removed.

Tip: Do the following before burying the cover over the crossover:
  - Anchor loop of slip knot by tying it to stationary object before starting to bury. You will then find you can use both hands and weight of body with more ease to bury the cover over the core and crossover (last two views in illustration).
  - Hold the crossover tightly and milk all the excess cover from point R to point X.

Flex and loosen the line at the crossover point during the final burying process. Hammering the cover at point X will help loosen strands. With larger ropes it is helpful to securely anchor a slip knot, attach a small line to the braided core at the crossover and mechanically apply tension with a block and tackle, capstan, come-a-long, or power winch. Tension will reduce diameter of core and crossover for easier burying (last view in illustration).

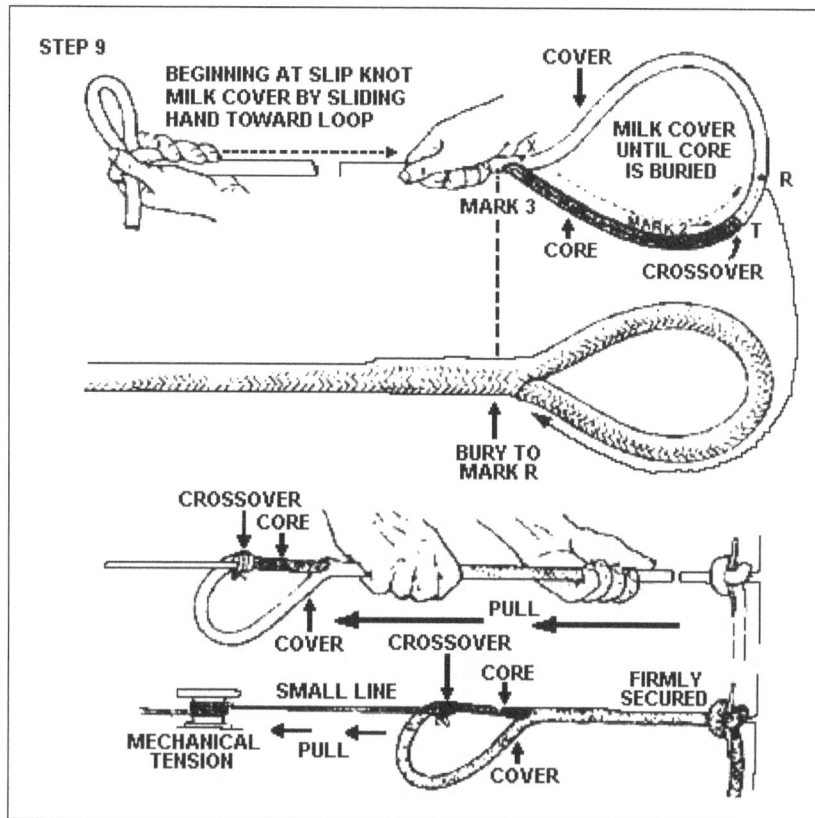

Figure 8-57. Burying the Exposed Core

● **Step 10. Finish the eye splice with lockstitch.** Lockstitch splices to prevent no-load opening due to mishandling. Use about one fid length of nylon or polyester whipping twine, about the same size as the strands in the line you are lockstitching. You may also use the same strands cut from the line you are lockstitching (Figure 8-58).

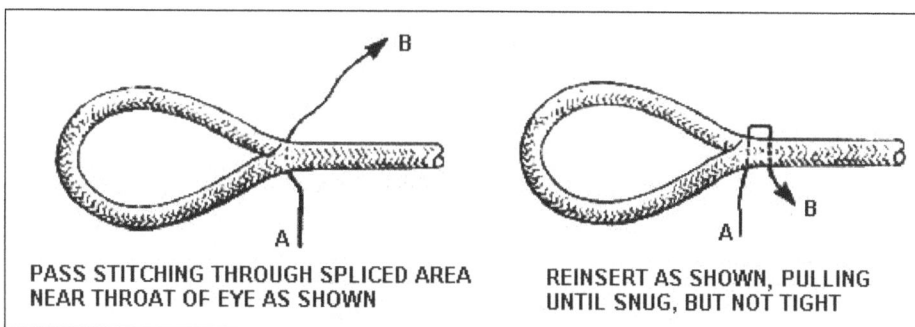

PASS STITCHING THROUGH SPLICED AREA
NEAR THROAT OF EYE AS SHOWN

REINSERT AS SHOWN, PULLING
UNTIL SNUG, BUT NOT TIGHT

Figure 8-58. Finishing the Eye Splice with Lockstitch

● **Step 11. Continue lockstitching.** Continue to reinsert as shown in Figure 8-59 until you have at least three complete stitches.

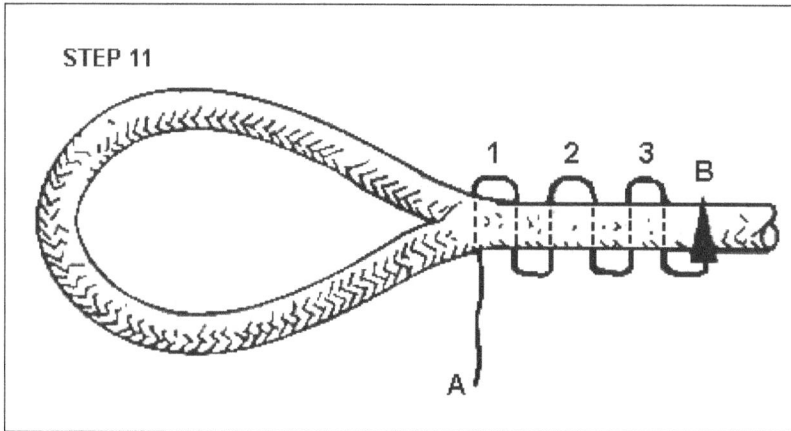

Figure 8-59. Continuing Lockstitching

- **Step 12. Complete lockstitching.** Rotate spliced part of line 90 degrees and reinsert end A into splice area in the same fashion as before. Make sure you do not pull stitching too tight. Complete last stitch so that end A comes out through the same opening in the braid as end B. Tie them together with a square knot and reinsert back ends into braid between cover and core as shown in Figure 8-60.

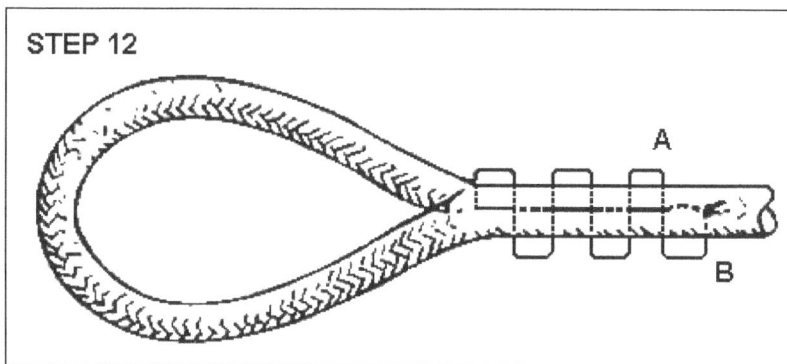

Figure 8-60. Completing Lockstitching

8-80. The splice will now be stitched on two planes perpendicular to each other.

END-FOR-END SPLICE

8-81. The Samson standard end-for-end splice can be done on new and used line (Figure 8-61). This is an all-purpose splice technique designed for people who splice used line as frequently as new line. It retains up to 85 percent average new line strength and up to 85 percent of the remaining used line strength.

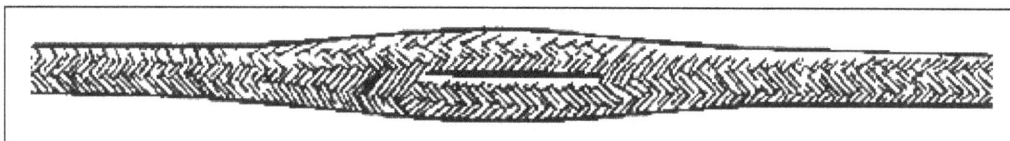

Figure 8-61. Standard End-for-End Splice

- **Step 1. Marking the measurements.** Tape the end of each line with one thin layer of tape. Lay two lines to be spliced side by side and measure one tubular fid length (two wire fid lengths) from end of each line and make a mark. This is point R (see Figure 8-62). From point R measure one short fid section length and mark again. This is point X where you should extract core from inside the cover. Be sure both lines are identically marked. Tie a tight slip knot about five fid lengths from point X. If you require the line with the finished splice to be a certain overall length, refer to Special Tools and Techniques (see page 8-36).

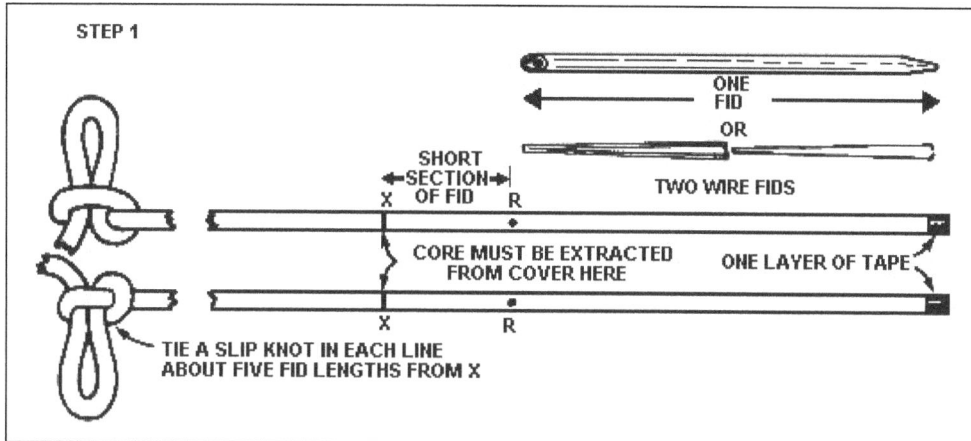

Figure 8-62. Marking the Measurements

- **Step 2. Extracting the cores.** Bend line sharply at point X. With the pusher or any sharp tool such as an ice pick, awl, or marlinespike, spread cover strands to expose core. First pry, then pull core completely out of cover from point X to the end of the line. Put only one layer of tape on end of core (see Figure 8-63). To be sure of correct positioning of Mark 1, do the following: Holding the exposed core, slide cover as far back towards the tightly tied slip knot as you can. Then, firmly smooth cover back from the slip knot towards taped end. Smooth again until all cover slack is removed. Then, mark core where it comes out of cover. This is Mark 1. Do this to both lines.

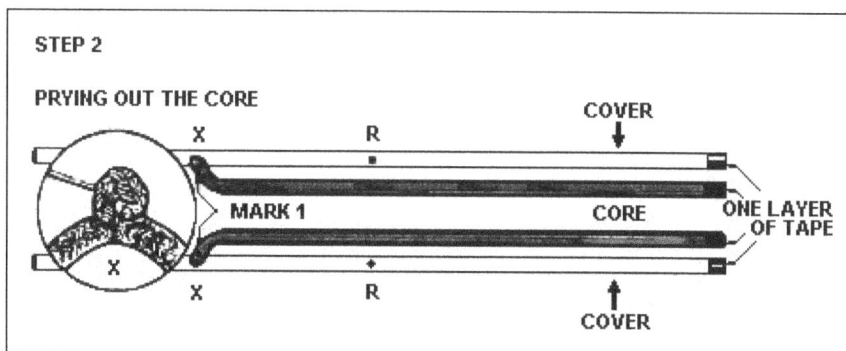

Figure 8-63. Extracting the Cores

- **Step 3. Marking the cores.** Hold one core at Mark 1 and slide cover back to expose more core (see Figure 8-64). From Mark 1 and following the core towards point X, measure a distance equal to the short section of fid and make two heavy marks. This is Mark 2. Measure one fid length plus another short section from Mark 2 in the same direction and make three heavy marks. This is Mark 3. Mark second core by laying it alongside the first and using it as an exact guide.

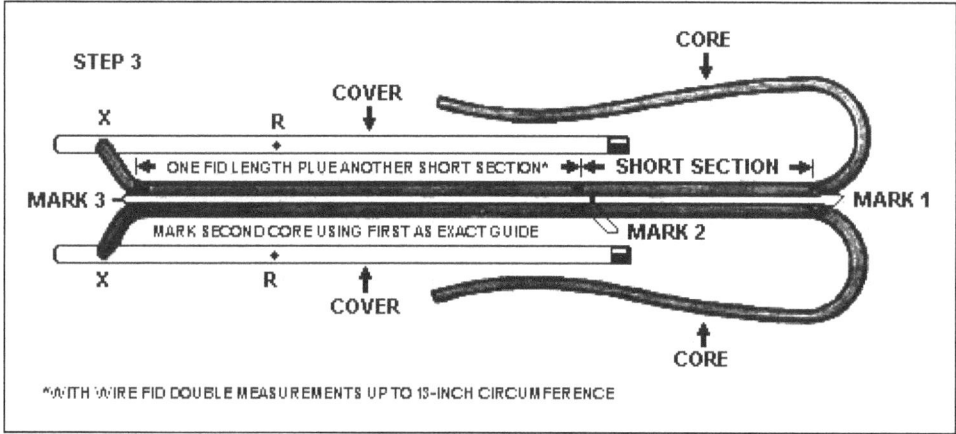

Figure 8-64. Marking the Cores

- **Step 4. Marking the cover for tapering.** Note nature of the cover braid (Figure 8-65). It is made up of strands. On inspection, you can see that half the strands revolve to the right around the line and half revolve to the left. Beginning at point R and working toward the taped end of cover, count eight consecutive pairs of cover strands, which revolve to the right (or left). Mark the eighth pair. This is point T (see Figure 8-65 insert). Make Mark T go completely around cover. Starting at point T and working toward taped cover end, count and mark every second right pair of strands for a total of six. Again, starting at point T, count and mark every second left pair of strands for a total of six (see Figure 8-65 insert). Mark both lines identically.

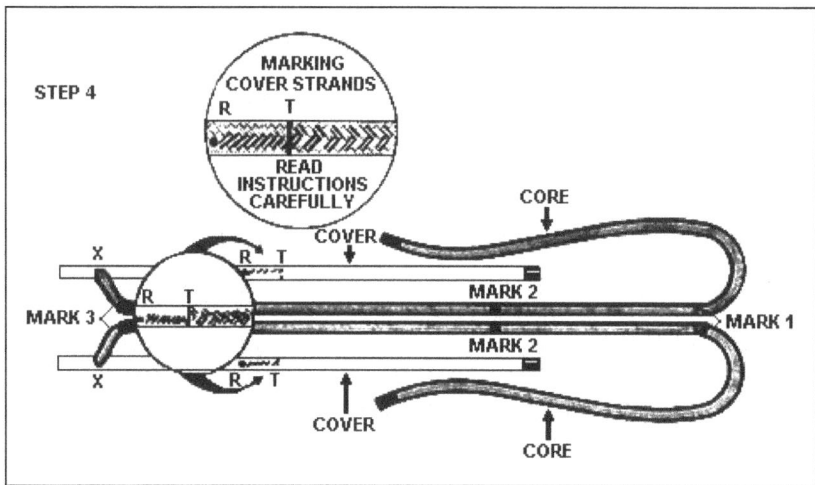

Figure 8-65. Marking the Cover for Tapering

- **Step 5. Performing the taper.** Remove tape from cover end. Starting with last marked pair of cover strands toward the end, cut and pull them completely out (see Figure 8-66 insert). Cut and remove next marked strands and continue with each right and left marked strands until you reach point T. Do not cut beyond this point (see Figure 8-66 insert). Re-tape tapered end. Cut and remove marked strands on the other marked cover, again stopping at point T. Re-tape tapered end.

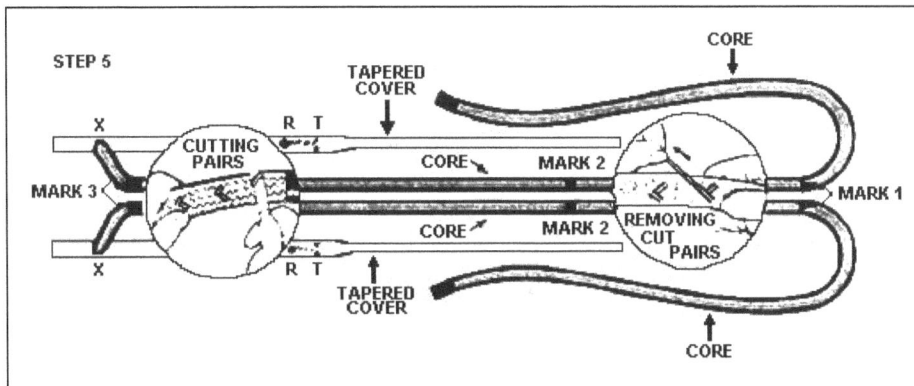

Figure 8-66. Performing the Taper

- **Step 6. Repositioning the lines.** Reposition lines for splicing as shown in Figure 8-67. Note how cover of one line has been paired off with core of the opposite line. Avoid twisting.

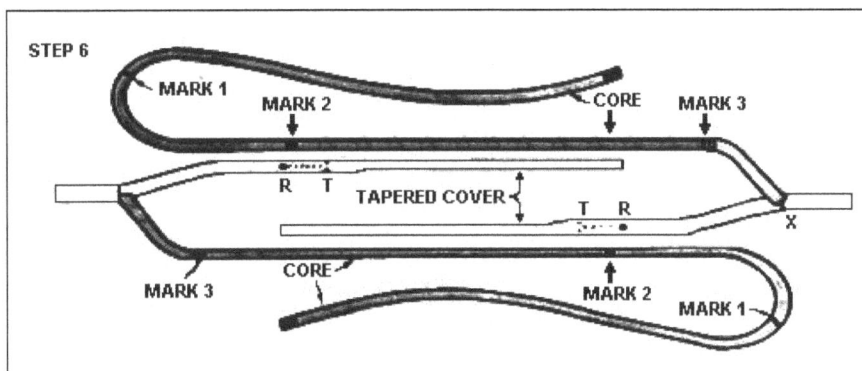

Figure 8-67. Repositioning the Lines

- **Step 7. Putting the cover inside core.** Insert fid into one core at Mark 2 and bring it out at Mark 3. Add extra tape to tapered cover end and jam it tightly into hollow end of fid (see Figure 8-68 insert). Hold core lightly at Mark 3, place pusher point into tapered end, pushing fid with cover in it from Mark 2 out at Mark 3. When using wire fid, attach fid to cover. Then pull fid through from Mark 2 to Mark 3. Pull cover tail through core until Mark T on cover meets Mark 2 on core. Insert other cover into core in same manner.

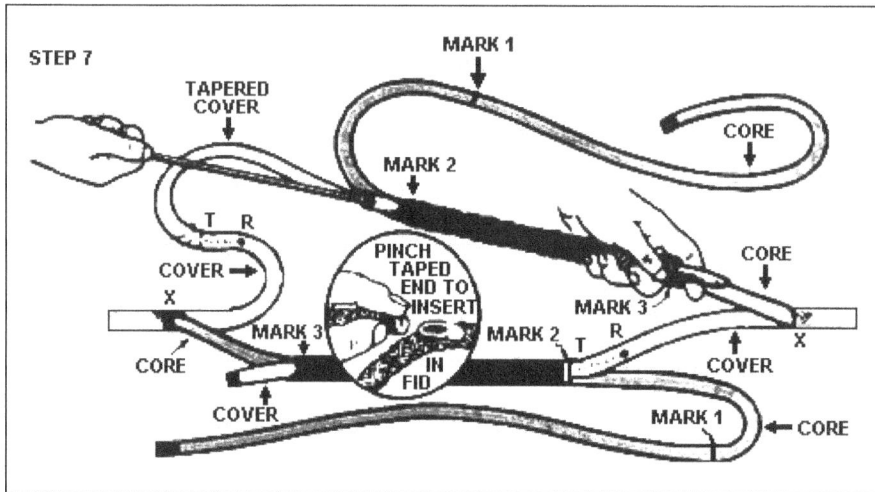

Figure 8-68. Putting the Cover Inside Core

- **Step 8. Reinserting the core into cover.** Now put core back into cover from point T to point X (see Figure 8-69). Insert fid at point T, jam taped core tightly into end of fid. With the pusher, push fid and core through cover, bringing out at point X. When using wire fid, attach fid to taped core. Then pull fid and braid through from point T to point X. Do this to both cores. Remove tape from end of cover. Bring crossover up tight by pulling on core tail and on tapered covered tail. Hold crossover tightly, smoothing out all excess braid away from crossover in each direction. Trim end of tapered cover on an angle to eliminate blunt end. Tapered cover tail will disappear at Mark 3. Cut core tail off at an angle close to point X.

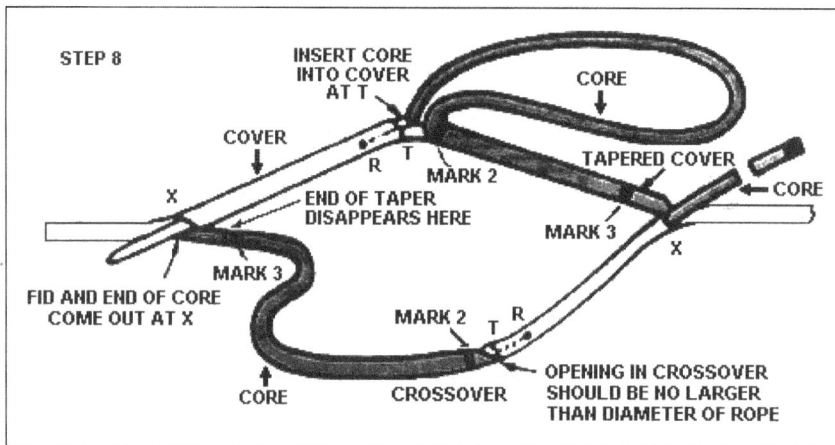

Figure 8-69. Reinserting the Core Into Cover

- **Step 9. Burying the exposed core.** Hold line at slip knot and with other hand milk cover toward the splice, gently at first and then more firmly (see Figure 8-70). The cover will slide over Mark 3, Mark 2, the crossover, and point R. Repeat with the other side of the splice. Continue burying until all cover slack between the knot and the splice has been removed.

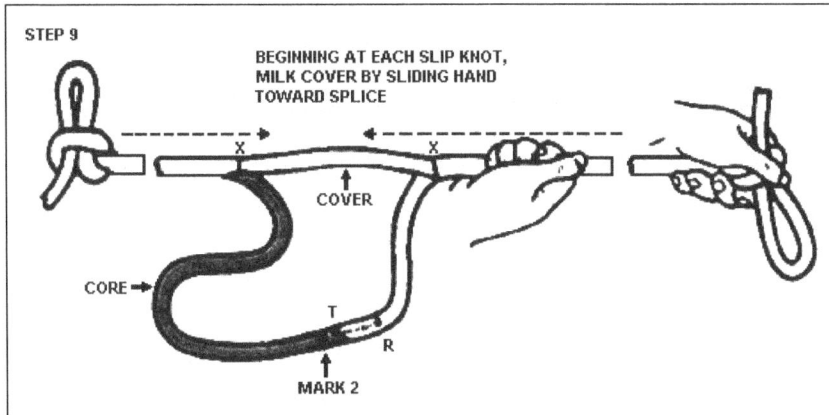

Figure 8-70. Burying the Exposed Core

- **Step 10. Finishing the splice.** The splice is done when all cover slack has been removed and there is an opening in the splice about equal in length to the diameter of line (see Figure 8-71). If one side of the splice at the opening is noticeably longer than the other side, something is wrong. Check steps 1 through 9 and remake if necessary. Now untie the slip knots.

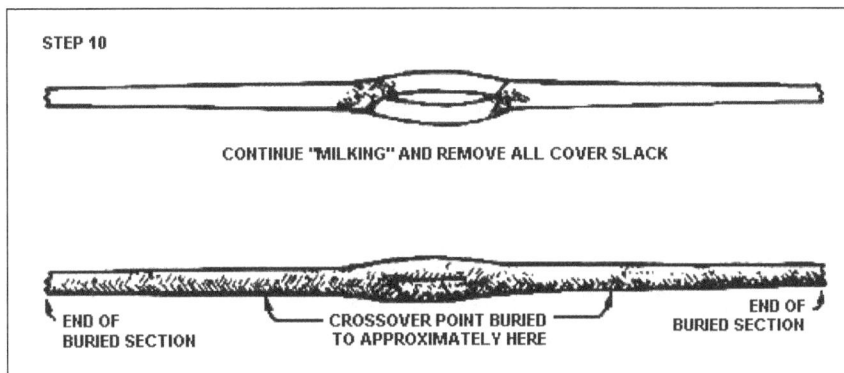

Figure 8-71. Finishing the Splice

This page intentionally left blank.

Chapter 9

Deck Maintenance

INTRODUCTION

9-1.   Vessel maintenance includes inspecting, cleaning, servicing, preserving, lubricating, and adjusting (as required). It can also require minor parts replacement within the capability of the crew. As a watercraft operator, you must take an active part in keeping your vessel at its peak operating condition. This is not an easy or simple task because you are constantly battling against the corrosive effects of salt water and salt air. The wind and sea also subject a vessel and its engines to strong stresses and strains. It takes day-by-day work and watchfulness to cope with all of these conditions; maintenance never ceases. This chapter covers the procedures and tools currently used for preventive maintenance and the required maintenance aboard ship. It should be used as a guide for all watercraft personnel responsible for shipboard maintenance.

PREVENTIVE MAINTENANCE

9-2.   These are the routine daily tasks that must be done aboard ship to prevent, or at least to minimize, the formation of rust or deterioration of the ship's equipment. The first and most important step in proper maintenance is keeping the vessel clean. This is necessary in order to maintain good health and efficient operation.

- **Decks.** Wash and scrub decks often to prevent tracking dirt throughout the vessel. If it can be obtained, canvas or cocoa matting can be laid on the deck wherever people walk. Scuppers must be kept clean and open so water can flow overboard freely and not leak into spaces below.
- **Topsides.** Topsides and superstructure must be washed often, using fresh water when possible. A small amount of washing soda can be added to the wash water to aid in cleaning. Parts washed with soda and water must be given a final wash-down with fresh water, if possible, or salt water.
- **Interior.** See that quarters are cleaned daily, giving close attention to dark corners and spaces blocked by lockers and other furnishings. Dirt collecting in these spaces results in unsanitary conditions where vermin can breed and rot can develop.
- **Bilges.** The rounded parts of a vessel's bottom, known as the bilges, collect water, oil, fuel, trash, etc. Keep them clean and well aired because dirty bilges are a fire hazard, produce disagreeable odors, and are harmful to vessels.
- **Cargo Holds.** Keep cargo holds clean. Stow and secure excess dunnage. Trace and eliminate sources of fumes and odors.

HAND TOOLS AND THEIR USES

9-3.   These tools must be cared for and used properly to get the most use from them. Safety in their use must also be stressed at all times.

9-4.   The following are the most commonly used hand tools found aboard ship:
- Chipping hammer.
- Wire brush.
- Hand scrapers.
- Portable electric grinder.
- Sandpaper.

9-5.   The use of each of these tools is described below.

- **Chipping Hammer.** Before letting anyone use this hammer, make sure they have been instructed on how to use only enough force to remove the paint. If a great deal of force is required to remove paint, the paint is still good and should not be chipped off. Feather the edges and paint.

- **Wire Brush.** This is a handy tool for light work on rust or on light coats of paint. It is also used for brushing around weld spots. When the surface is pitted, use a steel wire brush to clean out the pits.

- **Hand Scrapers.** These are more useful for removing rust and paint from small areas and from plating less than one-fourth inch thick, where it is impractical or impossible to use power tools.

- **Sandpaper.** Sandpaper can be divided into two types of abrasive materials: natural and artificial. The flint and garnet grits of ordinary sandpaper are natural abrasives. Emery and corundum are also used in the production of some of the cheaper grades of abrasive sheets. Artificial abrasives have largely replaced natural abrasives for use on metal. The two principal artificial abrasives are silicon carbide and aluminum oxide. The size of abrasive particles is indicated by code numbers ranging from 4 to 5/0 (or 00000). In garnet and artificial abrasives, 4 or 3 would be a very coarse abrasive (16-24 mesh); 2 1/2 to 1 1/2 would be coarse (30-40 mesh); 1 to 0 would be medium (50-80 mesh); and 2/0 to 5/0 would be fine (100-180 mesh). In flint paper or emery cloth, 3 to 1 would be coarse; 1 1/2 to 1/2 would be medium; and 0 to 3/0 would be fine. You will find sandpaper indispensable in cleaning corners. The usual procedure is to go over the surface first with coarse sandpaper and polish it with one of the fine grades. Do not polish any more than final finish requirements dictate, however, as paint bonds best to clean surfaces which are rough enough to provide "mechanical tooth." There is also a waterproof type of sandpaper. This usually consists of a better grade of garnet grit, bonded (made to stick on the paper) with a special resin. These sheets may then be used with water or oil for wet sanding. Ordinary sandpaper will disintegrate when used with liquids.

SHARPENING SCRAPERS AND CHIPPING HAMMERS

9-6.   Like other tools, scrapers and chipping hammers provide the best service when they are kept in good condition. Normally, this involves little more than sharpening the scrapers and hammers. The first step in sharpening a scraper is to square the end. Adjust the tool rest of the grinder so that it just clears the face of the wheel (see Figure 9-1, views 1 and 2). First, lay the scraper flat on the rest. Then, keeping the end of the scraper parallel with the shaft of the grinder, move the scraper back and forth across the face of the wheel. Grind across the entire width of the scraper. Use enough pressure to keep the wheel cutting out but not enough pressure to decrease its speed or overheat the metal. Keep a can of water handy while grinding and dip the scraper frequently into the water (this helps to prevent the scraper from overheating and drawing the temper from the metal). If the scraper has been chipped, grind away the edge until the chips disappear. With the end squared, begin to sharpen the scraper. Hold it in such a way that the original bevel lies flat against the face of the wheel (Figure 9-1, view 3). If the construction of the tool rest is such as to support it, hold your forefinger against the tool rest to serve as a guide as you pass the scraper back and forth across the wheel (Figure 9-1, view 4). Sharpening any tool in this manner causes the sharp edge to curl back or feather. The last step in sharpening is to remove the feathered edge. This may be done by lightly touching the flat side of the scraper to the side of the wheel, but a better method is to remove the feather with a file that has a fine surface.

9-7.   A chipping hammer is not sharpened like a cutting tool but rather like the blade of an ice skate. First, square the edge as described for scrapers. Then, as shown in Figure 9-2, grind away alternately on both bevels until the squared face is from one-sixteenth to one-eighth inch wide.

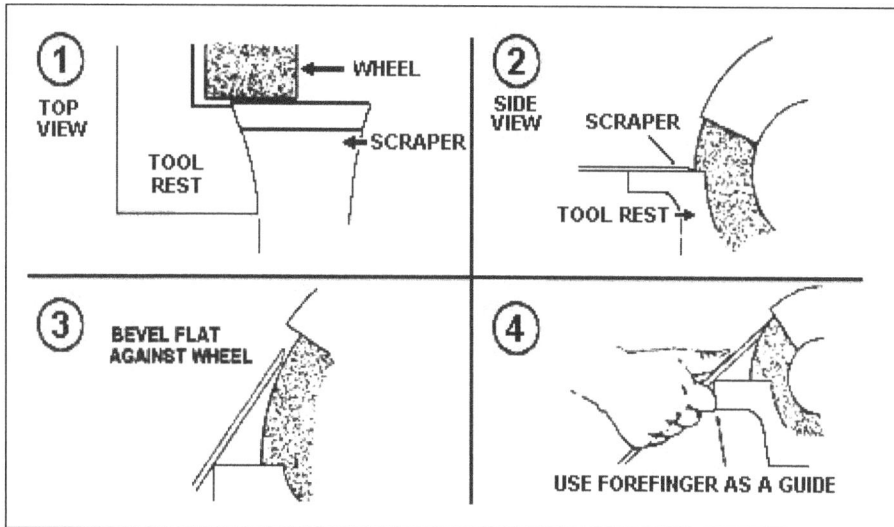

Figure 9-1. Sharpening the Scraper

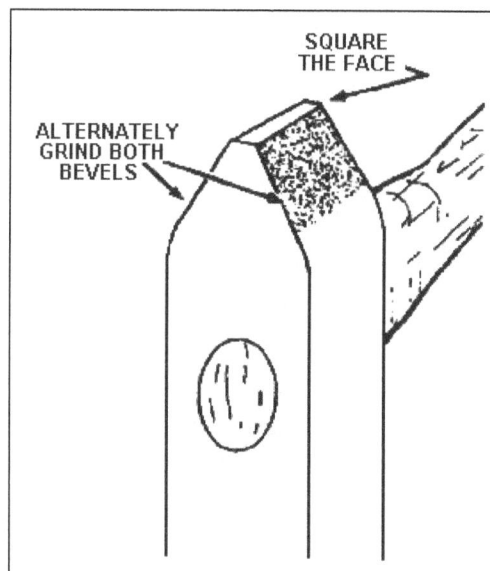

Figure 9-2. Sharpening a Chipping Hammer

POWER TOOLS

9-8.   The most useful power tool for surface preparation is the portable grinder (Figure 9-3). This usually comes equipped with a grinding wheel. This brush can be replaced with either the rotary wheel wire brush or the rotary cup wire brush. Light-duty brushes are made of crimped wire and heavy-duty brushes are made of tufts of wire formed by twisting together several strands of wire.

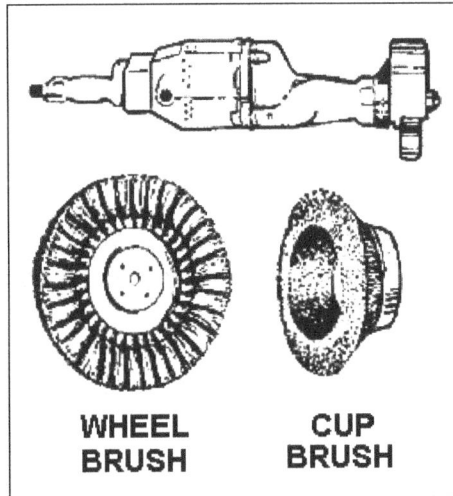

Figure 9-3. Electric Portable Grinder

9-9.   Scaling may be done by either of the tools shown in Figure 9-4. A chisel about 8 inches long and 1 1/4 inches wide is used with the pneumatic hammer. The hammer is held so that the chisel strikes the surface at an angle of about 45 degrees. Great care must be taken not to dent the surface. Denting forms low and high areas. This can lead to early failure of the thin paint film deposited on the high points.

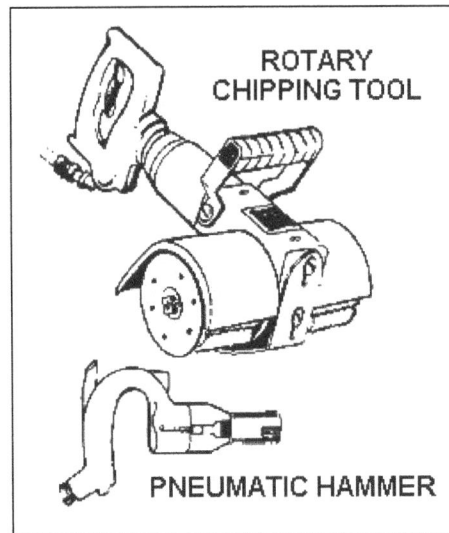

Figure 9-4. Power Scaling Tools and Wire Brushes

9-10.   The rotary scaling and chipping tool, sometimes called a "jitterbug," is electrically powered and has a bundle of cutters or chippers mounted on either side. Use it by pushing it along the surface to be scaled and letting the rotating chippers do the work. Replacement bundles of cutters are available. Also available is a larger, heavier model of this tool, designed especially for scaling deck. The electric disk sander is also a handy tool for surface preparation. Great care must be taken when using this machine. The disk should be moved smoothly and lightly over the surface. It should never be allowed to stay in one place too long because it will cut into the metal or wood.

SAFETY PRECAUTIONS

9-11.   Most electric tools are powered by 115-volt motors. Many people tend to regard 115 volts as not worthy of even moderate precautions. But make no mistake about it, 115 VOLTS CAN AND DOES KILL! All electric power tools are of the three-wire, grounded type. However, the operator can still receive a shock if the insulation on the wires becomes defective due to age, abrasion, or defective repairs; the ground circuit is not complete; or the operator becomes grounded.

WARNING

NEVER allow anyone to operate a power tool that is not functioning properly.

9-12.   Always make sure that personnel wear goggles when using power tools. This is particularly important with wire brushes because strands of wire frequently break off and shoot through the air like tiny arrows, which can penetrate a person's skin with ease. Insist that personnel give their full attention to the job and keep all parts of their bodies away from the working end of the tools. Keep nonessential personnel out of the area where power tools are in use. Always supervise work from behind the operators.

PAINT

9-13.   Paint is used primarily for preserving surfaces. It seals the pores of steel and helps to keep rust from forming. Paint also serves a variety of other purposes. It is valuable as an aid to cleanliness and sanitation because of its antiseptic properties and smooth, washable surface. Paint is also used to reflect, absorb, or redistribute light. For example, light-colored paint is used for the interior of the ship to distribute natural and artificial light to the best advantage. These same properties of reflection and absorption, incidentally, make camouflage painting possible.

9-14.   Paint consists of four essential ingredients: pigment, vehicle, drier, and thinner. To make any paint, the pigment is ground into the vehicle and the drier is added. Thinner is then added to make the paint the proper consistency for use by brush, roller, or spray gun.

- **Pigment.** The oldest of the opaque white pigments is white lead. White lead is no longer used in Army paint, although it is found in some commercial paints. It is made from acetic acid, carbon dioxide, and lead metal. The metal is corroded by the action of the other two ingredients until it becomes a fine, white powder. Linseed oil is usually added to this white lead to make a fine paste. It is then ready for use by the painter. Another white pigment is zinc oxide. Zinc oxide, by itself, makes a film that is too hard and brittle to withstand the extreme changes of outdoor temperatures, which cause it to crack and scale off. Therefore, because of its very fine texture it is usually mixed with titanium dioxide and other pigments for exterior work. Titanium dioxide and zinc oxide are now the principal white pigments in paints. Titanium dioxide is a white pigment with the highest known hiding power. Both titanium dioxide and zinc oxide are also considered "strengthening pigments" because they help increase the lasting quality of the paint in which they are used. Paint extenders, or inert pigments, are chemically stable and do not affect the color or durability of the vehicle. Extenders are used to—
  - Provide a less expensive base for certain kinds of colors.
  - Decrease the amount of chemically active pigments in the paint.
  - Reinforce the paint film.
  - Limit spreading power and increase the thickness of the paint film.
  - Make a good primer coat base for the finish coat.
  - Help prevent settling or caking in the container.

Note: Some of the more important extenders in common use are barium sulfate, calcium carbonate, whiting magnesium silicate or talc, and silica.

- **Vehicle.** The vehicle, usually referred to as the base, is the liquid portion of a paint which acts as a binder and brushing medium for the pigment particles. It wets the surface to be painted, penetrating the pores and ensuring the adhesion of the film formed by the drying vehicle. Until recently the base of most paints was oil (such as linseed oil). Today few paints contain raw oils of any kind. Some have bases of processed oils in combination with synthetic resins; others have vinyl bases. Some fire-retardant paints have chlorinated alkyd bases; some high-performance paints have two component epoxy or urethane bases. There are some that have water bases. Most oil-base vehicles dry partially by evaporation, partially by oxidation, and partially by polymerization. Polymerization is a process where two or more similar molecules combine chemically to form a larger molecule of a new substance. Older paints contained raw oils, had poorer physical properties when dry, and dried much slower than modern paints. For these reasons raw oils should never be added to paint. If the paint is thick and needs to be thinned, add some of the recommended thinner. Never add diesel oil, varnish, or other materials.
- **Drier.** Certain metallic compounds, when mixed with oil, add to the drying properties of paint. These are called driers. A paint drier acts as a conveyor of oxygen, taking it from the air and adding it to the oil. This speeds the oxidation of the paint. Without the drier, absorption of oxygen would be too slow a process, and you would have to wait too long for the paint to dry.
- **Thinners.** Thinners reduce the consistency of the paint to the proper degree for application by spraying or brushing. They also increase the penetration of the paint into the surface and help reduce gloss. The vehicle will become diluted if too much thinner is used. As you will remember, the vehicle is the binder. So if it is diluted too much, the durability of the paint will be affected. In flat paints the proportion of oil to thinner is deliberately reduced so that the paint dries without gloss. The most common type of thinner is mineral spirits, but the proper type to use depends on the base of the paint.

MIXING PRECAUTIONS

9-15.   Most paints you will be using will be ready-mixed. That is, when you draw them from the paint locker, they are ready for use. These paints have been carefully prepared to produce coatings that will be most satisfactory under the conditions in which the paints will be used. Certain paints require mixing immediately before use. These are zinc-dust, water-tank paint, aluminum paint, and high performance epoxy or urethane hull, tank, or nonskid deck paints (which contain more than one component). If the zinc-dust or aluminum paints are mixed and then stored, the heavier particles settle to the bottom. The zinc-dust or aluminum paste should be added in exactly the quantity needed, and the paints should be stirred often during use. Multi-component epoxy or urethane paints have a limited "pot life" after mixing and will thicken or harden if not used within that time. Aluminum and zinc-dust paints should always be freshly mixed just before use. If they are left standing any length of time after mixing, they lose the property of leafing. Leafing is the ability of the pigment to rise to the surface of the vehicle. In all cases, these paints should be used the same day as prepared. When kept in a sealed container, they have a tendency to become gaseous. The gases can rupture the container or blow the top off the can (presenting a danger to personnel). It can also result in a fire hazard as well as loss of the paint. So mix aluminum paint and zinc-dust paint only as needed, and use right away.

TYPES OF PAINT

9-16.   There are many different kinds of paint. For example, you cannot use the same type of paint on the deck topside and on the bulkheads in the captain's cabin. There is a different paint made for almost every purpose. The following describes some of the most important paints.

- **Primers.** Primers are paints specially prepared to adhere well to the surfaces for which they are mixed. A primer coat provides a good base for the finish coats of paint and, in the case of metal primers, includes chemicals that inhibit (hold in check) rust and other corrosion. Two common primers used are red lead and zinc chromate. Red lead is a general purpose primer used on all metals except aluminum. Use zinc chromate on aluminum. Always apply two primer coats on exterior, topside metal that has been stripped bare. Apply a third coat of the proper primer to the outside corners and edges. Allow at least 8 hours of drying time between coats.

- **Exterior Topside Paints.** Properly primed, vertical surfaces above the upper limit of the boot topping area are painted with two coats of gray. In general, horizontal steel surfaces are painted with two coats of deck gray. Refer to TB 43-0144, "Painting of Watercraft," for the exact color for each surface.
- **Bottom Paints.** Special paints have been developed for painting a ship's bottom. Because it is continuously underwater, the bottom is exposed to two dangers (corrosion and fouling). Either of these dangers can shorten the life of the ship if left unchecked. The part of the steel hull below the waterline will rust quickly from salt water if left unprotected. The steel hull can also become fouled with various types of marine growth. Two paint coverings that help overcome these problems are anticorrosive bottom paint and antifouling bottom paint.
- **Anticorrosive Paints.** These types of paint do not protect against fouling. Anticorrosive and antifouling paints are always used together on underwater hulls of active ships. The anticorrosive always goes on first. Vinyl and Formula 14N anticorrosive paints dry very quickly because the vehicles used evaporate rapidly. Because of this, you must apply anticorrosive with short, quick strokes and progress steadily over the area you are painting. Anticorrosive paint also contains heavy pigments that settle quickly. Because the beneficial effect of the paint depends largely on these pigments, you must stir the paint frequently.
- **Antifouling Paints.** These types of paint will prevent the fouling of the ship's bottom that result in loss of speed and increased fuel consumption. They contain copper oxide, the chemical most effective in preventing the attachment and development of marine growth. Remember that antifouling goes over an anticorrosive. They should not come in contact with the steel plating of the ship because they may pit the surface. Be sure that the total thickness and required coats of anticorrosive paint have been applied before putting on antifouling paint.
- **Deck Paints.** These types of paint are fairly standard. They do differ in color according to the compartment. For example, there is a dark green (interior) deck paint for decks in quarters and mess rooms, a gray deck paint for decks in most other living and working spaces, and a dark red deck paint for machinery spaces and workshops. Exterior steel decks and other horizontal surfaces may be covered in several ways. One system requires two coats of primer formula and two coats of exterior, gray deck paint. Where nonskid surfaces are required, nonskid decking material may be substituted.
- **Machinery Paints.** The usual coating for the parts of machines (which may be painted) is a gray enamel, silicone Alkyd, copolymer. When painting machinery, it is important to know what not to paint. DO NOT paint the following:
  - Start-stop mechanisms of electrical safety devices and control switchboards on machinery.
  - Bell pulls, sheaves, annunciator chains, and other mechanical communication devices.
  - Exposed composition metal parts of any machinery.
  - Identification plates.
  - Joint faces of gaskets and packing surfaces.
  - Lubricating gear, such as oil holes, oil or grease cups, zerk fittings, lubricators, and surfaces in contact with lubricating oil.
  - Lubricating oil reservoirs.
  - Machined metal surfaces of reciprocating engines or pumps.
  - Rods, gears, universal joints, and couplings of valve operating gear.
  - Ground plates.
  - Springs.
  - Strainers.
  - Threaded parts.
  - Zinc.
  - Working surfaces.
  - Hose and applicator nozzles.
  - Knife edges, rubber gaskets, dogs, drop bolts, wedges, and operating gear of watertight doors, hatches, and scuttles.

- Electrical contact points and insulators.
- Internal surfaces of ventilation ducts.

Note: A special heat-resistant paint is available for application on objects that are subjected to high temperatures.

- **Aluminum Paints.** These types of paint are composed of aluminum powder or paste and a varnish specially made for mixing purposes. The standard for practically all uses is 2 pounds of aluminum powder to 1 gallon of mixing varnish. Add the varnish gradually, increasing portions and stirring vigorously until a uniform mixture is obtained.

PAINT REMOVERS

9-17.   There are three types of paint removers in general use:
- **Flammable**. Solvent type containing benzol, acetone, and amyl acetate.
- **Nonflammable**. Containing chlorinated hydrocarbons.
- **Waterbase**. Alkali type containing caustic materials.

9-18.   Although all three are effective, their use aboard ship is limited because they are definitely hazardous. They must be used only in well-ventilated spaces, and all safety precautions, according to the type of remover in use, must be observed. Removers made to strip epoxy paints are extremely hazardous from both toxicity and skin burning standpoints. Alkali paint removers must not be used on aluminum or galvanized surfaces. Do the following regardless of the type of paint remover you use. Wet the surface with a smooth coat of the remover and let it soak in thoroughly until all paint or varnish is loosened. Then lift the paint off with a hand scraper. A film will form on the surface of the remover soon after it is spread on the object. Do not disturb or break this film until you are ready to lift off the paint. If you break the film, the remover will lose some of its effectiveness. After the surface is cleaned, wet it again with the remover and wipe it off with a rag. Then thoroughly wash the surface, with paint thinner or soap and water.

9-19.   Where sandblasting methods are not practicable, manual methods may be used. The surface should be roughened first with roughing tools such as chipping hammers and scrapers. Care must be taken to prevent nicking, denting, or scratching of the surface when using these tools. Nicks, dents, and scratches provide ideal starting points for early failure of paint films. The low portion of such surface irregularities is difficult to clean and becomes a source of rust. On the sharp-edged high points, only a thin film of paint will adhere causing early paint failure. Chipping hammers should never have a chisel-sharp edge. Thin plate (under three-sixteenth inch) should never be chipped, but should be scraped and wire brushed. When the bulk of rust and old paint has been removed by roughing tools, finishing tools should be used to complete the job. The main finishing tools are hand wire brushes, electric and pneumatic wire brushes, and sandpaper. The use of finishing tools without first using roughing tools results in a small amount of surface readied for painting or, with power wire brushing, the "glazing" rather than removal of the rust and old paint. The hand scraper is used on small jobs, where power tools are impractical, and in crevices and corners. The hand wire brush is a useful tool for light rust and for brushing around welds and in places not accessible to the power wire brush. Sandpaper and sanding discs are used where a smooth, clean surface is desired. Where old paint is removed in spots, the edge of the remaining paint film should be sanded so that it tapers (feathers) down to the bare metal. This will give a smooth appearance to the new paint film. Steel wool should not be used as a substitute for sandpaper. Small steel particles can become embedded in the paint and form a source of rust. After the finishing operation is completed, sweep the surface to remove paint chips, dirt, and dust, and apply a liberal coat of primer.

```
┌─────────────────────────────────────────────────────────┐
│                        WARNINGS                           │
│                                                           │
│  • NEVER use paint removers around an open flame because  │
│    some of them contain flammable ingredients.  DO NOT use│
│    them in confined spaces since some of them have dangerous│
│    anesthetic properties.                                 │
│  • DO NOT use paint and varnish removers if you have open cuts│
│    or sores on your hands (unless you are wearing rubber  │
│    gloves).                                               │
│  • Avoid letting the remover touch your skin (watch out   │
│    particularly for your face, eyes, and mouth).  If paint remover│
│    touches the skin and begins to burn, wash it off with cold│
│    water immediately and get medical assistance.          │
│  • NEVER use turpentine, spirits, or other thinners for cleaning│
│    your hands; they can be absorbed through the skin. Gasoline│
│    is also dangerous because it may contain lead.         │
└─────────────────────────────────────────────────────────┘
```

PREPARATION OF ALUMINUM SURFACES

9-20.   Sandblasting and wire-brushing aluminum surfaces are not recommended. Clean by brushing off dirt and dust and washing the surface with paint thinner. Then, wash with paint cleaner and water, rinsing with fresh water. Defective paint coatings on outside painted surfaces should be removed with paint remover. The cleaned surface must then be washed with paint thinner, scrubbed with paint cleaner and water, and rinsed with fresh water. Good paint on inside painted surfaces should not be removed.

9-21.   Flaking, scaling, or peeling patches may be removed with hand scrapers, being careful not to damage the aluminum surface. Dulling or rounding the corners of scrapers will help prevent nicking the aluminum. The edges of the good paint are flared with sandpaper. Do not sandpaper the aluminum surface. Wash the entire surface with paint thinner and again with paint cleaner and water and a fresh water rinse. Never use a chipping hammer or power tool on an aluminum surface.

PREPARATION OF GALVANIZED STEEL SURFACES

9-22.   Heavy blast-cleaning of galvanized steel shall be avoided. Chipping hammers will not be used on galvanized surfaces. Old paint shall be removed by means of a hand scraper and wire brushes. Never use sandpaper on this surface. After scraping and wire-brushing a galvanized surface, wipe the surface with paint thinner. As soon as the surface is dry, apply a liberal coat of primer.

SHIP BOTTOMS

9-23.   The condition of the ship's bottom has a considerable effect on steaming performance. Before applying paint to the bottom, be sure that it has been cleaned carefully. A special problem is involved near the waterline where oil and grease often accumulate. Paint applied over grease will not adhere or dry, so you must remove all traces of grease with a solvent. Gasoline was formerly used for this purpose but recently kerosene solvent-emulsion cleaners have been developed. They are inexpensive and efficient and do not constitute as great a fire hazard as gasoline. DO NOT REMOVE PAINT THAT STICKS AND IS FREE FROM FOULING. Remove blistered, flaked, or loose paint by sandblasting, hydro-blasting, or hand cleaning. DO NOT touch paint that adheres firmly and gives protection to the bottom. Clean antifouling paint, which is over 2 years old, to its original color and apply two additional coats. Where paint is completely removed and the metal is bare, replace all coats of the bottom system.

PIPING SYSTEM

9-24. Remember that you must not mar the surface when cleaning piping systems. The ordinary procedure is to remove loose paint from the pipe with a scaling tool; then go over the pipe with a wire brush to remove all loose particles. At some shipyards the pipes may be taken out and sandblasted while other machinery is being moved. One way the inside of a pipe is cleaned is with a tool called a vibrator. The vibrator has a long shank with a mallet-like head. The shank is inserted in the pipe and the vibration of the head removes scale and rust. Then an air hose is used to blow all the loose flakes out of the pipe.

ELECTRICAL EQUIPMENT

9-25. No attempt should be made to remove paint from electrical cables, fixtures, control enclosures, or switchboards. If you take the paint off a cable, you may injure the protective armor and watertight sheath directly beneath it. Damage to the sheath will allow moisture to enter and will result in grounding. Twisting or bending a cable to remove paint from it might destroy the watertight integrity of the packing in the bulkhead stuffing tubes. When scraping paint, sandblasting, or painting near electrical equipment, be sure that the equipment is covered to protect it from paint, dust, and sand particles. After your work is finished, clean the electrical equipment thoroughly, using a vacuum cleaner if you have one. Remember that paint dust is full of abrasive and semiconducting particles, which can seriously damage electrical equipment.

CAUTION

Note the flash point listed on container labels and take adequate precautions. Apply only when electrical equipment in the same and adjacent compartments is de-energized and surfaces to be painted are in a cold-iron condition.

PAINT PREPARATION

9-26. No matter how high the quality, paint will give poor service if not thoroughly mixed before applying. When paint stands for long periods of time, the pigment settles to the bottom of the container, and the vehicle rises to the top. Naturally, the paint must be remixed before use. The best system for mixing is to pour off most of the vehicle and mix the remainder thoroughly. Then add a small amount of the liquid at a time, until the entire vehicle has been added and the paint is uniform. To make sure the paint is thoroughly mixed, pour the paint back and forth between two cans (Figure 9-5). This process is called "boxing" and ensures a smooth and even mixture.

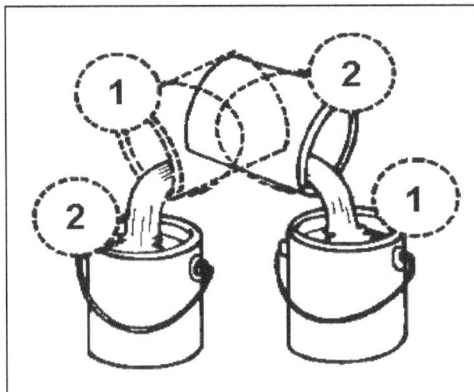

Figure 9-5. Steps in Boxing

9-27.   Some of the newer paints require special mixing procedures, including induction times (waiting periods) before use. For epoxy or polyurethane paints, carefully follow the manufacturer's instructions regarding mixing, induction time, and applying. When opening a paint can, you may find that a "skin" has formed on the surface of the paint. This must be removed carefully and thrown away. All particles of pigment, dirt, and skin should be taken out by straining the paint through a wire screen or cheesecloth. Straining should be done after the paint has been mixed thoroughly.

Note: During storage, turn paint containers "bottoms up" periodically (at least once every 90 days) to reduce the labor involved in mixing paint.

PAINTING BY BRUSH

9-28.   Smooth and even painting depends as much on good brush work as it does on good paint. There is a brush for almost every purpose, so be sure you use the right brush and keep it in the best condition. Table 9-1 lists the name and general use of the most frequently used brushes.

Table 9-1. Types and Uses of Brushes

Type	For Use On
Flat paintbrush	Large surfaces
Oval sash and trim brush	Small surfaces
Fitch brush	Small surfaces
Oval varnish brush	Rough work
Flat varnish brush	Medium work
French bristle varnish brush	High-grade work
Lettering brush	Small surfaces
Painter's dusters	Cleaning work

9-29.   The two most useful brushes are the flat brush and the oval sash and trim brush. A skillful painter using a flat brush can paint almost anything aboard ship. Flat brushes are wide and thick, carry a large quantity of paint, and provide a maximum of brushing action. Sash brushes are handy for painting small items and those hard-to-get places and for cutting in at corners. The most common used brushes aboard ship are shown in Figure 9-6.

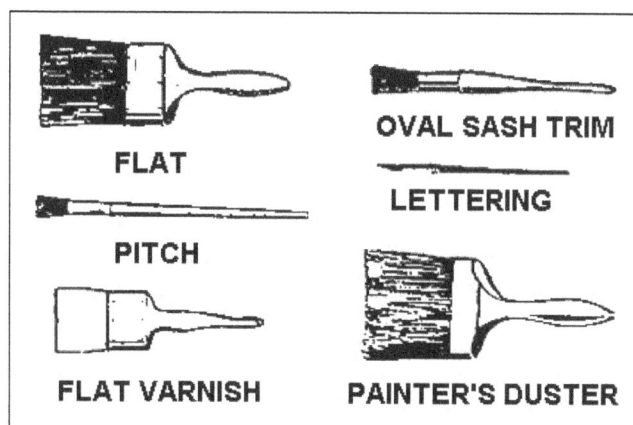

Figure 9-6. Types of Brushes

9-30.   Many of the brushes are made of horsehair and other natural bristles. More and more brushes are being made of synthetic bristles such as nylon. These brushes are much cheaper and provide comparable brushing action.

BRUSHES

CARE BEFORE USE

9-31.   Brushes are only as good as the care given them. The best brush can be ruined very quickly if not properly treated. If you follow the suggestions given, your brushes will last longer and give better service. When bristles of paintbrushes were set in wood, painters dampened the wood to cause it to swell and hold the bristles more tightly. However, almost all modern paintbrushes have bristles set in rubber or in some composition material. This means, of course, that wetting the end of the handle holding the bristle is useless. In fact, this practice will probably cause harm because it will tend to make the metal band (ferrule) rust faster.

9-32.   To make a new natural bristle brush more flexible and easier to clean, rinse it in paint thinner and soak it in boiled linseed oil for about 48 hours. Drain the oil from the brush before using. Wipe the bristles clean and wash them in a solvent or other oil remover. Synthetic bristle brushes do not require special treatment before use.

CARE AFTER USE

9-33.   Every paint locker should have a container with divided compartments for stowing different types of brushes (that is, paint, varnish, shellac, and so on) for short periods of time. The container should have a tight cover and a means of hanging brushes so that the entire length of the bristles and the lower part of the ferrule are covered by the thinner or linseed oil. The bristles must not touch the bottom because they eventually will become distorted, making it impossible to turn out an acceptable job with them. A simple brush keeper is shown in Figure 9-7. Drill a small hole through the brush handle and support the brush so that the ends of the bristles are allowed to soak in paint thinner or linseed oil. The keeper may be square or round, but it must have a tight lid to prevent evaporation and to avoid being a fire hazard.

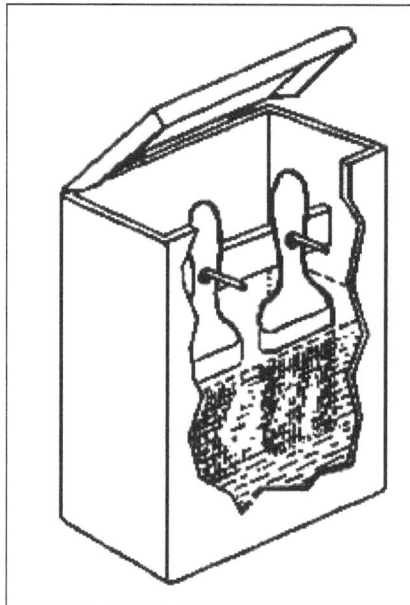

Figure 9-7. Small Brush Keeper

9-34.   Brushes to be used the following day should be cleaned in the proper thinner and placed in the proper compartment of the container. Brushes to be used later should be cleaned in thinner, washed with soap (or detergent) and water, rinsed thoroughly in fresh water, and hung to dry. After drying, they should be wrapped in waxed paper and stowed flat. Brushes should not be left soaking in water; the water causes the bristles to separate into bunches, flare, and become bushy. The proper cleaners for brushes used with different finishes are shown in Table 9-2. Remember that paint-soaked brushes should never be left in an open can of paint or exposed to the air. Good brushes are hard to get—take care of them. Clean them immediately after use; then store them properly.

Table 9-2. Brush Cleaners for Different Finishes

Finishes	Cleaners
Natural and synthetic oil-base paints and varnishes; chlorinated alkyd	
resin paint	Paint thinner or mineral spirits
Latex emulsion paints	Water
Chlorinated rubber paints	Synthetic enamel thinner or xylene
Shellac	Alcohol
Lacquer	Lacquer thinner

HOW TO USE A BRUSH

9-35.   There is an art to using a paintbrush properly. It is an art you will have to master if you are going to become a good painter. The following general hints will help you. Read them once to see how many mistakes you have been making. Then concentrate on each point separately until you are sure you have it mastered. Hold the brush firmly, but lightly, in the position shown in Figure 9-8. Do not put your fingers on the bristles below the ferrule. Hold the brush in a way that will permit easy wrist and arm motion.

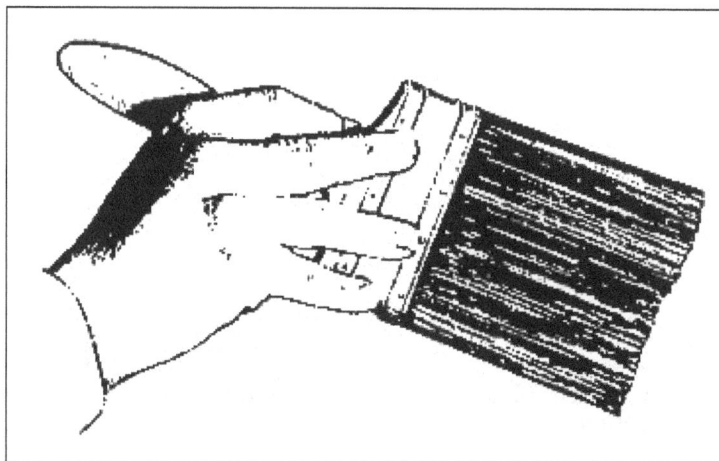

Figure 9-8. Correct Way to Hold a Brush

9-36.   Do not try to paint with the narrow edge when using a flat brush. That will wear the corners down and spoil the shape and efficiency of the brush. When using an oval brush, do not let it turn in your hand. An oval brush that has been revolved too much will wear to a pointed shape and become useless. Do not poke oversized brushes into corners and around moldings. Such use will ruin a good brush by bending the bristles. Use a smaller brush that will fit into such spots. Work the paint well into the brush before you start to paint. Hold the mixing paddle tightly over the rim of the bucket, dip the brush into the paint, and then wipe the brush clean across the edge of the paddle. Do this several times so you will be sure the brush is filled with paint. Dip slightly less than half of the bristles into the can when applying paint. Slap the brush lightly against the side of the can, and then apply it to the surface to be painted. To avoid paint from dripping off your brush, be careful not to overfill your brush. Hold the brush at right angles to the surface being painted, with the ends of the brush just touching the surface. Lift the brush clear of the surface when starting the return stroke. If the brush is held obliquely and is not lifted, the painted surface will be uneven, showing laps and spots and a "daubed" appearance. A brush that is held at too great an angle will soon wear away at the ends.

PAINT APPLICATION

9-37.   Use the "lay on" then "lay off" method to completely cover with paint (Figure 9-9). "Laying on" means applying the paint first in long, horizontal strokes. "Laying off" means crossing your first strokes by working up and down.

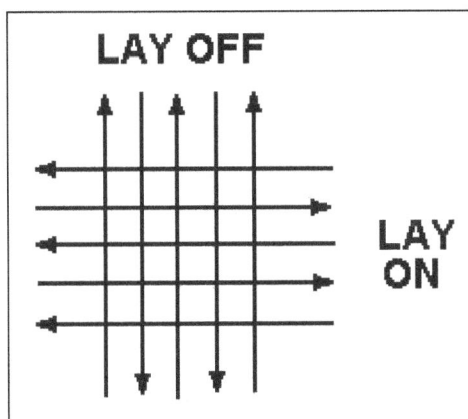

Figure 9-9. Laying On and Laying Off

9-38.   The laying-on and laying-off method distributes the paint evenly over the complete surface with the least amount of paint. A good rule to follow is to lay on across the shorter distance and lay off in the longer direction. When painting bulkheads or any vertical surface, lay on in horizontal strokes, lay off in vertical strokes. Always paint the overhead first and work from the far corner. By working the overhead first, you can keep the bulkhead free of drippings by wiping up as you go along. When painting overhead surfaces, paint strokes on the ceiling panels should normally be laid fore-and-aft, and those on the beams, athwart ships. But where panels contain many pipes running parallel with the beams, it is often difficult to lay off the ceiling panels fore-and-aft. In such cases, you will get better results by laying off the panels parallel with the beams. To avoid brush marks when finishing a square, stroke toward the last square finished gradually lifting the brush near the end of the stroke while the brush is still in motion. Every time the brush touches the painted surface at the start of a stroke, it leaves a mark. For this reason, never finish a square by brushing toward the unpainted area, but always end up by brushing back toward the area already painted. When painting pipes, stanchions, narrow strips, beams, and angles, lay the paint on diagonally. Lay off along the long dimension (Figure 9-10).

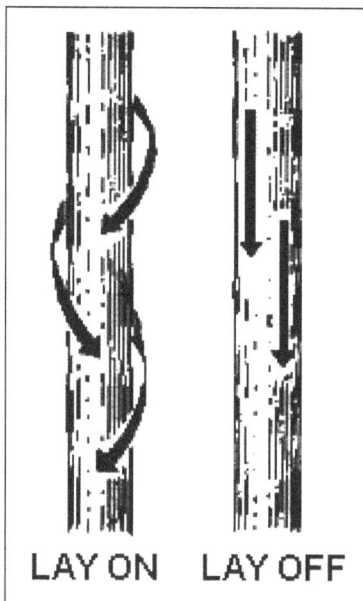

Figure 9-10. Painting Pipes and Stanchions

9-39.   Always carry a rag for wiping dripped or smeared paint.

FILM THICKNESS

9-40.   For interior painting, paint must be applied in the lightest possible coat that will cover the surface. Several reasons for this are:

- Heavy layers of paint constitute a fire hazard—the thicker the paint film, the more readily it will burn. Also, if paint is applied heavily, it is likely to entrap solvents and thinners that burn rapidly.
- Thick coats of paint tend to crack and peel. They are likely to be uneven and may show marks and scratches more readily than thin coats. Thick coats of paint do not penetrate as well as thin ones and do not dry as hard to a surface.

9-41.   If an interior surface has already had a total of four coats of paint (including primer) or if the total thickness of the existing paint amounts to 0.005 inch, the old paint should be removed before adding any more paint.

WORKING CONDITIONS

9-42.   Painting should not be attempted at a temperature below 32 degrees F. In cold weather, moisture condenses on surfaces and the paint will not stick. The thinner also evaporates too slowly, increasing the drying time. For best results, paint during warm weather (between 60 degrees F and 80 degrees F). In hotter weather, paint dries too rapidly and makes brushing and rolling difficult. Humidity and ventilation are also important considerations. High humidity may cause condensation on the bulkheads and make painting difficult. To reduce humidity inside, you can increase the temperature or improve the ventilation. Proper ventilation is also necessary to carry off the solvents and to furnish oxygen so the paint will dry properly.

STRIPING WITH MASKING TAPE

9-43.   Striping can be a relatively easy job if you use masking tape. You can use either a brush or spray gun with masking tape. There are two basic methods to follow, depending on whether the surface to be striped has been finished.

- **Striping Method Number 1.** If the surface is already painted and you do not want to do a complete repainting job, you can still add stripes without marring the finish. First decide on the position and width of the stripe; then apply masking tape to both sides of the stripe. Figure 9-11 shows how to apply the tape. It is a good idea to add a further protective covering on both sides, wide enough to prevent daubs or overspray from striking the rest of the surface. Placing newspapers or wrapping paper in the proper position before painting may also provide protection. The striping color is then sprayed or brushed on. When the paint has set, the masking tape is removed.

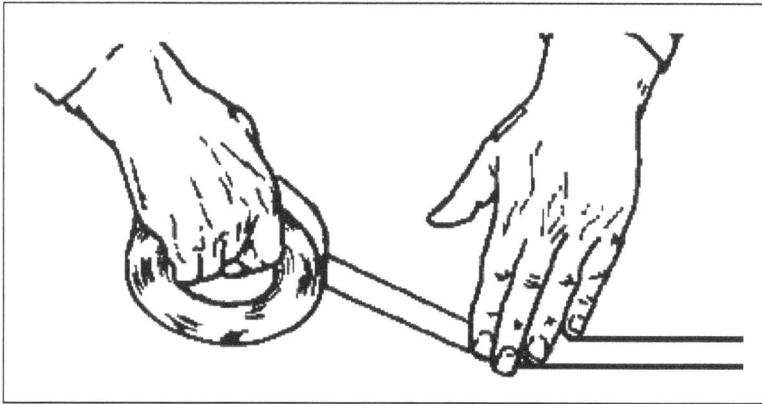

Figure 9-11. Applying Masking Tape

- **Striping Method Number 2.** If the surface is unfinished, the process of striping is a little different. First decide on the position and width of the stripe; then spray or brush the color on, allowing the paint to overlap the edge of the stripe a little on both sides. Allow the striping color to dry thoroughly, and then cover the exact area of the stripe with masking tape. Attach the tape firmly, but do not stretch it too much. Rub or roll it down to smooth out the wrinkles and make a tight, protective covering. Some painters recommend as the next step a light "fog" covering of the finishing material right over the tape. This will help to prevent the final coat of paint from sticking to the edges of the tape, cementing the tape to the surface. Now you are ready to spray or brush on the finishing coat. Do this right over the masking tape. When the surface coat has set, remove the tape to reveal a clean-cut stripe.

REMOVING MASKING TAPE

9-44.   There is a trick to removing masking tape so it will not mar the surface. The right way is to pull the tape off somewhat diagonally and back upon itself. The wrong way is to pull the tape directly away from the surface at a right angle. Figure 9-12 shows the proper angle. Work slowly, with your hands moving close and parallel to the surface.

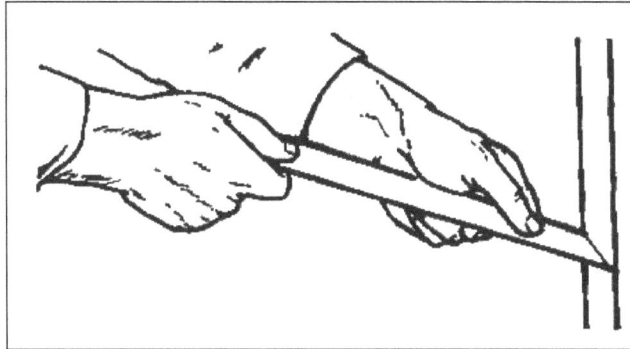

Figure 9-12. Removing Masking Tape

9-45. There may be a slight ridge along the edges of the stripe after you pull off the masking tape. If this is too noticeable, you can scrape it off after it has dried thoroughly and then rub it smooth with a rubbing compound.

STENCILS

9-46. All ships or boat companies should have adjustable stencil sets with locking edges. These sets are made of brass and include punctuation marks as well as the 26 letters of the alphabet and numerals from 0 through 9. The sets come in three sizes: 1/2 inch, 1 inch, and 2 inches. The edges of each piece are crimped allowing the use of a combination of letters, figures, and punctuation marks by slipping the edge of one piece into the edge of the adjoining piece.

9-47. Flat-ended brushes especially designed for stenciling are available, but an old toothbrush makes an acceptable substitute. Use the stencil paints available in general stores. After stenciling one surface, wipe off the back of the stencil before laying it on the next surface to be stenciled. Make sure the stencil does not slip while applying the paint. Stencils should be cleaned immediately after use—the brass ones with the proper thinner, the other type with only a clean soft rag.

CUTTING IN

9-48. A painter who has learned to "cut in" properly can do a job in less time than it takes another person to apply masking tape. Cutting in is not hard and anyone with a fairly steady hand can learn it in a short time. Suppose you have to cut in the angle between an overhead and a bulkhead. Start at one corner. Hold your brush at an angle of about 75 to 80 degrees from the bulkhead and about 10 degrees from the overhead. Run your brush along in fairly swift, long, smooth strokes. This is one job where working slowly will not produce better results. The slower your stroke, the wavier the line. If there is no definite break, such as the angle between bulkheads and overheads or decks, you should draw a line to follow. You can do this either with a straightedge or by snapping a chalk line. To snap a chalk line, first mark a couple of reference points, one at each end where the line will be. Then chalk the line and stretch it taut between the reference points. Have somebody pull the center of the line about 6 inches out from the surface and let it snap back against the surface. This leaves a neat, straight line. Cut in as already described. You may want to paint up close to the line and then cut in, but usually it is best to cut in first and paint out from that line.

PAINT ROLLERS

9-49. The dip type of paint roller consists of a replaceable, knotted Dynel, plush fabric roller having a solvent-resistant paper core which rotates on the shaft of a corrosion-resistant steel frame. Large areas, such as decks and ship's sides, free of rivets, bolts, cable, pipes, and so on, can be quickly covered with paint by using rollers. In order to get uniform coverage, always try to pick up the same amount of paint with your roller, and paint the same size area. A 7-inch roller filled with paint will cover about a square yard; a 9-inch roller, of course, will cover slightly more. Dip your roller in the paint at the lower end of the tray and roll it lightly toward the raised end. Repeat this process as necessary to fill the roller evenly. Then quickly apply it to the surface to be painted, using the same lay-on, lay-off technique used when brush painting. A moderate amount of pressure must be applied to the roller to ensure that the paint is worked into the surface. If pressure is not applied, the paint will not adhere and will peel off. The fabric cylinder should be stripped from the core after use, cleaned in the solvent recommended for the paint used, washed in soap and water, rinsed thoroughly, and replaced on the core to dry. Combing the pile of the fabric while damp will prevent matting.

SPRAY GUNS AND THEIR USES

9-50.   A spray gun is a precision tool in which paint is sprayed out through a nozzle by air pressure. The mixing area may be outside or inside the gun's spray cap.

CLASSES OF SPRAY GUNS

9-51.   Spray guns are classed according to where the air and paint are mixed (external-mix, internal-mix), how the air is controlled (bleeder, nonbleeder), and how the paint is fed to the nozzle (suction-feed, pressure-feed).

- **External-Mix Spray Gun.** In an external-mix gun, the air and paint are mixed outside and in front of the air cap as shown in Figure 9-13. This type of gun requires high air pressure, thereby using more cubic feet of air per minute than does an internal-mix gun. Atomization of the paint is extremely fine, however, and the size of the spray pattern can be controlled. There is no wear on the air nozzle. With different nozzles, an external-mix gun works with both suction and pressure feeds.

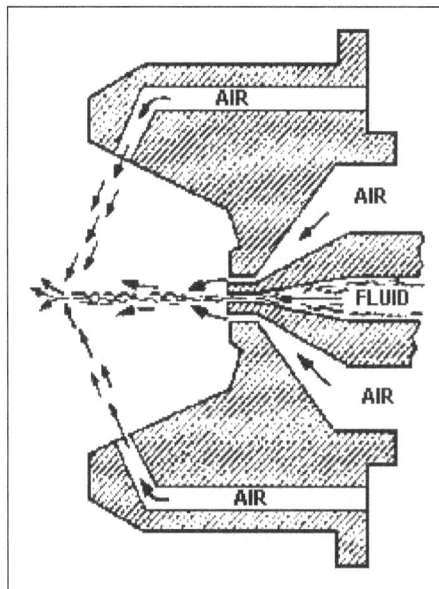

Figure 9-13. External-Mix Air Cap

- **Internal-Mix Spray Gun.** In an internal-mix gun, air and paint are mixed within the gun as shown in Figure 9-14. In this type of gun, atomization of the paint is coarse, and the spray pattern is fixed. This gun works only with a pressure-feed, but the pressure is lower and the amount of air used is less than for the external-mix gun. Because atomization of the paint is coarse, more paint is applied on each pass.

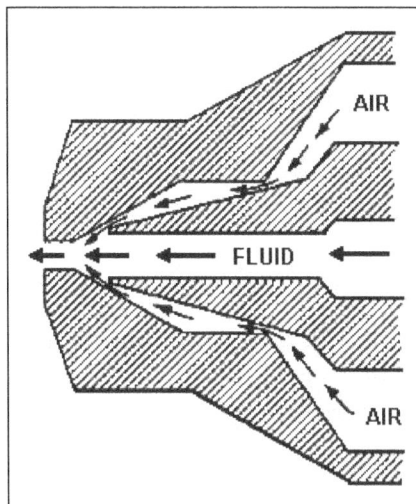

Figure 9-14. Internal-Mix Air Cap

- **Bleeder and Non-bleeder Spray Gun.** The bleeder type of gun is one in which air is allowed to leak or bleed from some part of the gun to prevent air pressure from building up in the air hose. In this type of gun the trigger controls only the fluid. It is generally used with small, air compressing outfits that have no pressure control on the air line. The non-bleeder gun is equipped with an air valve that shuts off the air when the trigger is released. It is used with compressing outfits having a pressure-controlling device.
- **Suction Feed Spray Gun.** In a suction-feed gun, the air cap is designed to draw the fluid from the container by suction (Figure 9-15) in somewhat the same way that an insect spray gun operates. The suction-feed spray gun is usually used with 1-quart (or smaller) containers.

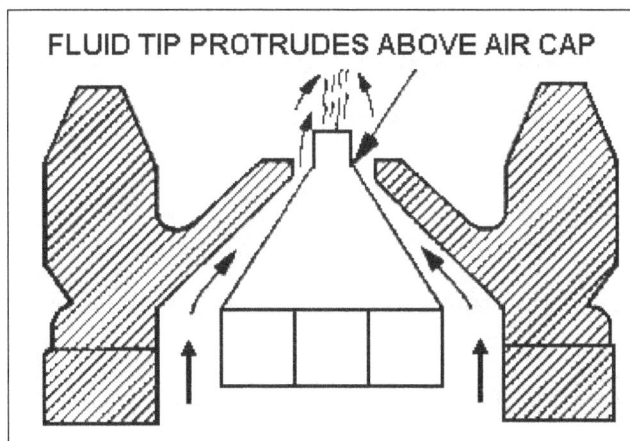

Figure 9-15. Suction-Feed Air Cap

● **Pressure-Feed Gun**. A pressure-feed gun operates by air pressure (Figure 9-16) which forces the fluid from the container into the gun. This is the type used for large-scale painting.

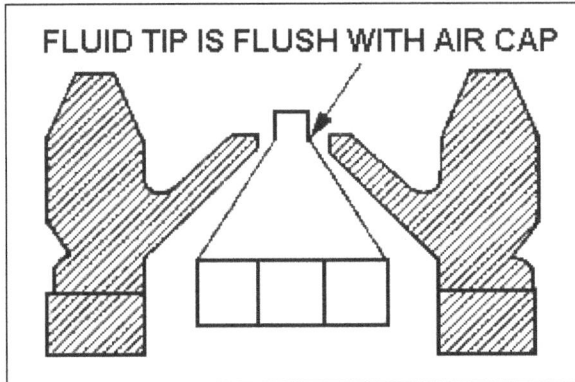

Figure 9-16. Pressure-Feed Air Cap

PARTS OF THE SPRAY GUN

9-52.   The two main assemblies of the spray gun are the gun body assembly and the spray head assembly. Each of these assemblies is a collection of small parts, all of which are designed to do specific jobs. Figure 9-17 shows the principal parts of the gun body assembly. The air valve controls the supply of air and is operated by the trigger. The spreader adjustment valve regulates the amount of air that is supplied to the spreader horn holes of the air cap. This will vary the paint pattern. It is fitted with a dial, which can be set to give the desired pattern. The fluid needle adjustment controls the amount of spray material that passes through the gun. The spray head locking bolt locks the gun body and the removable spray head together.

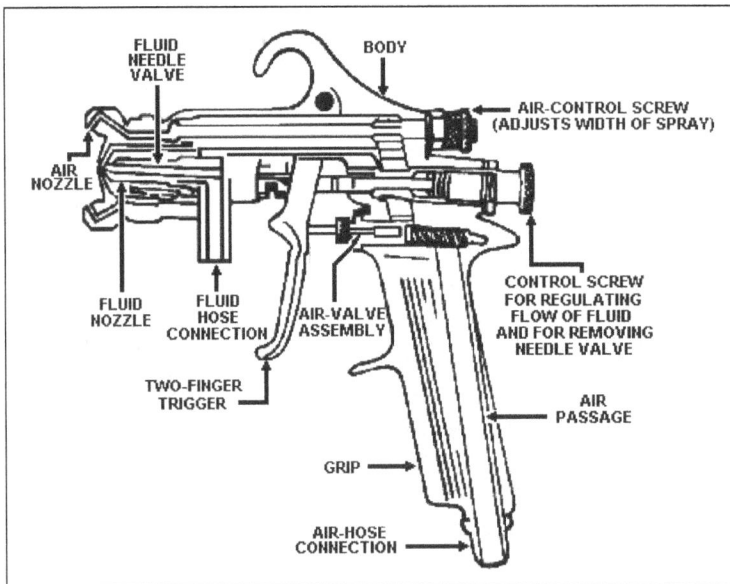

Figure 9-17. Cross-Section of a Spray Gun

9-53.   Most guns are now fitted with a removable spray head assembly. This type has many advantages. The head can be cleaned more easily; the head can be quickly changed to use a new color or material; and, if damaged, the head alone can be replaced, using the old gun body. The principal parts of the spray head assembly are the air cap, the fluid tip, fluid needle, and spray head barrel (Figure 9-18).

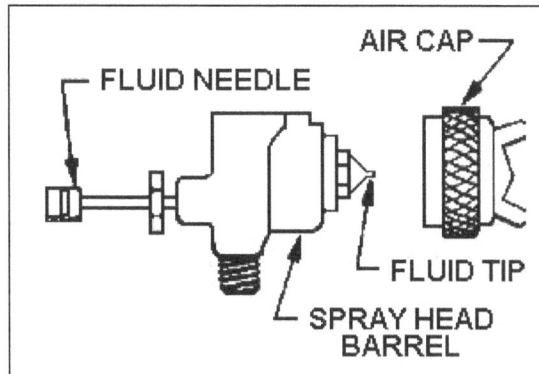

Figure 9-18. Principal Parts of the Spray Head

9-54.   The fluid tip regulates the flow of the spray material into the air stream. The tip encloses the end of the fluid needle. The spray head barrel is the housing, which encloses the head mechanism.

Containers

9-55.   The cups or tanks, which hold the spray material before delivery to the gun, are called containers. The job to be done determines which one of several kinds of containers to use.

*   Suction-feed cups are used for small quantities of lightweight and medium weight spray materials, such as lacquers.
*   Gravity-feed cups are also small and are attached directly to the top or side of a gun. They normally are used only on artist's and decorator's guns or on small touch-up guns.
*   Pressure-feed cups (Figure 9-19) are considered best for handling small quantities of enamels, plastics, or other heavy materials on jobs that need fine adjustments and speed of application.

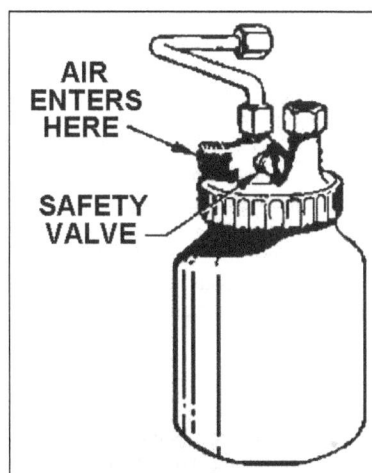

Figure 9-19. Pressure-Feed Cup

9-56.   Pressure tanks are large containers with a capacity of 2 to 60 gallons. Figure 9-20 shows a common type of pressure tank.

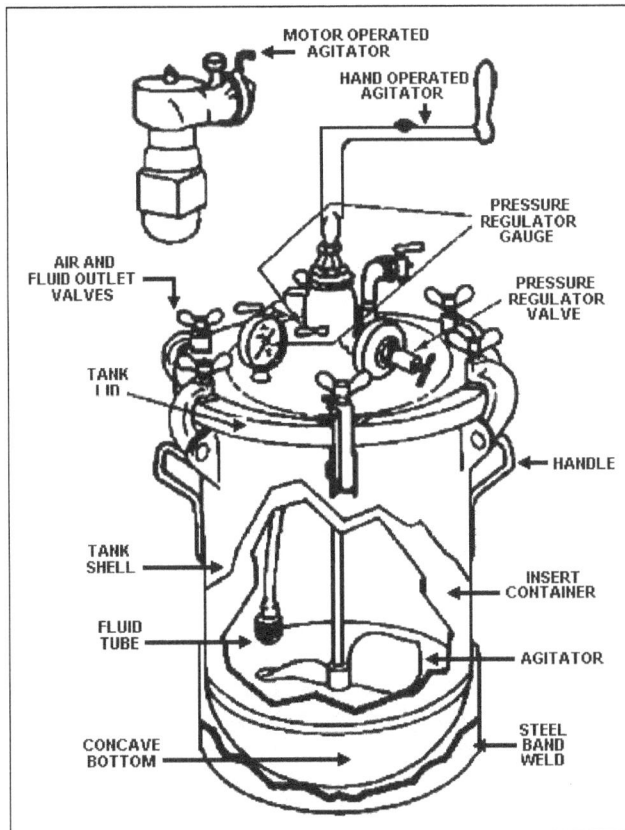

Figure 9-20. Pressure Tank

9-57.   There are two general types, the regulator type and the type that uses the equalized pressure tank. The equalized pressure tank is equipped only with a safety valve and a release valve. The regulator type is equipped with one or two regulators, a safety valve, release valve, and pressure gauge. It may also have one or two hand-operated or motor-operated agitators. If there is only one regulator, it regulates the fluid pressure in the tank only. If there are two regulators, one regulates the fluid pressure in the tank while the other regulates the air pressure from the container to the gun. Each regulator operates independently of the other. The pressure tank shown in Figure 9-20 is equipped with air outlets and fluid outlets. The fittings, pressure regulators, and gauges permit the use of more than one spray gun at the same time. Sometimes, instead of pouring the material directly into the tank, you can put a separate container (called an insert container) into the tank. With this type of container, it is possible to make quick changes of color or material without having to clean the tank. You can also mix your materials ahead of time and have them on hand.

Hose Lines

9-58.   Spray gun hoses are of two varieties—one kind to handle air and another to handle liquids. Air hoses are usually made of braid covered tubing, with either one-braid or two-braid construction. Fluid hoses are made of a special, solvent-resisting material.

AIR SUPPLY

9-59.   Spray guns are operated by compressed air, which may be supplied by either portable or installed compressors. However, aboard ship, guns using pressure tanks are usually connected to the low-pressure ship's service air line. Pressure on this line is usually from 100 to 125 psi, but this is cut down to spraying pressure at the tank by a pressure regulator valve. The manufacturer's instructions for the operation of air compressors must be followed exactly. If you intend to use air from the low-pressure line for long periods, it is a good idea to inform the engineering officer of the watch. To properly spray paint, the air should be dry and free from dust. All air, in varying amounts, contains moisture and dust and some means must be provided to remove both. An air transformer (Figure 9-21) is usually used to remove moisture and dust. The air transformer is also called an air separator or air regulator.

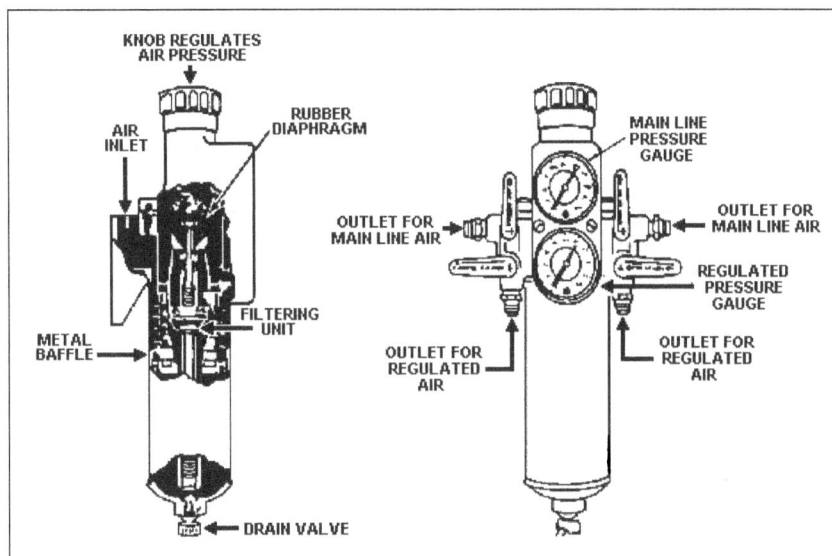

Figure 9-21. Air Transformer

9-60.   Air enters through an air inlet, passes through a series of baffles and a filter chamber, and then through a regulator diaphragm which adjusts the pressure. Normally, the transformer should be drained daily. If the weather is damp, it should be drained several times daily. You do this by turning a drain cock at the bottom. The packing and filtering material should also be changed at regular intervals.

OPERATION OF THE SPRAY GUN

9-61.   Squeezing the trigger of the spray gun opens the air valve, admitting compressed air through the air inlet. The air passes through the gun body to the spray head. In the most common (external-mix) type of spray head, the air does not come in contact with the paint inside the gun, but is blown out through small holes drilled in the air cap. Paint is shot out of the nozzle in a thin jet, and the force of the air striking it breaks the jet into a fine spray. You can control this spray, making it into various patterns, by setting the air-control screw that regulates the spreader adjustment valve. Turn the screw clockwise for a round spray. For a fan spray turn it counterclockwise. Turn the fluid-control screw clockwise to increase the flow. To obtain the same coverage over the wider area, the flow of paint must be increased as the width of the spray is increased. The handling of a spray gun is best learned by practice, but here are some tips. Before starting to spray, check adjustments and operation of the gun by testing the spray on a surface similar to that which you intend to coat. There are no set rules for spray gun pressure or for distance to hold the gun from the surface because pressure and distance vary considerably with the nozzle, the paint used, and the surface to be coated. The minimum pressure necessary to do the work is the most desirable, and the distance is normally from 6 to 10 inches. Always keep the gun perpendicular to and at the same distance from the surface being painted (Figure 9-22). Start the stroke before squeezing the trigger, and release the trigger before completing the stroke (Figure 9-23). If the gun is not held perpendicular or is held too far away, part of the paint spray will evaporate and strike the surface in a nearly dry state. This is called "dusting." Failing to start the stroke before starting the spray or spraying to the end of the stroke will cause the paint to build up at the end of the stroke, and the paint will run or sag. Arching the stroke makes it impossible to deposit the paint in a uniform coat.

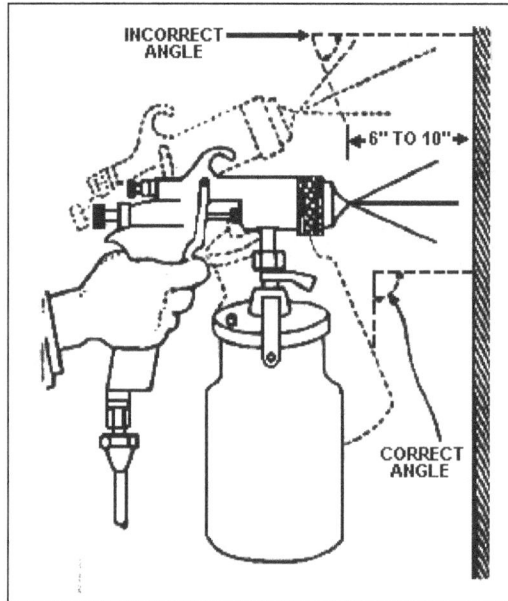

Figure 9-22. Hold Spray Gun

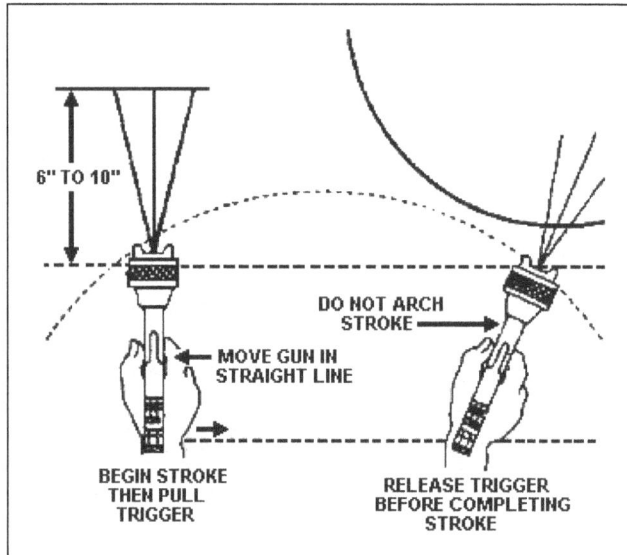

Figure 9-23. Proper Spray Gun

PERPENDICULAR TO SURFACE STROKE

9-62.   When spraying the inside and outside corners, stop 1 or 2 inches short of the corner. Do this on both sides, then turn your gun on its side and, starting at the top, spray downward, coating both sides at once (Figure 9-24).

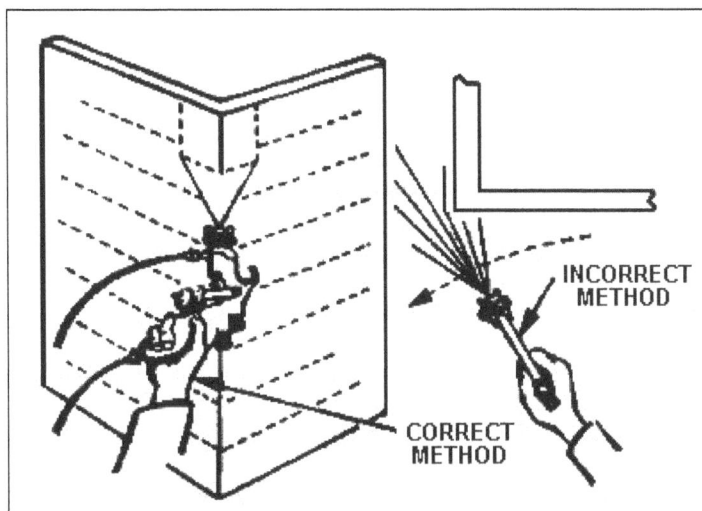

Figure 9-24. Correct and Incorrect Methods of Spraying Corners

9-63.   If you are spraying a large area from which small parts and pieces protrude, first lightly coat those items. Then go over the whole surface. For example, if you are painting a compartment, first spray the hatch coamings, door frames, rivets, exposed bolt heads, and all small items secured to the bulkheads. Then do the entire compartment. This eliminates a lot of touching up later.

COMMON SPRAYING DEFECTS

9-64.   The most common defects in sprayed-painted coats are "orange peel," runs and sags, pinholes, blushing, peeling, and bleeding.

- **Orange Peel.** This is a general term used to describe a painted surface that has dried with a pebble texture resembling an orange peel. This may be caused by using improper thinners, a spray which is not fine enough, holding the gun either too far or too close to the surface, improper mixing of the material, drafts, or low humidity.
- **Runs and Saps.** These are usually the result of using material that is too thin. Sags result from too much material. Runs and sags can also be caused by allowing a large lap in spraying strokes and by poor adjustment of the spray gun or pressure tank. Dirty or partially clogged passages for air or fluid will also cause uneven distribution.
- **Pinholes.** These may be caused by the presence of water or excessive thinner in the paint or by too heavy of an application of quick drying paint. Either way, small bubbles form and break in drying, leaving small holes.
- **Blushing.** This resembles a powdering of the paint. The cellulose material in the paint separates from its solvent and returns to its original powder form. Water is usually the cause of this, either moisture on the sprayed surface or excessive moisture in the air. When blushing occurs, you will have to remove the defective coating because the moisture is trapped within the material and will remain there unless the coating is removed.
- **Peeling.** This is almost invariably due to carelessness in cleaning the surface. Before any spraying is attempted, the surface must be absolutely clean. Cheap spray materials sometimes will give poor adhesion, but you will not have this trouble if you always use standard paints.

- **Bleeding.** This occurs when the color of a previous coat discolors the finish coat. Paint containing a strong aniline dye (synthetic organic dye) will do this when another color is sprayed over it.

CARE OF THE SPRAY GUN

9-65.   Spray guns (as well as paint containers and hoses) must be cleaned thoroughly after they are used. Figure 9-25 shows the steps in cleaning a pressure-feed gun. First, back up the fluid, needle-adjusting screw and release the pressure from the pressure tank by means of the release valve. Hold a cloth over the air cap and pull the trigger (this forces the spray material back into the tank). Now remove the fluid hose from the gun and run a solvent through the hose. There is a special hose cleaner made for this purpose. Dry out the tip and clean the tank. Soak the air cap in solvent. If the holes are clogged, use a toothpick to clean them. Put all clean parts back in place, and the gun is ready for use again.

9-66.   Figure 9-26 shows how to clean a container-type gun. First, remove the container. Then hold a cloth over the air cap and pull the trigger. Empty the container and pour in a small quantity of solvent. Attach the container to the gun and spray in the usual way. This process cleans out all passageways. Clean the air cap by soaking it in a solvent and then replace it. Some spray gun troubles, their possible causes, and their remedies are listed in Table 9-3.

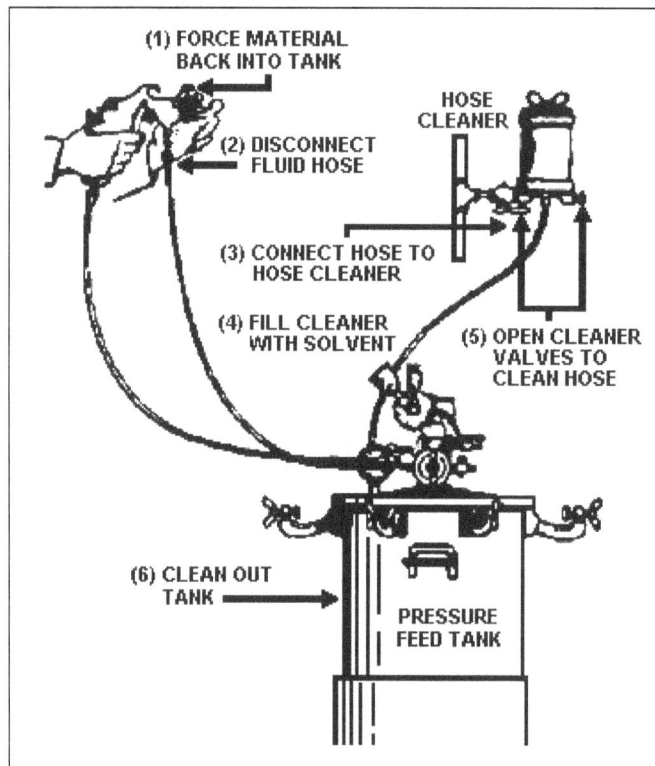

Figure 9-25. Steps in Cleaning a Pressure-Feed Gun

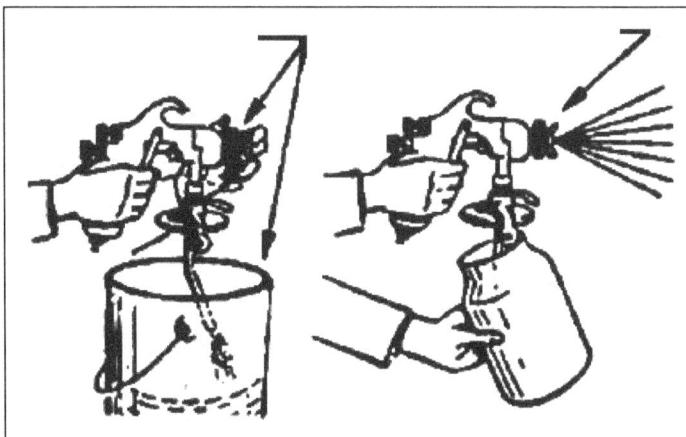

Figure 9-26. Steps in Cleaning a Container-Type Gun

Table 9-3. Spray Gun Troubleshooting Chart

Trouble	Possible	Causes Remedies
Air leaks from front of gun	Foreign matter on valve seat Worn or damaged valve seat Sticking valve stem Bent valve stem Packing nut loose	Clean Replace Lubricate Replace Adjust
Fluid leaks from front of gun	Worn or damaged fluid tip or needle Foreign matter in fluid tip Packing nut too tight Wrong size needle	Replace Clean Replace Adjust
Jerky or fluttering spray (both suction- and pressure feed)	Insufficient material in container Tipping container to excessive angle Obstructed fluid passageway Loose or cracked fluid tube Loose fluid tip or damaged tip seat	Refill Take greater care Clean Tighten or replace Tighten or replace
Jerky or fluttering spray (suction-feed only)	Too heavy a material Clogged air vent in container lid Loose or damaged coupling nut or cup lid Fluid tube resting on bottom	Change to pressure feed Clean Tighten or replace Use proper fluid tube
Defective spray pattern	Air cap horn holes partially plugged Dirt on air cap or fluid nozzle	Rotate air cap one-half turn and spray another pattern. If defect is inverted, fault is on/in air cap. If pattern is same, fault is on/in fluid nozzle. Clean proper part.

LUBRICATION OF THE SPRAY GUN

9-67.   Your spray gun also needs a little lubrication. The fluid needle packing should be removed occasionally and softened with oil. The fluid needle spring should be coated with grease or petrolatum. Figure 9-27 shows the parts and the oil holes in which you occasionally should put a few drops of light oil.

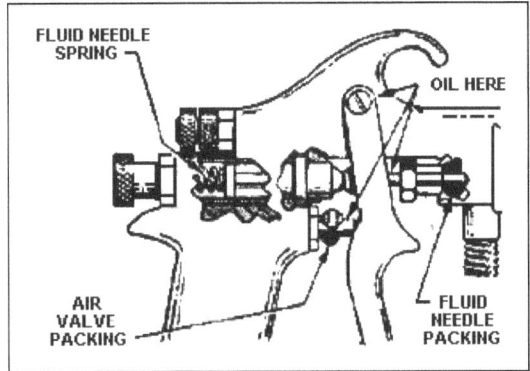

Figure 9-27. Lubrication Points of a Spray Gun

HOW TO REMOVE THE SPRAY HEAD

9-68.   To clean, to repair, or to change paint color, you may have to change the spray head. First, remove the gun from the air and fluid hose lines. Holding the gun in the left hand, pull the trigger all the way back and loosen the locking bolt with the wrench provided for the purpose. Push the trigger forward as far as possible and pull the spray head forward (Figure 9-28). To replace the head, push the trigger forward and insert the spray head. Then hold the trigger back and tighten the locking bolt.

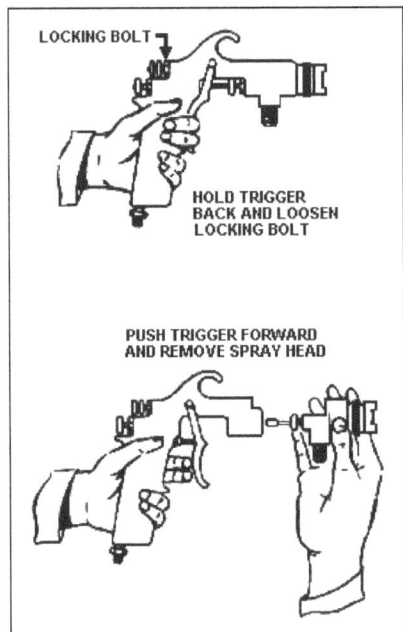

Figure 9-28. Removing the Spray Head

WHAT NOT TO USE IN YOUR SPRAY GUN

9-69.   As a general rule, paints, enamels, lacquers, synthetics, varnishes, and shellacs are suitable for spray work with ordinary equipment. Except in an emergency, material containing small gritty particles (such as alkaline coverings, rubber hose paints, plastics, and mastic paints) should never be used in standard equipment. They will damage the ordinary machines; therefore, use only the special outfits designated for use with those paints.

RESPIRATORS

9-70.   Spray painting breaks up the paint into a fine spray that releases fumes, pigment, and vehicle into the air. If you breathe them or otherwise absorb them into your body, these fumes and particles can cause injury. BE CAREFUL. Always wear a respirator when spraying or in the vicinity of spray work. Respirators are specially designed to give you maximum protection. Here are the most common types:

FILTER RESPIRATOR

9-71.   The filter respirator (Figure 9-29) is equipped with filter pads. It can be used for spraying, grinding, or dust blowing when dust and fumes are not too severe.

CARTRIDGE RESPIRATOR

9-72.   The cartridge respirator is designed for more severe conditions than those that can be met by a filter respirator. It uses a filter pad and a large purifying cartridge made of chemically treated charcoal.

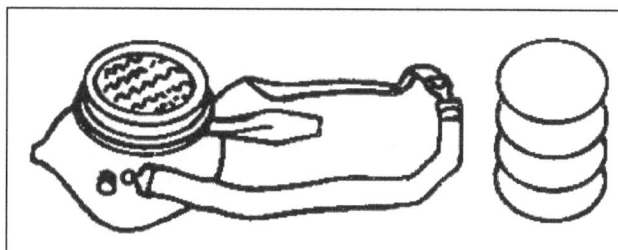

Figure 9-29. Filter Respirator

DUST RESPIRATOR

9-73.   The dust respirator (Figure 9-30) is one of the most common types of respirators. It contains a replaceable cartridge and its lightweight makes it easy to wear.

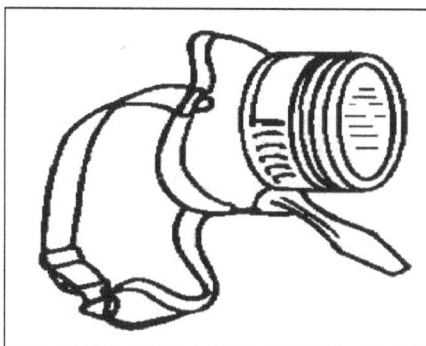

Figure 9-30. Dust Respirator

AIR SUPPLY RESPIRATOR

9-74. The air supply respirator (Figure 9-31) provides you with complete protection when working in old, tank interiors, and other areas where no ventilation is possible. This type is supplied with fresh air through a compressed-air line, purified by a charcoal cartridge, and then fed to the breathing compartment of the respirator.

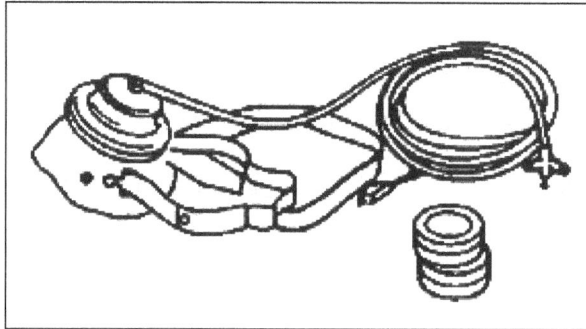

Figure 9-31. Air Supply Respirator

HOOD RESPIRATOR

9-75. The hood respirator (Figure 9-32) consists of a flameproof hood, a headgear of fiber with a metal eyepiece, and an air hose. The neck cloth at the bottom of the hood ties snugly around your neck to prevent entrance of fouled air.

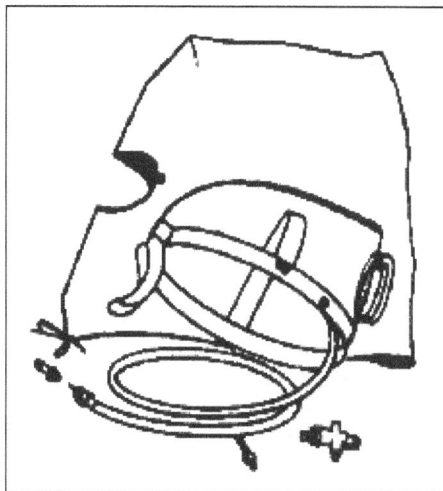

Figure 9-32. Hood Respirator

9-76. The opening in front of the hood is the only outlet for the constant flow of air entering from a hose attached to the back of the hood. Foul air cannot enter because pressure inside is slightly greater than pressure outside. The generous opening permits maximum range of vision.

SAFETY PRECAUTIONS

9-77. The application of paints, varnishes, lacquers, enamels, wood bleaching liquids, and other flammable liquids by the spray process is more hazardous than brush or roller applications. This is due to the volume and concentration of fumes and particles as well as the production of a flammable residue and deposits, which are subject to spontaneous ignition. Health hazards from potentially harmful substances such as lead, benzene, and silica may also be present in paint-spraying operations. To ensure immediate removal of vapors and paint dust from spraying operations, complete ventilation of the compartment is essential. A system balanced to supply fresh air as well as to exhaust vapors is recommended. Ordinarily the ventilation necessary for the health and comfort of the operators is also sufficient to remove flammable vapors.

9-78.   Personnel using spray guns should wear clothing that fits tightly at the ankles, neck, and wrists. Approved respirators must be worn, and parts of the body not protected by clothing should be covered with petrolatum (Vaseline). Smoking, open flames, welding, grounding of spray equipment, chipping, and other spark-producing operations are prohibited in compartments where spraying is in progress. Explosion-proof portable lights should be used. Pre-caution should be taken to ensure that wires do not become exposed from dragging and pulling. Bulbs must not be replaced in a compartment or tank being painted until flammable or explosive vapors have been removed.

ESTIMATING PAINT JOBS

9-79.   To plan the work of your ship more competently and, particularly, to make the most effective use of your spray painting teams, you should develop the ability to estimate the number of man-hours and amount of paint required to do the ordinary, shipboard painting jobs. It is difficult to list more than a few guidelines for estimating painting requirements. To lay down any hard and fast rules is impossible because of the many variables involved (type of compartment, skill of the team, type of paint to be used, and so on). Keep notes on the jobs that your personnel perform. These records will help you with future estimations. Note such things as the number of square feet a gallon of each type of paint will cover when applied by different methods (see Table 9-4). Also how much time is required to ready the equipment for spraying, how many square feet of surface a team can paint in 1 hour, the number of gallons of paint required for each compartment, and so on.

Table 9-4. Paint Coverage per Gallon

1. Enamel – 400 ft2 (Brush)
2. Enamel – 500 ft2 (Spray)
3. Haze Gray – 500 ft2 (Brush)
4. Haze Gray – 500 ft2 (Roller)
5. Gray Deck – 500 ft2 (Brush)

MAINTENANCE OF SHIP'S RIGGING AND DECK MACHINERY

9-80.   Wire rope must be lubricated properly to ensure long life and safety. The internal parts of the wires move against each other wherever the rope passes over a sheave or winds on a drum. Each wire rotates around its own axis, and all wires slide against one another.

CLEANING AND SLUSHING STANDING RIGGING

9-81.   A slush of specially prepared grease is used to prevent rust on standing rigging. It is very easy to handle and creates a minimal amount of drippings if applied sparingly. This is very important in regards to our protection of the environment. Remember that excessive use of cable lubricant will result in runoff and eventual water pollution. Galvanizing metal gives it a very thin coating of rustproof zinc. However, this coating eventually wears off in places, or the elements penetrate below it. Therefore, even though much standing rigging is galvanized, it should be greased with slush periodically. The man going aloft should take a steel scraper and a wire brush to slush down standing rigging. Make sure that safety goggles and harnesses are worn. Any scale on the wire must be chipped or scraped off, and the wire brushed down, either to the bare metal or to a good hard coat of slush put on previously. New galvanized metal must be rubbed down with a rag soaked in vinegar before slush or paint is applied.

RUNNING RIGGING

9-82.   A vessel's running rigging consists of all the guys, tackles, whips, blocks, boat falls, and so on, used to control the motion of the ship's movable gear or to handle cargo.

- **Blocks.** Blocks on cargo davits and rescue boat davits must be periodically inspected and lubricated. Failure to do so could result in equipment failure at a critical time.
- **Slush Down Running Rigging**. Modern Army watercraft has very little, if any, running rigging. A thorough understanding of the care of this equipment is still necessary since it may be encountered. Wire in running rigging is protected from wear and the weather by being greased at regular intervals with "Crater C" lubricant grease. On older break bulk types, a ship's rigging was cared for from the boatswain's chair, or the rigging may be unreeved and slushed while it is on deck. This slush is applied with a rag and it must be handled carefully to avoid getting spots on the deck, awnings, or paint work. The wire pendants or stationary supports on the ends of the tackles of cargo guys are also slushed because the pendants also need to be lubricated. Remember that excessive use of lubricant will result in deck run off that leads to water pollution.

WARNING

Before slushing any running rigging, be sure the winch motor is de-energized.

SHACKLES AND TURNBUCKLES

9-83.   Particular attention must be paid to protecting the threads of shackles and turnbuckles. They are the parts that will be eaten away first if not cared for properly. Turnbuckles should be opened out frequently, the threads should be brushed well, and the parts lubricated with graphite grease.

BOAT DAVITS

9-84.   Boat davits should be inspected at least once a week. The regular lubrication of the mechanical components, as outlined in the individual manufacturer's manual, should be carried out. The wire rope of the hoisting slings should be coated entirely with grease. As an alternative, grease may be applied only to those rope areas where saltwater will form a pocket, adjacent to shackles, buttons, or clamps, and around the thimble.

WINCHES, CRANES, AND ANCHOR WINDLASSES

9-85.   The maintenance and lubrication of heavy deck equipment (such as winches, cranes, and anchor windlasses) are performed by personnel of the engineering department. You need to keep in mind that you must work with this equipment. Therefore, for your own protection, you should assist as much as possible in the maintenance and lubrication of this equipment.

Chapter 10

Beaching and Retracting Operations

INTRODUCTION

10-1.   The term "landing craft" implies that the craft can safely land on a beach or shore. The most important phase of landing craft operations and the most severe test for the crew is beaching. Poor seamanship during beaching can risk life and property. The first task of the coxswain or master is to get past any obstructions that may be between deep water and the shore. The craft must be beached in a way that will not cause it to broach to and which will permit troops or cargo to be unloaded quickly and safely. A craft is broached to when it is thrown broadside to in heavy surf, heavy seas, or on a beach. After the craft is firmly on the beach, it must be retracted safely off the beach to be of further service to a unit. Like beaching, retracting from the beach requires skill in boat handling and seamanship.

RULES FOR LANDING OPERATIONS

10-2.   During land operations, the landing craft coxswain or master must remember (even when he/she has the right of way) the importance of doing everything possible to avoid a collision. If there is immediate danger of a collision, his/her prime responsibility is to save the crew, passengers, cargo, and boat. Heavy surf, fog, smokescreens, and similar hazards met during landing operations call for special precautions and good judgment. Some general traffic rules for landing operations are as follows:
- Heavily loaded boats have right of way over lightly loaded or empty boats.
- Boats in tow have right of way over all boats.
- Retracting boats, which have their bows toward the beach and are in the surf zone, have right of way over empty or loaded boats.
- Boats, after clearing the surf zone, should keep clear of inbound craft. They should also make for the designated flank before continuing back to the transport.

SURF ACTION

10-3.   To better understand this chapter, you should know what causes surf and also the definitions of the terms. Definitions are as follows:
- Breaker. A single breaking wave.
- Breaker line. The outer limit of the surf.
- Comber. A wave on the point of breaking. A comber often has a thin line of white water on its crest.
- Crest. The top of a wave, breaker, or swell.
- Foam crest. The top of the foaming water that speeds toward the beach after the wave breaks.
- Surf. A number of breakers.
- Surf zone. The area between the first break in the swells and the shoreline.
- Swell. A broad, rolling movement of the surface of the water.
- Trough. The valley between waves.

10-4.   Surf is caused by the swells as they move in toward the beach (Figure 10-1). As this movement approaches shore, it is confined between the rising ocean floor and the surface of the water. The more confined the water becomes, the more the crests peak up in the form of combers. Combers usually, but not always, form into breakers. A sandbar or reef between the outer surf (or breaker) line and the beach sometimes causes two (more or less) well-defined surf belts.

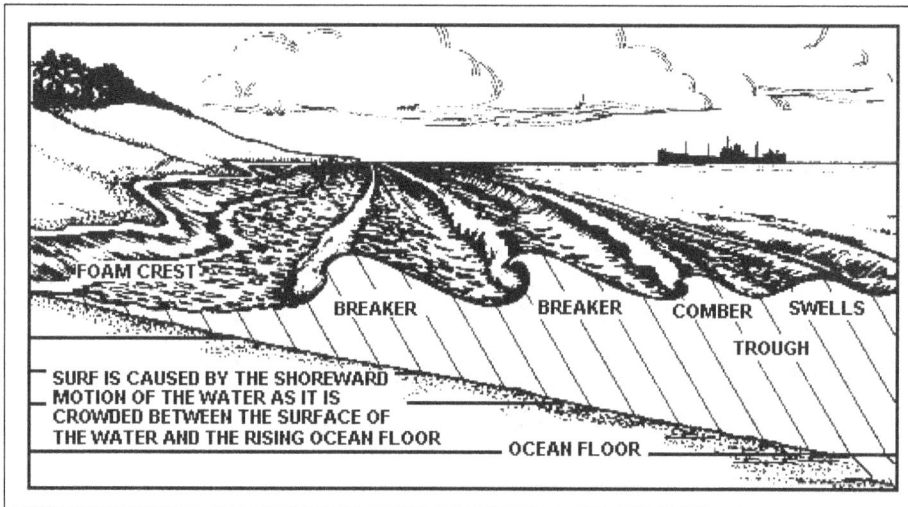

Figure 10-1. Cross Section of the Surf

10-5.   Breakers vary in size and sometimes may follow a sequence for a short interval (such as a large breaker following a certain number of smaller ones). There is no regularity to the pattern, so do not count on, for example, every seventh breaker being larger than the six preceding ones. However, the interval between breakers is fairly constant, tending to stay the same for several hours. Swells causing surf are created by winds far out to sea. The distance the swells travel from their origin, which may be several hundred miles, determines the interval between swells. The important points to remember about surf are that you must not be lulled into expecting the surf to be consistent, you must respect it, and you must learn how to make it work for you while beaching and retracting.

PREPARING TO HIT THE BEACH

10-6.   As stated earlier, the most important phase of landing craft operation and the severest test for the landing craft crew is beaching. Poor seamanship during this operation can jeopardize life and property. A number of important rules to be observed by the master or coxswain when hitting the beach are as follows:

*   Check to see that all equipment and cargo are properly secured.
*   Ensure that all personnel are wearing life jackets.
*   Make sure that each crew member is in his/her place and ready for the run.
*   Prior to entering the surf zone, check the ground swell and attempt to gauge the nature of the surf. Surf will appear only half as high as it actually is when looking from seaward to the beach. Therefore, what appears to be a 3-foot surf is actually a 6-foot surf.
*   Cross the surf line at a right angle to the advancing waves (Figure 10-2). Waves are not always parallel to the beach.
*   Approach the surf line at reduced speed and take each wave carefully. Pick out an object on the beach as a guide (range markers are desirable), select a stern wave of considerable size and, as the wave gets under the craft, increase speed to that of the wave, and ride in just behind the crest. If impossible to keep up, select another wave and repeat the process. However, be careful to ensure that the surf is kept perpendicular to the stern to prevent broaching to. Once inside the surf line, the course should not be changed, and the craft must be kept lined up with the object on the beach.
*   If a bar stops the forward progress, reduce speed, wait for flotation, and proceed. If necessary to unload at the bar, check the depth of water prior to debarking troops or equipment. Do not debark troops or equipment into water too deep to ford.

- Hit the beach at the fastest speed possible. Keep the engines ahead, and use the rudders and engines to keep the craft on the beach.
- Constantly check the seawater strainers and discharge to prevent engine failure at this critical time. When the ramp is down, it will help keep the boat on the beach.

WARNING

DO NOT drop the ramp without first checking underneath to ensure that all is clear.

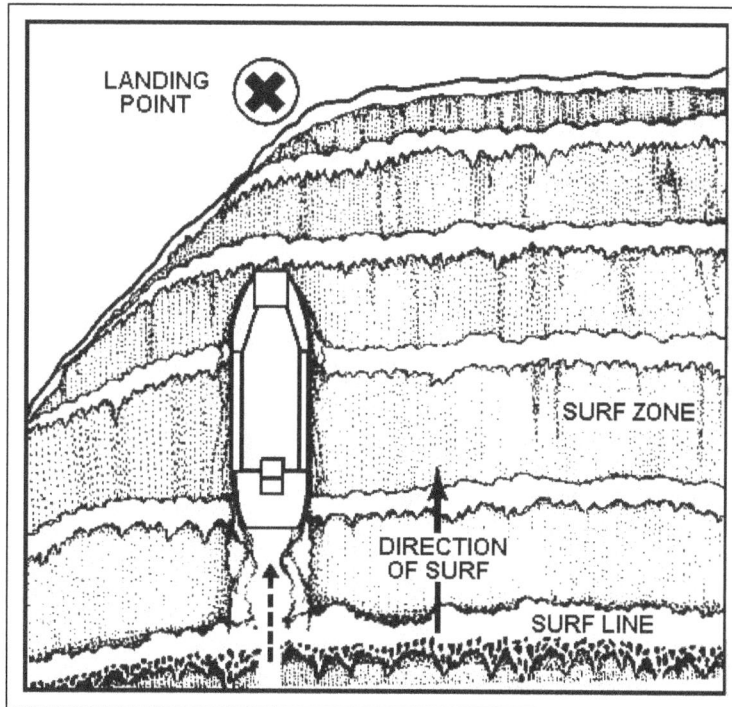

Figure 10-2. Crossing the Surf Line

BEACHING HAZARDS

10-7.   There are a number of hazards that can be encountered. Be prepared to take the appropriate safety measures to safely maneuver your vessel.

SANDBARS

10-8.   Sandbars encountered on the run to the beach should be hit by slowing down your craft. In many cases, the boat's momentum and the following wake will be sufficient to carry the craft over the obstruction. However, if the forward motion of the craft is stopped, the engines should be slowed immediately to idling speed. Then, when the stern wash comes under the craft, the engines are run at half throttle until the craft works free. If this method fails, it is still possible to get the craft over by use of the propeller streams. The starboard or port engine is reversed and set at half throttle, and the other engine is held at half ahead. The screw current from the reversed propeller will then wash the sand away from the side of the keel, cutting a small channel through the bar. If the freeing of one side is not enough to get the boat free, the engines can be shifted to dig the sand away from the other side of the keel. Turning the wheel from side to side helps to work the craft free. As soon as the boat becomes free and begins to move through the bar, both engines are put ahead until the craft is completely clear. In some cases, it is possible to work a boat clear by moving its cargo, combining the shift in trim with lift of the waves, to provide the necessary flotation. When the craft is free, it is run into the beach just behind the crest of a wave. As soon as the craft beaches, the engines are throttled down until the wash or a wave provides additional flotation, at which time the engines are accelerated to move the boat up on the beach. In calm surf, the engines are slowed when about two boat lengths off the beach and then accelerated when the stern wash lifts the craft. The sudden stopping of the boat when it touches on a bar or beach can cause serious injury to personnel in closed spaces. All personnel aboard should therefore be kept out of the engine room during the last stages of the run.

REEFS

10-9. Two general types of reefs are fringing and barrier. Fringing reefs are attached to the land, while barrier reefs are separated from the land by a body of water called a lagoon. The problem of crossing a reef in a landing craft is largely a question of water depth. At high water, the minimum depth over a reef should be about equal to the tidal range. Under certain conditions, the rising and falling action of tidal waters usually cuts passageways through the reefs. Whenever possible, use the channels when approaching or clearing a beach.

SHELVING BEACHES

10-10. A shelving beach is one where the shore gradually slopes toward the sea. Where there is a long, shelving, sandy beach it may be necessary to run the craft a long way over the bottom to reach the beach. In this case, it is best to continue ahead by taking advantage of the waves for added flotation whenever possible to aid in carrying the craft onto the beach. When small sandbars are encountered, the craft can be freed by using the propeller streams as described earlier. If the craft cannot be freed by this method, lower the ramp and use a boat hook or sounding pole to check the depth of water immediately ahead. If shallow enough, one man should be sent in with a pole to test the depth all the way to the beach before troops or cargo are unloaded.

BROACHING TO

10-11. The most difficult task of the coxswain or master, in heavy weather, is to keep the craft from broaching to (or turning broadside to the sea or wind so as to risk capsizing). This occurs when the bow stops in still water and the stern is thrown around by the impact of the next wave. The same thing can happen, however, when the bow stops on the beach and the stern swings in toward the beach. In the first case, the craft probably will be thrown up on the beach before the coxswain can regain control. In the second case, the craft is already on the beach with the seas pounding it, and the boat may fill with water. In moderate surf, a broached landing craft sometimes can be freed under its own power. If the stern lies to port, it may be brought around by accelerating the port engine in reverse (causing the port screw current to wash sand from under the craft), and idling or accelerating the starboard engine ahead with the rudders hard left. The procedure is reversed if the stern lies to starboard. During this operation, the engine saltwater discharge lines must be visually checked to make sure the saltwater pumps are functioning properly. The five measures listed below are designed to aid in avoiding situations that might cause broaching to.

- The coxswain or master must watch the seas carefully and maintain all possible headway during the run to the beach.
- Breaking seas must be kept dead astern and perpendicular to the craft at all times, even if it means hitting the beach at an angle.
- If the stern starts to swing, the engine opposite the swinging side should be accelerated and the wheel put over sharply in the direction of the swing.
- As incoming waves float the boat, the engines should be accelerated in forward gear to force the craft well upon the beach so that as much of the keel as possible is on the beach.
- The stern anchor on LCUs should be used as an anti-broaching aid.

BEACHING PROCEDURES

BEACHING OPERATIONS

10-12. During beaching operations, the operator must stand by the helm to hold the boat securely on the beach. This is true especially when the craft is on a steep beach with its stern partially afloat or while it is being loaded or unloaded. If the stern begins to swing around, the anti-broaching procedure must be followed. In every case, the stern must be held directly into the seas or it will broach to immediately. The operator should apply forward throttle in varying amounts on each engine, depending on the particular situation, and also use the rudders as needed. The thrust of the propellers from the LCM will build up a sandbar in back of an average beach. On a steep beach, however, this problem will not occur because of the depth of the water at the stern. The engines must get enough seawater for cooling while the craft is beached. If the engines show signs of overheating or if there is inadequate seawater discharge, the craft must be retracted sufficiently to allow the seawater suction pump to draw in more water. The tides must be carefully checked during a prolonged stay on the beach and the craft moved in and out with the tide so that the stern stays in the water while the bow remains properly beached. In some localities, circumstances may require that the craft be beached prior to low tide and left resting on the bottom while the tide is out. In this case, the operator should select a position where the bottom is clear of rocks and obstructions.

CAUTION

Boats must never be left on the beach unattended or unwatched.

BEACHING AN LCU

10-13. When beaching an LCU, the crew should be alerted as to the intended beaching operations and crew members should be at their assigned stations. The following procedures should be used:

- Before entering the surf line, select a stationary object on the beach to help you establish your angle of approach.
- At all times, keep the LCU at right angles (90 degrees) to the surf line, even if it means hitting the beach at an angle.
- Begin the approach to the surf zone at a reduced speed.

RETRACTING PROCEDURES

RETRACTING AN LCU

10-14. The following procedure will be used when retracting an LCU:

- Have personnel at their assigned stations.
- Put the rudders amidships.
- Have the chief engineer start up the anchor windlass engine.

CAUTION

The chief engineer must keep up the speed of the anchor windlass to take up all the slack in the anchor cable. This will prevent the danger of the LCU overriding the anchor cable and fouling the propellers.

- Put the port, center, and starboard engines half speed astern to break the LCU free from the beach.
- Stop the engines once the LCU is free of the beach.

- If sandbars are encountered, put one engine half astern and the other half ahead. Alternate the outboard engines until a channel has been cut through the sandbar.
- House the anchor.

Note: The next step is accomplished after the anchor has been housed and the surf line has been cleared. The direction of turning will depend on the wind. Use the wind to your advantage.

- Put the rudders hard left, starboard engine full ahead and the port engine half astern. Reverse the procedure if turning to starboard.

Note: The turn is made at the crest of a wave as rapidly as possible and the turn is completed in the trough so that the LCU can meet the next sea head on.

CAUTION

Never turn in the surf zone. Boats retracting which have their bows toward the beach and are in the surf zone have the right of way over empty or loaded boats coming into the beach. Boats beyond the surf zone and leaving the beach will keep clear of inbound boats.

RETRACTING AN LCM

10-15. The procedures for retracting or backing the boat off the beach are generally the same for all landing craft except the LCU, in which case the stern anchor and winch are employed. Before retracting, the coxswain or master ensures that the rudders are not turned. All engines are backed down and used to steer until the craft is in deep enough water to permit proper rudder control. Landing craft are normally retracted at low speeds for control through the propellers. If bars are encountered during retracting operations, the procedure described earlier is partially reversed. One engine is backed down at half throttle and the other is put ahead about half speed, alternating until a channel has been cut through the bar. Both engines are then put in reverse and the boat is backed straight out. When completely clear of the surf line, the craft is turned to starboard by putting the starboard engine in reverse, the port engine in forward, and the wheel hard over right (this procedure is reversed when turning to port). The maneuver should be started on the crest of a wave and completed in the trough so that the craft will meet the next wave head on. When the breeze is fairly strong, the coxswain should use the wind to aid the start of the turn by letting the bow swing off to leeward. When putting about, the coxswain must make sure that the turn will leave the craft clear of incoming traffic. Turning in the surf zone must be avoided. When going against the waves, headway is maintained between waves but reduced somewhat when larger crests are encountered. This reduces the craft's resistance to the sea and allows it to ride over without danger of taking on excessive water.

CAUTION

The coxswain must make sure to pause a few seconds in neutral before engaging the opposite gear to allow the propeller time to stop turning. Failure to do so will result in transmission slippage plus an expensive and time-consuming repair job.

10-16. Retracting from a shelving beach in bad surf poses special problems for the landing craft crew because, in most cases, the boat has been landed well up on the beach. The best solution is to rely on the seas and the power of the engines. By this method the engines are accelerated in reverse each time a wave washes under the stern and the process is repeated until the craft works free. When completely cleared, the boat is turned seaward. Retracting from steep beaches presents no special problems under normal conditions, but because of the depth of the water just off the beach the surf has a more direct and immediate effect on the operation in heavy weather. This situation calls for careful handling and rapid response in getting the craft off the beach.

SALVAGE PROCEDURES

10-17. Experienced salvage boat crews never lose sight of the fact that their main objective is to keep the beach clear for incoming waves of boats. They never become so involved in freeing a disabled boat that they impede the progress of other landing craft en route to the beach. On the other hand, salvage boat crews must free broached and stranded landing craft as quickly as possible for the safety of the crew of the disabled boat. For their part, the men in the operable boat must do all they can to keep the craft shipshape. The engines must be kept running at all costs. Once the engines fail, the landing craft is helpless, even if free from the beach. In light surf, the salvage boat may back in far enough to pass the towline to members of the crew of the stranded boat who wade out to receive it. Whenever possible, it is better for the salvage boat to remain outside the breaker line and let another inbound landing craft carry the towline to the beach, from which it may be passed to the stranded boat. Another possibility exists if the beach is flat and the surf is breaking well out. Under such conditions, the salvage boat may be beached and the line then passed to the disabled boat. But if the surf is breaking close inshore or if the beach is steep, the salvage crew may approach the weather (windward) side of the broached boat and throw a heaving line so that the heavier towing line may be hauled aboard. Approaching from windward enables the heaving line to be thrown more effectively. The distance between the boats also decreases, as the salvage boat is set toward the stranded boat by the wind. In heavy surf it may be necessary to remain outside and use the line-throwing gun. At other times, it may be better seamanship to anchor the salvage boat by the bow outside the breakers and pay out the anchor line so the salvage boat drops astern close enough to pass a line to the helpless boat. The strain on the anchor line will hold the bow of the salvage boat to the sea. This procedure is generally undesirable, because it is difficult for the salvage crew to haul in the anchor line with sufficient speed when their boat begins to move seaward in the direction of the anchor. Once the towline has been passed, the crews of both boats should keep in mind the following procedures (see also Figure 10-3):

- Both the salvage boat and the towline should be perpendicular to the waves.
- A bridle is always used when freeing an LCM.
- The towline must not foul the screws of either boat.
- A broached boat is never towed by one quarter. Such a tow would be both dangerous and inefficient.
- The salvage boat never attaches the towline to its own bow, but to maintain maneuverability, the towline must be secured well forward of the screws and rudder.
- After the salvage boat has moved out beyond the breaker line, a steady strain is put on the towline. Slack must be taken up smoothly. Do not use full throttle until all slack is removed. The stranded boat should come off the beach a few inches at a time as each sea raises her. The broached boat may not break free immediately, so a steady pull should be maintained until ordered otherwise.
- The coxswain of the disabled craft should keep it in forward gear. As the engines are gunned forward while a wave is receding, the discharge current blasts the sand away from the rudders and skegs. This prevents rudder damage and enables the boat to draw off the beach without digging into the sand. Likewise, keeping the engines in forward gear adds to the strain of the disabled craft. When the stern is broken free, the engines are reversed to assist the salvage boat. Once freed from the beach, the boat is towed clear of the surf. The towline is cast off unless the tow is crippled.

10-18. The foregoing of general rules serves only as examples of common procedures followed in typical situations. There are no hard-and-fast rules for salvaging. No two salvage jobs present exactly the same problems; each must be solved individually. Below are a few examples that illustrate procedures carried out in more unusual situations.

- When a boat is stranded and lying almost parallel to the beach, attaching the towline to the bow may be desirable. Sometimes the boat can be swung around with its stern serving as a pivot. In such cases, slip the towline under the bow, bring it up and around on the shoreward side, and fasten it to a forward bit or cleat. When freed, the boat is towed out to sea, bow first.

- On a steep beach, made treacherous by a heavy backwash and current, the best salvage approach may be for the salvage boat to beach at some distance from the stranded craft. The towline may then be carried across the beach by hand and secured. This method of passing the line lessens the danger of the heavy backwash carrying the towline into the screws of either or both boats.
- In the foregoing situation, if the salvage boat cannot draw near the shore, attach a light line to a life ring and let the life ring float in with the surf. The crew of the stranded craft can use the line to haul in the heavier towline.

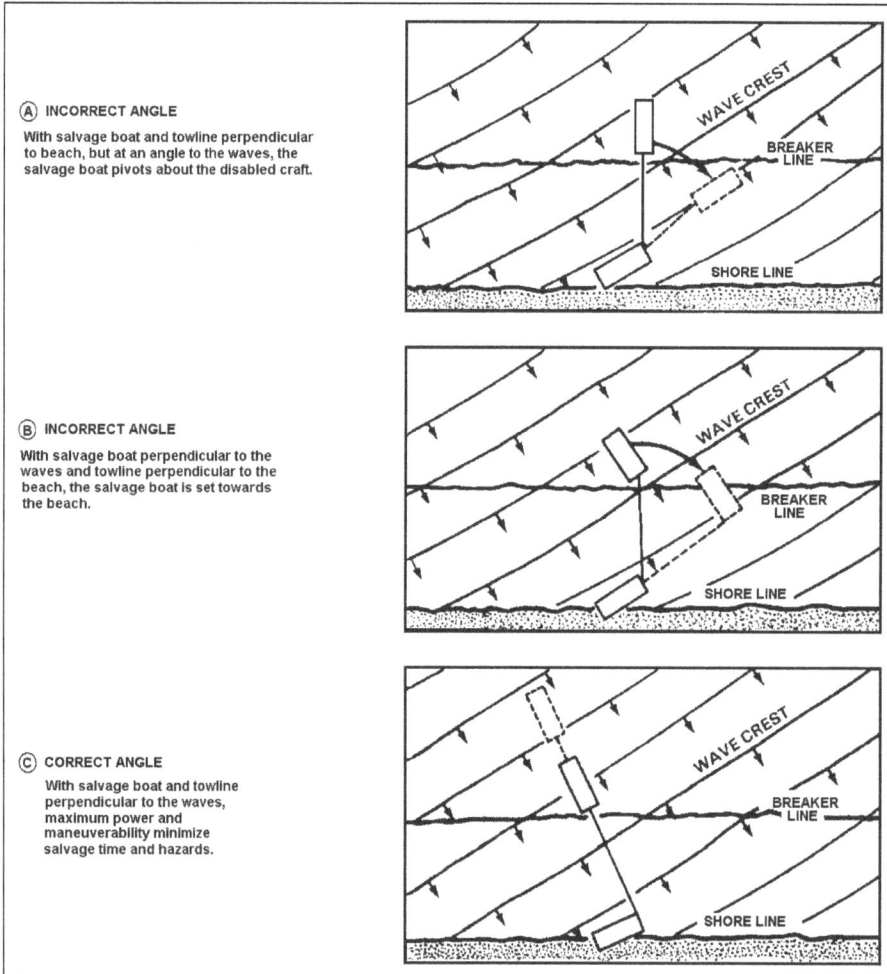

Figure 10-3. Correct and Incorrect Angles for Towing Broached Boat Clear of Beach

Chapter 11

Landing Craft Operations

INTRODUCTION

11-1.   Two principal types of landing craft operations are administrative and tactical. If the deployment is such that it is undesirable for watercraft units to move under their own power, various types of oceangoing vessels can transport them overseas.   An administrative operation is one during which no enemy interference or contact is anticipated. Emphasis is on economy; that is, maximum use of the transport capability.   An overseas administrative move will be documented according to DoD Regulations.   A tactical operation is different in that personnel, supplies, and equipment are loaded so that they may be unloaded easily and rapidly in accomplishing the tactical mission. Here, maximum use of the transport capability is secondary to successful accomplishment of the mission.

LOADING FOR MOVEMENT OVERSEAS

11-2.   The LCM-8 and LCU are usually transported aboard a heavy lift ship. When conventional ships are used, landing craft must be loaded or unloaded by cranes.

BOAT GROUPS

11-3.   The boat group (under a boat group commander) is the basic naval organization of landing craft and support vessels. It is composed of the numbers and types of landing craft and support vessels required to land a particular troop unit (normally a battalion landing team). The boat group is organized into waves. Each wave will consist of the landing craft that will beach simultaneously. Each landing craft is assigned to load troops or supplies from a certain ship. Troops load at debarkation stations marked by colored squares and numbers. Debarkation numbers run forward to aft, with odd numbers to starboard and even ones to port. The sequence of colors, also forward to aft, is red, white, blue, yellow, green, and black. At night, single-cell flashlights with appropriately colored lenses mark the debarkation stations.

CALLING BOATS ALONGSIDE

11-4.   Flag hoists are used to call landing craft alongside naval cargo ships discharging cargo during tactical operations. These flag hoists are flown from either the main yardarm or from a special yardarm at the stern of the vessel being discharged.

- **Day Signals.** Flag hoists flown from the starboard yardarm call boats to the starboard side. Hoists flown from the port yardarm call the landing craft to the port side (see Table 11-1).

Table 11-1. Day Signals

Top Flag	Type of Craft	
Papa	LCVP	
Lima	LCPL	
Six	LCM-6	
Eight	LCM-8	
Bottom Flag	*Station Number*	
(color)	*(stbd)*	*(port)*
Red	1	2
White	3	4
Blue	5	6
Yellow	7	8
Green	9	10
Black	11	12

- **Night Signals.** At night, boats are called by three vertically aligned lights. These lights are coded as shown in Table 11-2.

Table 11-2. Night Signals

Top Light	Side of Ship	
Green	Starboard	
Red	Port	
Center Light	*Type of Craft*	
Red	LCVP	
Amber	LCPL	
Blue	LCM-6	
Green	LCM-8	
Bottom Light	*Station Number*	
(color)	*(stbd)*	*(port)*
Red	1	2
White	3	4
Blue	5	6
Yellow	7	8
Green	9	10
Black (no light)	11	12

LANDING CRAFT WAVES

11-5.   Landing craft are formed into groups of six to eight boats. These groups of boats are called waves. The number of craft in each wave depends on the landing plan. The convoy commander is on the lead boat in the convoy. Control craft are stationed on the port and starboard flanks; salvage and maintenance boats are in the rear. When a boat in a scheduled wave is loaded, it is given a paddle with two numbers on it--the first indicating the number of the wave, and the second, the boat's position in the wave. After loading, a boat proceeds to the rendezvous area, falls in with its wave, and commences to circle slowly. The first wave of landing craft circles clockwise, and so on. Moving out of the rendezvous area, the boats proceed in a column, and when clear of the transport area, form into a wedge with odd-numbered boats to starboard and even-numbered boats to port of the boat carrying the wave commander. Before crossing the line of departure (LOD), the boats are formed in line abreast. The distance between boats is usually 50 yards. The coxswain of each craft should have sufficient charts and navigational aids aboard to enable him/her to travel alone in emergencies. The coxswain should be briefed on such details as follows:

- Approach and landmarks of the location.
- Currents and tides that prevail in that specific area.
- Suitability of the location for anchoring, depth of the water, and type of bottom.

● Facilities for procuring fuel, freshwater, and supplies.

TYPES OF FORMATIONS

11-6.   Three basic reasons to run in formation are to keep in order and maintain contact and control between craft, facilitate landing large numbers of troops or amounts of supplies in a designated place of concentration, and present a difficult target for enemy fire. The following paragraphs discuss four different types of formations.

● **The Closed-V Formation.** This formation enables craft to maintain closer contact than the straight column. The closed-V formation is used primarily in moving from the rendezvous area to the Line of Departure (LOD). It is also the preliminary step in forming the open-V formation.

● **The Open-V Formation.** This formation is used primarily in assault beaching operations. It permits a large number of troops and supplies to be landed in one place with good control between craft and a minimum of vulnerability to air attack. In this formation, each craft is 50 yards astern and 100 yards abeam of the craft ahead.

● **The Line-Abreast Formation.** This formation is used during and after crossing the LOD. Such a formation is vulnerable to flanking shore fire and more difficult to control. The number of craft in a wave depends on the width of the beach. The craft are normally stationed 50 yards apart.

● **The Straight Column.** This is a simple formation in which the craft operate in a straight line at intervals of 15 to 50 yards (depending on visibility). This formation is used when leaving the beach, when in a rendezvous area, or when operating in a noncombat situation.

LANDING CRAFT VISUAL SIGNALS

11-7.   Special visual signals used in directing watercraft formations during both day and night are an absolute necessity in certain situations (Figure 11-1). Every watercraft operator should know how to send and receive signals. These signals must be given carefully and distinctly.

● **Day Signals.** Hand and arm signals are used between landing craft by day when visibility permits. A signal flag may also be shown in the hand to permit recognition of the signal.

● **Night Signals.** Two methods are used to transmit signals at night. Hand and arm signals are given using a flashlight equipped with a red lens, or a signal light may be used to transmit by Morse code signal the type of formation required. Arm and hand signals are used when radio silence is in effect and/or when a radio is inoperative. All crewmembers should know arm and hand signals. A receiving vessel should pass the signal to the vessel astern.

Figure 11-1. Arm and Hand Signals

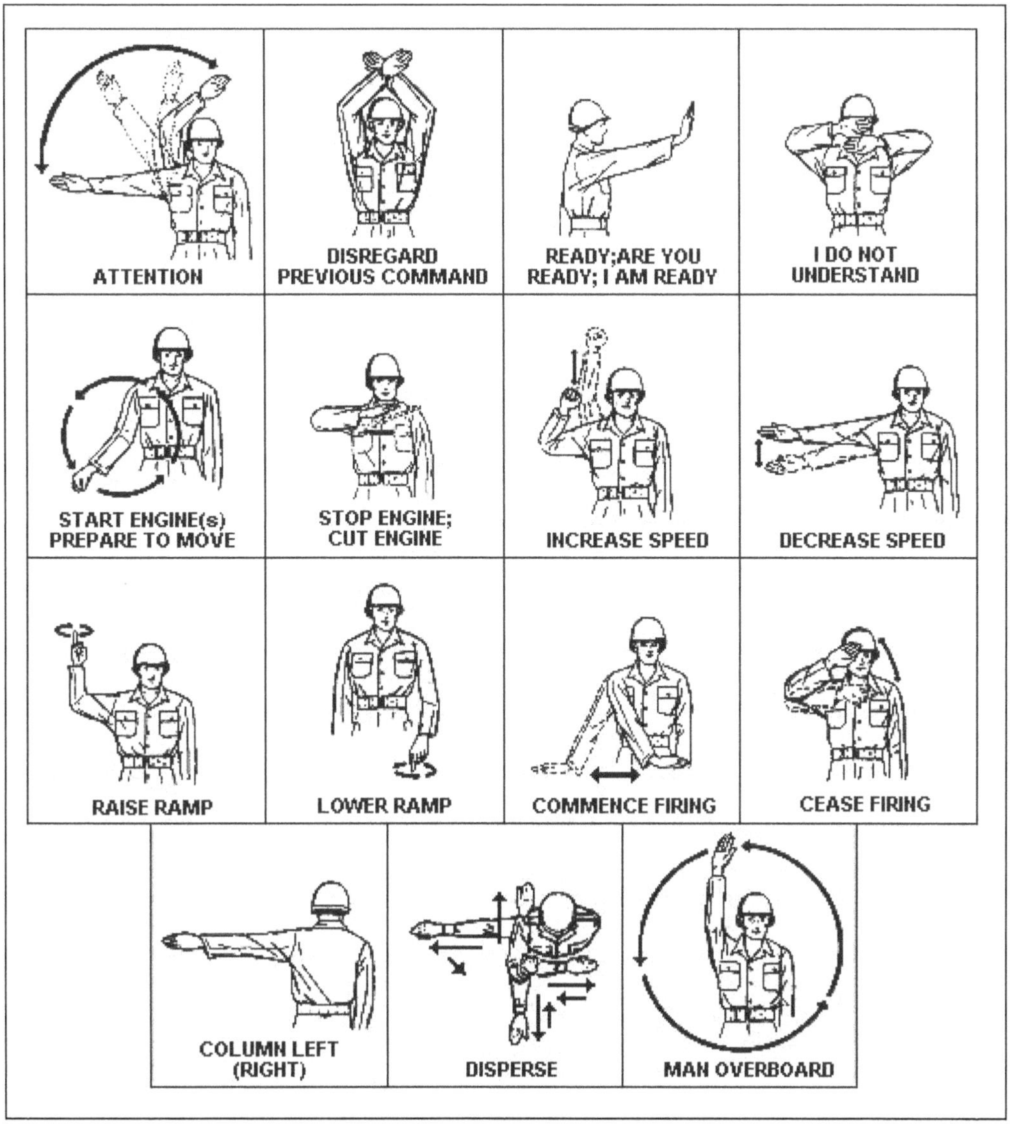

Figure 11-1. Arm and Hand Signals (continued)

MANEUVERING SIGNALS

11-8.   These signals (see Table 11-3) may be transmitted by signal flag or blinker.

Table 11-3. Maneuvering Signals

Command	Signal Flag	Signal Light (Morse Code)
Attention	Mike	− −
Assemble in a column	Alpha	· −
Cease firing	Hotel	· · · ·
Commence firing	X-Ray	− · · −
Execute	Echo	·
Forward	Foxtrot	· · − ·
Increase speed	Kilo	− · −
Line abreast	Romeo	· − ·
Man overboard	Oscar	− − −
Stop	Delta	− · ·
V-formation	Uniform	· · −
Close the V-formation	Uniform/Tango	· · − / −
Open the V-formation	Uniform/Romeo	· · − / · − ·

HYDROGRAPHIC AND BEACH MARKERS

11-9.   The naval beach party is landed early in the assault. When they reach the beach, they proceed with their duties of marking channels and hazards to navigation.

> *Note*: The U.S. Navy Beach Master Unit is responsible for setting up and maintaining beach markers. These markers are used only during a tactical operation.

11-10.   During the process of beach organization, debarkation points for various categories of supplies and equipment are selected on each beach where they best support the tactical plan. Shore party personnel erect beach markers and debarkation points as soon as possible after the initial assault of the landing site. These debarkation markers along with lights for night operations are set up to indicate to the boat crews where the various types of cargo are to be landed. Beaches under attack are given a color designation such as red beach, green beach, and so forth, and beach markers are constructed in corresponding colors. The center of a beach is marked by a large square of cloth with the color side facing seaward. The left flank of the beach, as seen from the sea, is denoted by a horizontal rectangle of the same color, while the right flank is marked by a vertical rectangle, also of the same color (see Figure 11-2).

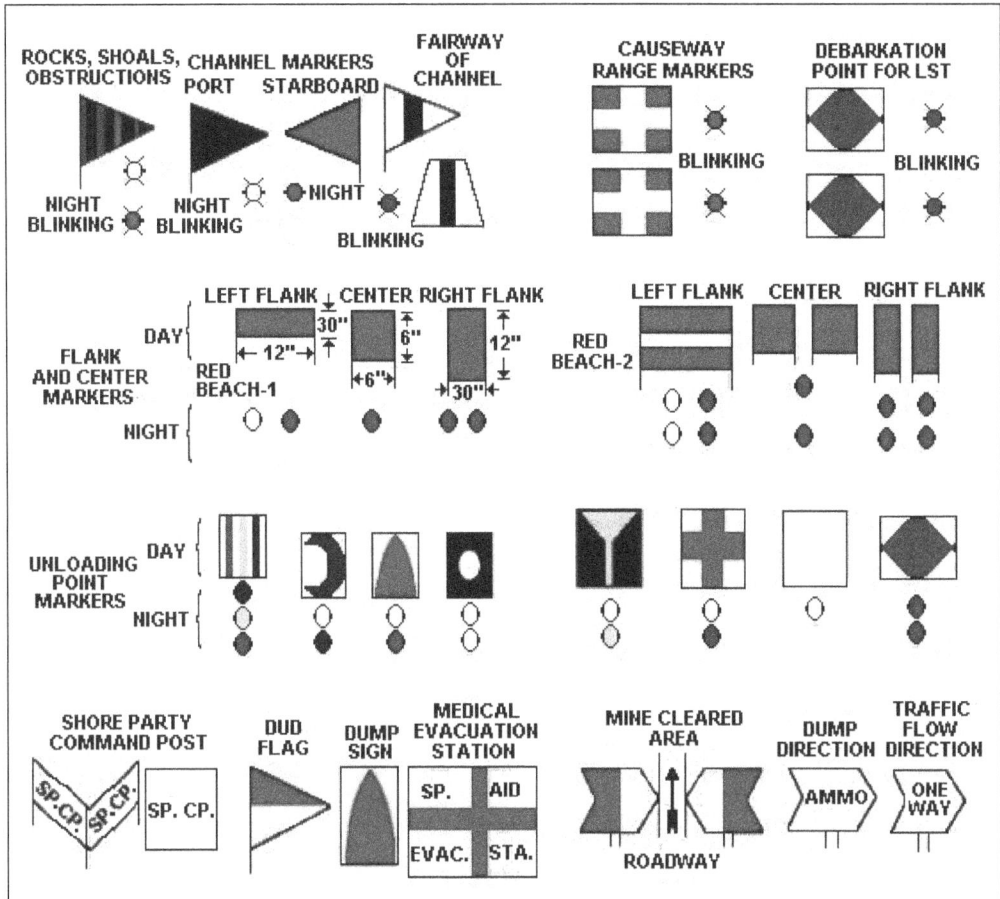

Figure 11-2. Hydrographic and Beach Markers and Signs as Seen From Seaward

GENERAL UNLOADING PHASE

11-11. During the general unloading phase, loaded boats do not maintain a formation on the trip to the beach, although several of them may be required to move as a unit. On the way to the beach, they must stop for orders at the primary control ship and at the boat group commander's boat. The type of cargo in a boat is indicated by the color of special flags flown. These flags are described as follows:

- Red flag. Denotes bulk cargo, which needs manpower for unloading.
- Yellow flag. Shows the load is such that a prime mover is required.
- Blue flag. Denotes self-propelled cargo.
- Red burgee. Shows that the boat is a fuel boat.
- Green flag. Shows that a boat belongs to a floating dump.
- Numeral flag. May be flown under it to indicate the type of cargo carried.

Note: For information on Joint Logistics Over-the-Shore (JLOTS) Operations, refer to Joint Publication 4-01.6.

CARGO DOCUMENTATION

11-12. To provide a record of cargo handled by each link in the loading and unloading chain, the coxswain or master will be given a DD Form 1384 (Transportation and Movement Control Document) for each unit of cargo loaded. The coxswain keeps one copy which is used for the log entry of tonnage hauled, gives the required number of copies to the cargo checker at the discharge point, and then delivers the required number of copies to the control officer. The exact number of copies and their distribution will be prescribed in the unit SOP. Detailed procedures for tallying cargo and the use of this form in accounting for cargo are discussed in DOD Regulations.

CARGO LOADING OPERATIONS

11-13. In a LOTS operation where you become involved in a resupply situation, accountability and condition of cargo are of utmost importance. When transporting cargo aboard landing craft, the crew must make certain that it is properly stowed and secured.

TIPS ON SECURING CARGO ABOARD LANDING CRAFT

11-14. When securing cargo aboard a landing craft, there are a number of things that can be done to reduce the risk to the safety of the ship or the health or safety of any person on board.
- Properly load, stow, and secure all cargo.
- Properly pack and secure cargo within the containment.
- Correctly load and transport heavy cargo or cargo with abnormal physical dimensions to reduce the risk of damage to the ship's structure.
- Check the strength of securing points and lashings.

DUNNAGE

11-15. Aboard landing craft, dunnage usually consists of 1-x 6-inch x random length lumber and timbers. Dunnage with dimensions of 4 inches x 4 inches x random length or larger is called timber. During cargo operations, dunnage is carried by all landing craft.

PALLETIZED CARGO

11-16. Dunnage is not required for palletized cargo; the pallets serve that purpose. Pallets can be loaded directly on deck.

MILVANS AND CONTAINERS WITHOUT CHASSIS

11-17. Lay a dunnage floor in a landing craft before MILVANs and containers without chassis are loaded. The bottom layer of dunnage should be laid athwartship to allow any water in the well deck to spill off into the bilge. If seas are heavy or considerable water is expected to come aboard, a second layer of dunnage is laid—this time, in a fore-and-aft direction. This will serve two purposes. First, it will protect the bottom of the containers from the water, and second, it will allow even distribution of weight over the deck of the landing craft. It will also protect the deck from being gouged and torn up in general. When loading MILVANs or containers without chassis, lashing is not required. Their weight and size will hold them in place.

MILVANS AND CONTAINERS ON CHASSIS

11-18. Dunnage should be laid only under the wheels and the front stands. The MILVAN or container and chassis must be lashed down. Four lashings are required—two forward and two aft.

CAUTION

Make sure that the MILVAN or container on chassis is loaded with the front (where the tractor connects with the chassis) facing the bow.

PECK AND HALE QUICK RELEASE

11-19. Several methods and types of lashings are used and in the system. The peck and hale quick release is only one of the systems used. This tie-down is used primarily for securing vehicles, but similar cables are incorporated into nets and used to secure cargo on deck. If such gear is unavailable, wire rope and turnbuckles can be used to prevent movement of the chassis.

WHEELED VEHICLES

11-20. These vehicles can be loaded directly on deck. ALWAYS LOAD THE VEHICLE WITH THE ENGINE OF THE VEHICLE FACING THE BOW. Then the vehicle can be driven straight off the landing craft. Once the vehicles are loaded and spotted on deck, make sure that the brakes are set, that vehicles are left in gear, the wheels are chocked, and that lashings are used to secure the vehicles to prevent them from shifting.

TRACKED VEHICLES

11-21. The same principles for securing wheeled vehicles are used when securing tracked vehicles. One exception that you must remember is that a double layer dunnage floor must be laid. When laying the dunnage floor, keep the cloverleaf's or tie-down rings clear so that you will be able to tie down. For tanks and tracked vehicles of this size, use 8- x 8-inch or larger timbers for effective chocking.

PETROLEUM, OILS, AND LUBRICANTS (POL) IN DRUMS

11-22. Drums are stowed on their side in a fore-and-aft direction and stowed bilge-to-bilge. The drums are tiered in pyramid fashion and stowed no higher than three tiers. If only a few drums are to be loaded, then they are stowed in an upright position with their bungs up. They must be stowed on dunnage to prevent sparking and to reduce the possibility of shifting. Drums stowed in an upright position must be lashed in place.

WARNING

DO NOT accept any drums that are leaking.

LOADING TROOPS

11-23. Before loading troops aboard landing craft, passenger lists must be prepared. Embarkation personnel should prepare these lists ahead of time. The passenger list should show the full name, grade, serial number, and unit of all troops to be loaded. At loading time, troops should be loaded in list order and at that time the accuracy of each entry verified. One copy should go to the embarkation officer loading the vessel, or if from a port area, to the harbormaster, and one copy to the coxswain or master of the landing craft. When taking troops aboard from the pier or a troop ship, crew members on the landing craft should do the following:

- Make sure each passenger is wearing a life jacket.
- Assist passengers in boarding and getting into the well deck.
- If passengers are without life jackets when boarding, issue life jackets from the ship's supply, if available.
- Take life jackets back from the passengers when they debark.

11-24. At all times, the vessel master or coxswain is responsible for the safety of passengers carried aboard the craft. Each crew member is responsible to the master or coxswain to help ensure the safety of the passengers.

Note: DO NOT allow passengers on the ramp, catwalks, upper decks, engine compartment, or in the crew's quarters.

Chapter 12

Search and Rescue

INTRODUCTION

12-1.   Due to the nature of harbor craft operations (working in coastal waters, inland waters, and harbor areas) you are apt to become involved in a search and rescue (SAR) operation. This chapter describes a limited SAR operation. This type of SAR occurs when a crewman is lost overboard, a small craft is lost, or when only your vessel is involved in the search.

PERSONAL SURVIVAL

12-2.   With a man overboard, time is critical. All crew members must be fully aware of what is being done and how it is done. In the event you are the victim or whether it is one person or an entire crew, survival depends on three things: courage, training, and time. Courage is your mental attitude - DO NOT GIVE UP! Do what you were trained to do to survive if you fall in the water. Time is of the essence. The ship's response to the situation is critical. Table 12-1 gives you an estimate of survival times in various water temperatures. This table is only a guideline to emphasize the need for fast action and not a means of setting an arbitrary limit on the search effort.

Table 12-1. Survival Times in the Water

Water Temperature		Survival Time (Average Duration)
Centigrade	Fahrenheit	
Less than 2 Degrees	Less than 34 Degrees	Less than 45 minutes
2 Degrees to 4 Degrees	34 Degrees to 40 Degrees	Less than 90 minutes
4 Degrees to 10 Degrees	40 Degrees to 50 Degrees	Less than 3 hours
10 Degrees to 15 Degrees	50 Degrees to 59 Degrees	Less than 6 hours
15 Degrees to 20 Degrees	59 Degrees to 69 Degrees	Less than 12 hours
Greater than 20 Degrees	Greater than 70 Degrees	Indefinite (depends on physical condition)

COLD WATER SURVIVAL AND HYPOTHERMIA

12-3.   Any time the sea temperature is below 70 degrees Fahrenheit (F), the water is considered to be cold. To survive in cold water, two things must be prevented: drowning and hypothermia. Jumping into cold water will put a severe strain on your entire system. It can cause you to gasp for breath. If your head is below the surface of the water, you will drown. By wearing a life jacket, you will keep your head above water. That is why the wearing of a life jacket is so essential during heavy weather operations, drills, or shipboard emergencies.   The life jacket will take care of the first problem, which is drowning. Hypothermia is defined as subnormal temperature of the body. In this case it is the lowering of the central body temperature. As the body temperature decreases it causes the person to become irrational, lose consciousness, and finally drown. To prevent hypothermia you must slow down the rate of "core" or central body cooling. Get out of the water as soon as possible. At all times keep your head up and as dry as possible. Your head is the greatest heat-loss area of your body. If you have a hat or cap on—keep it on.

When wearing a life jacket in cold water, you can protect the high heat-loss areas of your body. This includes the head, neck, sides, and groin. There are two ways to do this (see Figure 12-1).

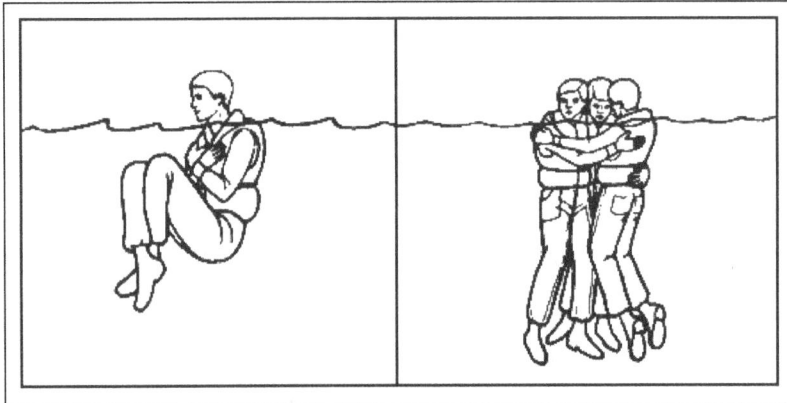

Figure 12-1. Retaining Body Heat

12-4.   If you are alone, hold your upper arms against your sides with the wrists placed over your chest. Draw your legs up as close to your chest as you can and cross your ankles. If there are other persons in the water, group together. All of the people face one another with their chests and sides as close together as possible and their arms about one another. Either one of these actions will increase survival time up to 50 percent. However, even partial covering of the sides, neck, and groin will cut down the heat loss and extend the survival time.

WARNING

DO NOT exercise or swim - this will only speed up the rate of body heat loss.

SECTOR SEARCH PATTERN FOR ONE SHIP

12-5.   SAR covers many situations and many types of operations. This paragraph describes a sector search pattern for one ship (Figure 12-2). It is used when the position of the search target is known and within close limits. With a man overboard, the ship returns immediately to the datum; or, if the search target is once sighted and then lost, the ship heads for the datum. All turns are 120 degrees to the starboard. Each leg of the pattern is approximately 2 miles. The search pattern will always start from the datum point. This pattern gives a very high probability of detection close to the datum point and spreads the search over the probable area quickly. Upon completion of the search pattern, re-orient the pattern 30 degrees to the right and research a new pattern (see the dashed lines in Figure 12-2). This procedure can be repeated three or four times until either the victim is found or the search has been called off.

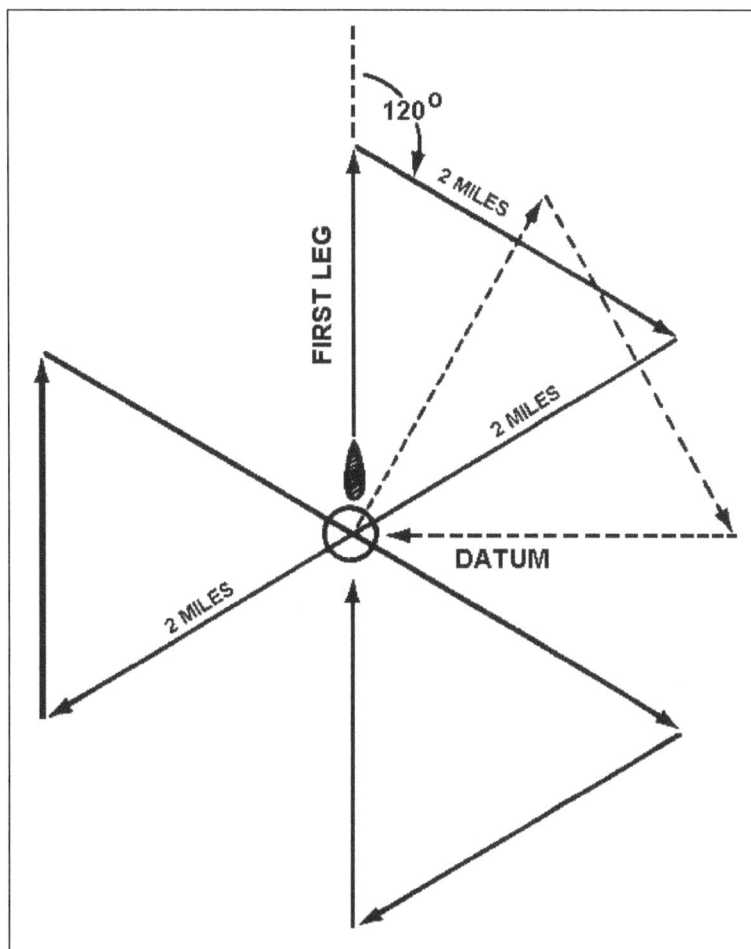

Figure 12-2. Sector Search Pattern—Ship

This page intentionally left blank.

Chapter 13

Towing

INTRODUCTION

13-1.   Even though towing is a routine task for tugs, it is still one of the most dangerous operations Army mariners must perform. The practice of good seamanship is necessary to prevent endangering the crew, tug, or tow. The tug master is responsible for the entire operation, but the boatswain and leading seaman are responsible for preparing, making up, and rigging the tow. You should know how to make up a tow, the different types of tows, and how to assemble and rig for the different types of tows. This chapter is intended to improve your knowledge of these procedures.

TYPES OF TOWS

13-2.   The use of tugs and their connection to the tows can vary. The following are some basic configurations:
- Single tug, single unit tow.
- Single tug, multiple unit tow.
- Multiple tugs, single unit tow.

SINGLE TUG, SINGLE UNIT TOW

13-3.   This consists of a tug and tow. Several methods are used in connecting the tug and tow. The single leg and bridle or stern tow is used for long distance towing in open waters. The alongside or hip tow is used where maximum control and maneuverability are required. The push tow is the most effective use of the tugs power and gives the best control of the tow. The single leg and bridle is made when the towing ship passes the towline, which is shackled to a flounder plate at the apex of the bridle. Each leg of the bridle consists of chain or wire rope passed through the bow chocks and secured on the tow's deck to pad eyes or bitts. Towing alongside (hip tow) is most often used in congested waters. Towing alongside offers excellent control; it is not recommended for the open ocean. For alongside towing, the tug generally secures to one side of the tow with her own stern abaft of the stern of the tow to increase the effect of her screw and rudder. The side chosen depends on how much the towing ship must maneuver with the tow.

SINGLE TUG, MULTIPLE UNIT TOW

13-4.   This consists of one tug and several tows. The connection and makeup of the tows can vary. The following are the four versions used:
- Tandem rig (for congested waters where control is required).
- Honolulu rig (for short distance towing).
- Christmas Tree rig (for long distance in open ocean towing).
- Modified Christmas Tree rig (restricted to protected waters).

MULTIPLE TUG, SINGLE UNIT TOW

13-5.   It may be desirable to use more than one tug for only one tow. Greater power, increased towing speed, and better control may be obtained in a multiple tug tow. This type tow is generally used in towing large ships, deep-draft, large-displacement dry docks, or deep-draft barges.

Note: When a multiple tug, single unit tow is to be made up, the senior tug master is in charge and is responsible for the tow and its makeup.

DESCRIPTION OF TOWING EQUIPMENT

13-6.   All deck crew members should know the basic terminology of towing gear, its function, and what to check for in safety and rigging procedures. Most of the towlines used today are made up of either nylon or one of the new synthetic polyester fiber lines. Based on size and weight, they provide great strength. If properly cared for, they will last for a long time. Usually wire rope and chain are used for deep sea tows. Nylon or synthetic fiber towlines are used in inland waters and for coastal tows.

CARE OF NYLON TOWLINE

13-7.   The tensile strength of nylon line is approximately twice that of manila line of equal size. It will stretch under strain and return to its normal length. Nylon line resists rot and mildew. If properly maintained, it should last five times longer than manila line.

ABRASION

13-8.   Chafing gear should always be used at points of abrasion where there are sharp metal edges. If possible, ensure that all holding devices have smooth surfaces. If the yarn becomes abraded, cut away the chafed section and splice the ends. Glazed areas may appear where the line has worked against bitts or chocks. These areas do not affect the strength of the line. When reeling in, ensure that thimble and connecting links do not chafe or cut hawsers.

CLEANING

13-9.   If nylon becomes oily or greasy, scrub it with fresh water and a paste-like mixture of granulated soap. Then rinse with fresh water and allow to dry before stowing.

STRETCHING

13-10. Nylon line can stand repeated stretching. It thins out when under load and returns to normal if the stretch does not exceed 40 percent. Beyond that, the line may part. When the stretch becomes excessive, double the lines by passing the bight. This reduces the danger of snapback.

NEW HAWSERS

13-11. New cable-laid nylon hawsers may be stiff. Tension the cables at 30 percent extension for 20 minutes. When new lines are strained, they produce a sharp crackling sound. This is the result of readjustment of the line strands to stretching and should not be alarming. To uncoil new line, unreel it as you would wire rope. Do not pull the end through the eye of the coil.

COILING AND WINDING

13-12. When not using a powered reel, coil the line in a different direction each time. This prevents unbalancing the lay.

STOWAGE

13-13. Stow nylon rope away from strong sunlight, heat, and strong chemicals. Cover the rope with tarpaulins. If the line is iced over, thaw it carefully and drain it before stowing.

USE OF NYLON WITH OTHER ROPE

13-14. When using sets of lines in parallel (for example, purchases), use nylon line with nylon line only, not with wire or manila. Other kinds of line must not be used on the same chock or bitt with nylon. Always use a nylon stopper with nylon towlines under load.

CHARACTERISTICS OF NYLON LINE UNDER LOAD

13-15. Plain-laid nylon hawsers may elongate around the bitts under load. Minimize by taking a turn under the horn and crossing the line on itself before taking more turns. Sudden surges may occur when nylon hawsers are used on capstans for heavy towing or impact loading. Take six turns on the capstan. The standing part of the line is led to a set of bitts and made fast by figure eight. This provides a back up and is a precaution against surges.

ALONGSIDE TOWING

13-16. Make up forward and backing towlines as closely as possible. Take up slack in the relaxed line while the other line is under load. It may be necessary to reverse the tug's engines slightly when easing pull. This counteracts the elasticity of the nylon and prevents snapback.

SPECIAL PRECAUTIONS

13-17. Snapback may occur after a 50 percent stretching. Ensure that personnel are away from the direct line of pull when applying heavy loads. Nylon line around bitts may slip when eased out under heavy load because its coefficient of friction is below that of manila. This may cause injury to personnel who have not been thoroughly instructed in the peculiarities of nylon line. Take two or three turns on the bitt then make fast by figure eight; this provides closer control. Stand well clear of the bitts.

NAMES OF TOWLINES

13-18. Towlines may be called spring line or towing line, bowline or backing line, stern line or turning line, and bow breast line (see Figure 13-1).

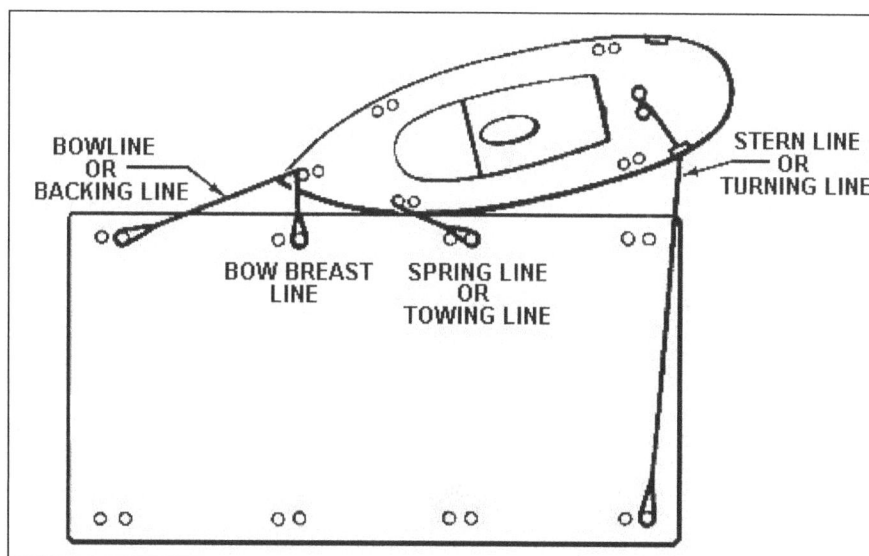

Figure 13-1. Display of Towlines

STERN TOWLINES

13-19. The description given here is for typical towlines that are used for stern tows over long distances and in heavy weather. For a deep sea tow you may carry as much as 2,100 feet of 2- x 6- x 37- inch high-grade, galvanized plow steel, fiber-core wire rope. Wire rope is primarily used for long distance ocean tows. For short coastal or light displacement tows, you may carry 2,400 feet of 7- to 9-inch circumference nylon towline.

CARE OF WIRE ROPE TOWLINE

13-20. Wire rope towline is very expensive. It is important to keep it in excellent condition, protect it against excessive wear, and inspect it regularly.

LUBRICATION

13-21. Wire rope, like a machine, is made up of many moving parts. The steel fibers slide independently and must be protected by adequate lubrication against the effects of stress. Corrosion damage is also a danger. There is no way of estimating the exact loss of strength resulting from corrosion. During manufacture, wire rope is thoroughly lubricated. However, this lubrication is lost during operation and must be renewed periodically and after each use. Use relatively fluid oil of light viscosity (such as linseed oil). If the oil is stiff, it must be heated. Apply the oil so that it penetrates through all the strands to the core. Apply the oil by swabbing, using a brush, or by pouring the oil on the rope. If wire rope is to be stored for some time, lubricate it with heavy oil (such as crude petroleum) to which a small amount of graphite has been added.

WINDING ON A DRUM

13-22. Ideally, only one layer of wire rope should be wound on a drum. Where this is impractical, the rope must be carefully wound so that each layer spools evenly over the preceding layer. The drum must be the correct size for the rope size.

DISTRIBUTION OF WEAR

13-23. A towline can be reversed occasionally to redistribute the wear, or one end can be cut. These are complicated operations because of the construction of wire rope. They must be done in a shipyard.

ACCELERATION

13-24. Suddenly applying a load to wire rope by rapid acceleration causes stress much greater than the weight or resistance of the tow. Avoid such strain on the rope by gradual acceleration.

STOWING

13-25. When not in use, stow wire rope in a dry place where there is no acid. Keep the outer layer lubricated.

BREAKING IN NEW ROPE

13-26. New wire rope is broken in by using it first with a light load then gradually working it up to its working load.

INSPECTION

13-27. Inspect the rope thoroughly as it is being wound after each use. Look for broken strands in the outer layer. When one-half the diameter of the fibers of the outer layer is worn away, the rope should be replaced. Worn wire rope is an indication of chafing. The cause of the chafing should be investigated. Ensure that the rope is properly lubricated and stored after use and inspection.

Note: Destroy damaged wire rope rather than discarding items that have been judged defective. They might be used again by someone not aware of the hazard or defect.

SPECIAL TREATMENT OF WIRE ROPE EXPOSED TO SEAWATER

13-28. After using a wire rope towline in seawater, wash it with fresh water. Then lubricate it with a thorough coat of environmentally friendly grease. Use a high pressure wire rope grease machine that will impregnate grease all the way to the core of the hawser.

MISUSE OF WIRE ROPE TOWLINES

13-29. It is extremely important that wire rope be used and maintained properly because of the great expense involved. It should not be subjected to any of the following common abuses:

- Chafing.
- Rope of incorrect size.
- Drum of inadequate size.
- Improper winding on drum.
- Improper or insufficient lubrication.
- Exposure to acid fumes.
- Lack of protection against moisture and saltwater.
- Kinks.
- Sudden acceleration.
- Touching the seabed.

Note: Simple shipboard action can prevent damage to wire rope. Adequate use of chafing gear, use of correct size fairleads; and timely change of the nip will greatly increase the life of wire rope.

CAPSTAN

13-30. Power capstans are provided on the deck of a tug for taking a strain on line-towing hawsers. The capacity, characteristics, speed, and location vary from ship to ship.

Note: Towlines should never be made fast on the capstan or a cathead.

CHAFING GEAR

13-31. Chafing gear is used to prevent wear. It is additional protective material that is placed over and around a line or wire rope to protect it from damage through rubbing. An old fire hose cut into 4- to 6-foot lengths and then split lengthwise makes excellent chafing gear. It is wrapped around the hawser or towing cable to protect it from wear due to constant rubbing (Figure 13-2). Chafing gear can be made up of the following:

- Old canvas and burlap wrapped around and secured to the towline at the point of chafing.
- Four- to six-foot lengths of old fire hose split and then wrapped and secured around the towlines.
- Metal chafing boards fabricated from split steel pipe or channel iron and clamped to the cable for ocean towing.
- Commercially available Kevlar chafing sleeves for line hawsers.

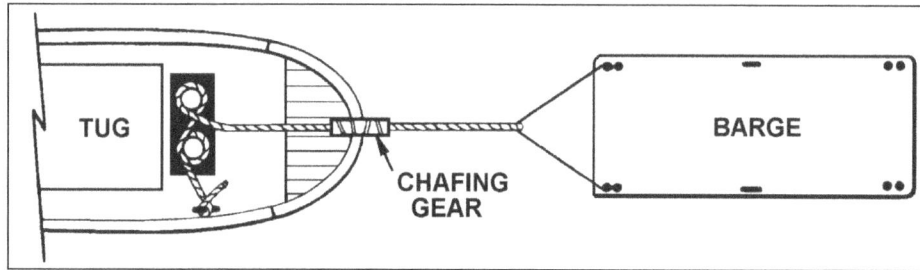

Figure 13-2. Chafing Gear

FLOUNDER PLATE (FISH PLATE)

13-32. This is a triangular steel plate used as a central connecting point for the tow bridles and towline.

BRIDLES

13-33. Stern tows should be fitted with a two legged towing rig of wire or chain attached to towing pad eyes or bitts on the tow. At the apex of the bridles is the flounder plate. Each bridle leg must be capable of full towline strength. A good way to determine bridle length is that the length of each leg of the bridle must be at least one and a half times the width of the tow. For a barge with a beam of 50 feet, each bridle leg would be at least 75 feet long.

RETRIEVING LINE

13-34. This is a wire rope or fiber line that is connected to the flounder plate and usually led to the tow. The retrieving line should be longer than the distance from the tow to the flounder plate. This prevents it from taking any load. The legs of the bridle form two sides of an equilateral triangle. An imaginary line between the towing pad eyes or bitts on the tow forms the triangle base.

PENDANTS

13-35. Pendants are made up of either wire or chain and they can be used to connect the towline to the flounder plate. A pendant can be a length of wire used as an under rider wire in a "Christmas tree" rig. A pendant or single leg rig is the simplest and most straight forward towing rig used for open ocean towing of ships with fine bows, sonar domes, and bulbous bows or when the tow is most stable in this configuration.

PLATE SHACKLE

13-36. Plate shackles are frequently used in salvage and offshore towing operations because they are simple, efficient and easily fabricated to fit a specific application. Once fabricated, they should be load tested so their safe working load (SWL) is known. See Figure 13-3.

HANDLING TOWLINES

13-37. Three ways can be used for making up when towing alongside or towing astern. When towing alongside, towlines can be made up as a single towline or the towline can be doubled up. The method used will depend on the situation.

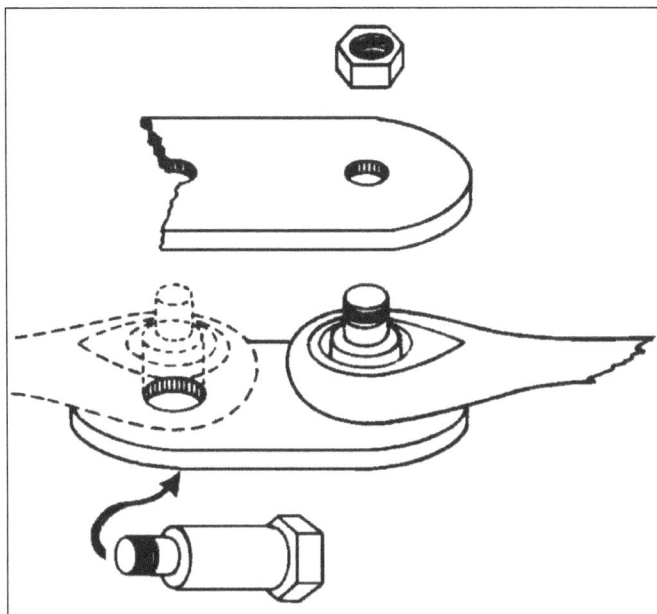

Figure 13-3. Small Plate Shackle

SINGLE-LINE LEAD

13-38. When leading out a towline, lead it out between the towing bitts and make the eye fast to the bitt on the tow that is nearest to you. Then take in all the slack and secure the line with figure eight or round turns.

DOUBLING UP A TOWLINE

13-39. In this method (see Figure 13-4) the eye splice of the towline is put over one of the bitts on the tug, and the bight of the line is then led around the bitt on the tow. The bitter end of the tow line is made up on the same bitt as the eye. The bitter end is lead from the outboard side, and one or two round turns are taken on the bitt, making a figure eight of the line on the bitts.

- Doubling the lines gives added strength.
- When releasing the tow, you slack off on the line, cast off the eye from the bitt on the tug, and take in the line. This eliminates the need of having to put an individual aboard the tow to release the line.

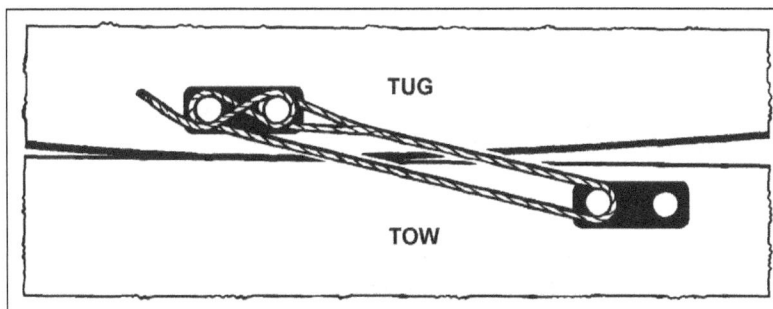

Figure 13-4. Doubling-Up Towlines

RIGGING A STERN TOWLINE

13-40. To rig a stern towline, the towing hawser should be faked out in the fantail of the tug (Figure 13-5). This will ensure that the hawser will pay out without becoming fouled. The eye of the hawser is led back over the top of the "H" bitt, over the shoulder of the horn, and back through the legs of the bitt (Figure 13-6). Then the hawser is payed out. When you get close to the point where you are going to secure the tow, take a full round turn and cross the line back onto itself. Then take two or three additional round turns before you figure eight the line on the bitts, and finish it off with two or three turns on the arm of the bitt.

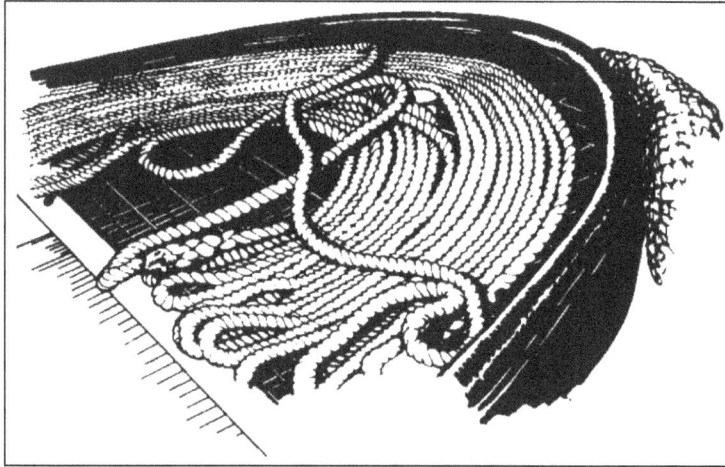

Figure 13-5. Hawser on Fantail

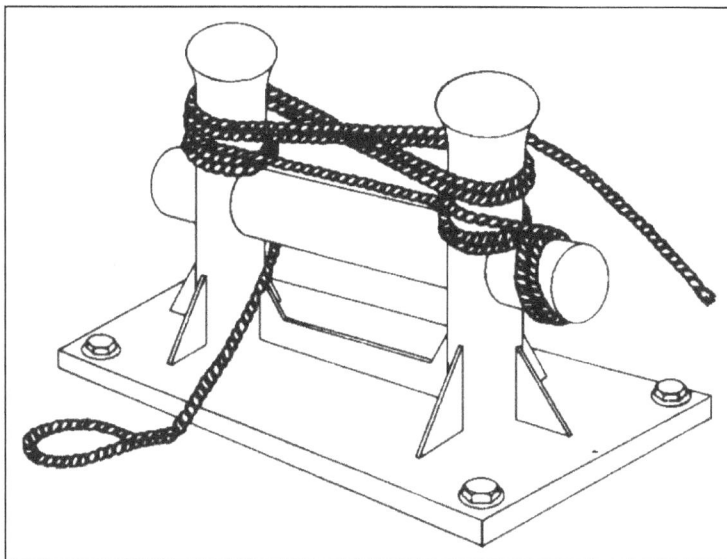

Figure 13-6. Leading the Hawser Over the "H" Bitt

WARNINGS

1. ALWAYS face your work.

2. NEVER step over a line lying on the deck. Either lift it up and walk under it, or step on top of it and cross over. NEVER straddle or step in the bight of a line.

3. When towlines are coming under or are under a strain, work fast. Get the turns or figure eight on as quickly as possible. When surging or slacking off on a line that is under strain, keep your hands clear of the bitts.

4. Know where the fire axe is located.

TOWING ALONGSIDE (HIP TOW)

13-41. For tugs working inside harbors and for towing short distances or in confined areas where constant control is required, towing alongside or the hip tow is the preferred method. A hip tow can be made up on either the port or starboard side of the tug (see Figure 13-7). There are various ways of making up a hip tow; however, some standard requirements must be met.

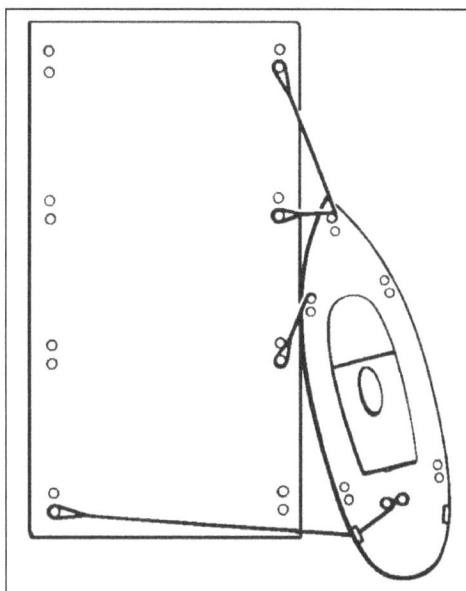

Figure 13-7. Making Up a Hip Tow

TOWLINES

13-42. Three lines should be used: the spring line, the bowline, and the stern line. For large or heavy tows, you may want to double up on the towlines and also use a bow breast line. Before the tug goes out to make up for a hip tow, the towing lines must be inspected and made ready. Inspect the lines for signs of severe chafing and the eyes and the eye splices for fraying or breaks. Check the towlines for wear and breaks. If you find a line damaged or one that you have doubts about, point it out to the boatswain or mate. When selecting the lines to be laid out, the usual procedure is that the best line is used for the spring line. This serves as the towline and takes the greatest strain. The second best line is used for the bowline, and the third best line is used for the stern line. The lines are then faked down (they are laid out so that they are free of kinks and obstructions). They can then be paid out rapidly when they are needed.

DETERMINING WHICH SIDE TO MAKE UP TO

13-43. The tug secures to one side of the tow with her own stern abaft of the stern of the tow. This will increase the effect of the tug's screw and rudder. The side chosen depends on how much the tug must maneuver with the tow. If all turns are to be made with the tug's screw going ahead, she will be more favorably placed on the outboard side of the tow-- the side away from the direction toward which the most turns are to be made. If a sharp and difficult turn is to be made under headway, the tug should be on the side toward which the turn is to be made. Here she is properly placed for backing to assist the turn, because as she slows, the tow's bow will turn toward the side the tug is on. If a turn is to be made under no headway, the tug is more efficient on the starboard side of the tow. When the tug backs to turn, the port send (side force) of her screw will combine with the drag of the tow to produce a turning effect greater than that which could be obtained with the tug on the port side. The best position for a long back in a straight line is to have the tug on the port side. Then the drag of the tow tends to offset the port send of the backing screw.

SECURING THE TOWLINES

13-44. The towing line or spring line, usually a 6-inch (or larger) hawser, is led from the forward towing bitts on the tow side of the tug to the aft set of bitts on the tow. This line is secured first. Next, the bowline or headline is paid out thru the closed chock in the bow (commonly called the bull nose) to a bit on the forward end of the tow. Next the bowline or backing line is paid out over the outboard side of the bow stem or king post and lead to a bit on the forward end of the tow. Once the bowline is secured on the tow, all the slack is taken in and the bowline secured. This will bring the tug into proper position, slightly bow-in to the tow. When backing down, the bowline becomes the towline. The stern line or turning line is lead from the tug's stern to the outboard side of the tow's stern. The purpose of this line is to keep the tug's stern from drifting out. The three lines, when properly secured and made taut, will make the tug and tow work as one unit.

> *Note:* If for some reason the stern line cannot be fair led and secured to the outboard side of the tow, it is then secured to the inboard bit on the stern of the tow.

13-45. A fourth line (optional), the bow breast line, can also be used for greater control when making up to a heavy tow. Check all the lines to ensure that they are as taut as possible. Perform this by easing the tug gently forward, then aft, to see that all the towlines are secure. The tug and the tow should be made up as a single unit.

CAUTIONS

1. When securing these towlines, remember to NEVER secure the line so that it cannot be thrown off quickly and easily.

2. Areas of the harbor subject to wave action should be avoided whenever possible. The tug and tow seldom pitch in the same tempo. When both start pitching out of harmony, the lines take a heavy strain and may part. Size and loading of the tow may obstruct the view of the tug's conning officer. In that case, a lookout is stationed aboard the top who keeps the conning officer fully informed of activity and hazards in the blind area.

SHIFTING THE TOW TO THE OTHER SIDE

13-46. Occasionally it may be necessary to shift a tow from one side to the other. One method of doing this is shown in Figure 13-8.

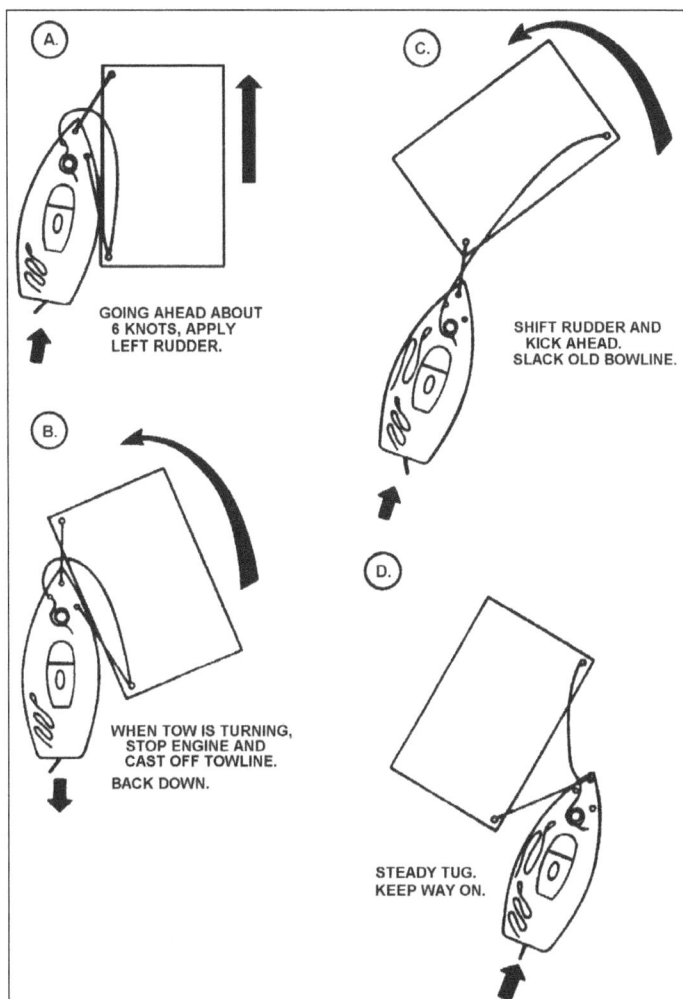

Figure 13-8. Shifting a Tow From One Side to the Other

TOWING TWO BARGES ALONGSIDE

13-47. Two barges may be towed alongside. Figure 13-9 shows the makeup for alongside tow.

PUSH TOWING

13-48. A vessel that normally pushes its barges or tow ahead is referred to as a towboat or a pusher tug. This type of towing typically happens in inland waters by a vessel fitted with push knees. It is the most effective use of the tugs power and gives the best control of the tow.

RIGGING A PUSH TOW

13-49. The cables from the towboat to the barge are called face wires, and are normally made up on winches, either hand cranked or mechanically driven. For large or heavy tows, you may want to double up your facing wires to the tow. A line leading from the bow of the towboat to a bit in the center of the barge or to bits on the corners of the barge are called jockey lines and prevent the knees from shifting when the towboat is turned hard over. One of the hazards of push towing is the possibility of parting a face wire. If this should happen while the tug is pushing at high speed, it could result in the towboat being tripped (capsized) by the other face wire. The best thing is to have lines rigged from the corner bits of the barge to the midship bits of the tug. These lines are called preventers or safety lines. The barge would then drag the tug along bow first while you could regain control and rig another facing wire.

Note: Tug and/or tow must be capable of being rigged for a stern tow in the event that it becomes too rough to safely push tow.

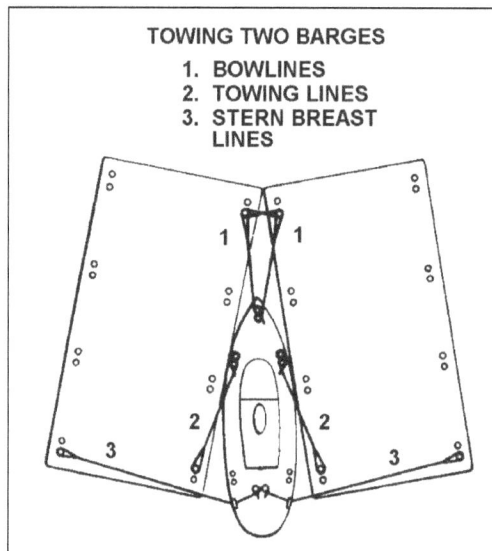

Figure 13-9. Towing Two Barges Alongside

TOWING ASTERN (INLAND WATERS)

13-50. There are many variations of a stern tow. Different towing materials are also required. Some factors that must be considered in planning for a stern tow are as follows:
- Whether the tow is being made in inland waters, bays, coastal waters, or overseas.
- Weather and sea conditions.
- Size and horsepower of the tug.
- Size of tow.
- Number of tows involved.

SHIFTING THE TOW FROM PUSHING AHEAD OR ALONG SIDE TOWING TO ASTERN

13-51. Shifting usually is necessary when a tug is to tow a barge from port to port where rough waters may be encountered along the transit route. The tow is taken along side or pushed within the harbor and shifted astern outside. The shifting procedure is simple. The towing hawser is connected to the fish plate and towing bridles before getting underway. Outside the harbor the lines used for towing alongside or pushing ahead are cast off, allowing the tow to drift away from the tug. Then, by slowly accelerating and carefully altering course and judiciously paying out the tow hawser, the tug gets underway with the tow and comes to the required course.

Note: When towing astern, you have limited control over the forward motion and turning of the tow. For this reason, stern tows are made in open waters. The longer the towline the less control you will have.

TOWING LINES

13-52. When towing in inland waters, the towing hawser is usually made up of nylon or synthetic aramid fiber lines such as Kevlar, Amsteel or Plasma. For their size, the lines are lightweight and have tremendous strength. They are easy to handle. For inland waters, the length or scope of the stern towline is relatively short. Seldom will it ever be longer than 600 feet. The area in which you are towing and the master's desire determine the length. Due to their relatively light weight, synthetic fiber towlines will float when they lie in the water. When a strain is taken on the line, it will rise up out of the water and stretch out. When using synthetic fiber towing lines there will be no catenary or dip in the towline. As the strain of the tow increases, the towline stretches out like a rubber band. As the line stretches, it will reduce its diameter by as much as 30 to 40 percent. The tow will surge forward because towing actually consists of pulling by jerks. Then the cycle starts over. The greatest danger in using synthetic fiber towlines such as nylon is that if the line should part when under strain, it will snap back its full length like a bull whip. The force of the snapback is tremendous depending on the strain that the line was under at the time it parted. There is no set pattern on how the line will whip back. It may snap back directly on itself or it may whip from side to side. There is no way to tell what it will do. If you see a synthetic fiber line under strain parting or beginning to part--DO NOT RUN—just fall flat down on the deck.

TOWING ASTERN (OPEN SEA)

13-53. Deep-sea towing places many more requirements on the deck crew members of the tug. Now we are talking of much heavier towing gear, a variety of equipment, and the added requirements for assembling the towing gear.

Note: The tug master is responsible for determining the size and length of the towing gear to be used. The mate is responsible for obtaining this equipment. The boatswain and crew must be familiar with it and know how to assemble and rig for the different types of stern tows.

INSPECTION OF TOWING EQUIPMENT

13-54. Before assembling the towing rig, the mate and boatswain will inspect each item of equipment. If there is any doubt of its serviceability, REPLACE IT! If there is a question of size, for safety's sake, go to the next larger size. Remember, if any one of these items fails you at sea, you stand the chance of losing the tow and even the life of a crew member.

SCOPE OF HAWSER

13-55. When underway, the tug and tow should be "in step"; that is, meeting and riding over the crests of waves at the same time (see Figure 13-10). Otherwise, the towline is alternately slack and taut, causing heavier than normal stresses. You can easily adjust the scope or length of the hawser when you have a towing machine. However, if you do not have the machine, it is almost impossible to make an adjustment. If you have the tow's anchor chain shackled to the hawser, you can let out or heave in the chain and hawser with the windlass. The scope of a hawser should be long enough to provide a good catenary, but not to the extent of having the towline drag on the bottom if in shallow water. A catenary absorbs shocks. The scope of the hawser should be no less than 200 fathoms to provide a good shock-absorbing catenary when towing a large vessel. You should not put stress on a towline to the extent of lifting it out of the water, but you can increase the catenary by reducing the tug's speed.

HAWSER WATCH

13-56. A hawser watch must be posted on the after deck to keep tow and gear under constant observation. Instruct the crew member, on watch, to immediately report the following:

- Too much tension is on the towline.
- The tow is not weathering properly.
- The bridles or other gear fail.

PREVENTIVE MAINTENANCE

13-57. The boatswain is responsible for maintaining the towing cable or line, bridles, and other gear efficiently at a minimum expense. Begin by having the cable or line properly wound and stowed. Then as rusted surfaces appear, have them scraped with wire brushes and greased with a high pressure wire rope grease machine. Using chafing materials at points where the cable contacts the tug, tow, and bitts reduces wear and tear. In addition to chafing gear, continued monitoring of the towline's condition is necessary and important. Stern rollers and other fairleads must be properly lubricated and all possible points of line wear protected. Canvas, hose, line, wood, split steel pipe or channel iron, Kevlar sleeves, or other materials should be used for chafing gear as required. Chafing must be eliminated or reduced on board the tow and the tug as much as possible. Freshening the nip is the process of lengthening or shortening the tow wire. It should be done at least every day in moderate weather and more often during heavy seas to prevent the towline from chafing where it comes in to contact with the stern of the tug.

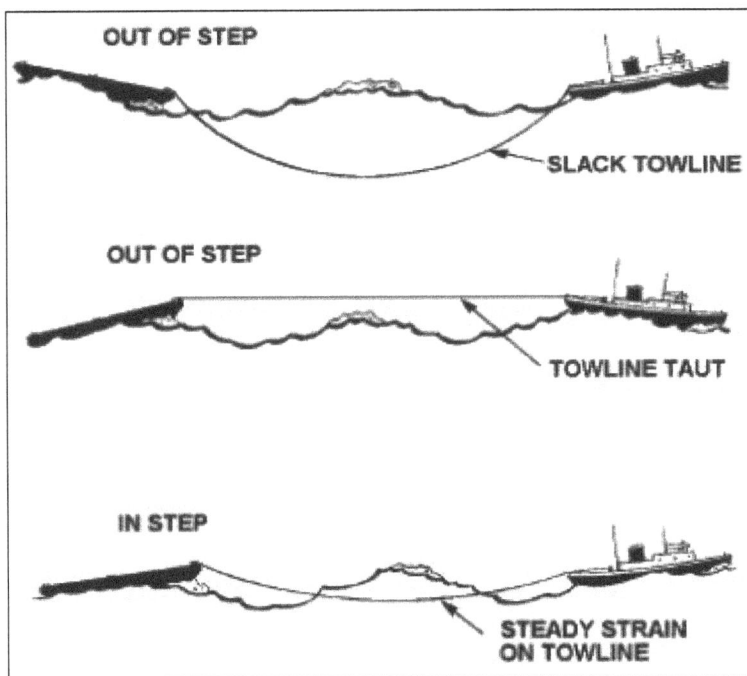

Figure 13-10. Keeping a Tow in Step

TOWING IN TANDEM

13-58. When towing more than one barge astern, it is referred to as tandem towing. In a pure sense, tandem means one behind the other. Within the tandem rig are three other methods called the Honolulu rig, Christmas Tree rig, and Modified Christmas Tree rig.

TANDEM RIG

13-59. In this method, the tug is connected to the first tow. The first tow connects to the second, and so on if additional units are towed (see Figure 13-11). The intermediate hawser, connecting the first tow to the second, must be streamed and allowed a proper catenary depth. The surging action must be eliminated between tug and first tow and between first tow and second tow.

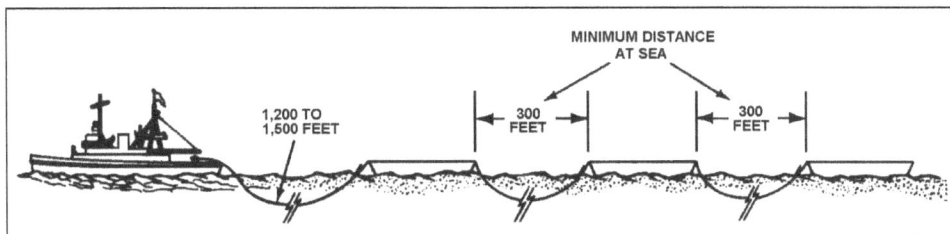

Figure 13-11. Tandem Rig

HONOLULU RIG

13-60. In this method, the first tow is connected to the main tow wire. The second tow is connected, with an auxiliary tow wire, to the bitts on deck (see Figure 13-12). The Honolulu rig allows independent connection of the two tows. Disconnecting and control are readily workable.

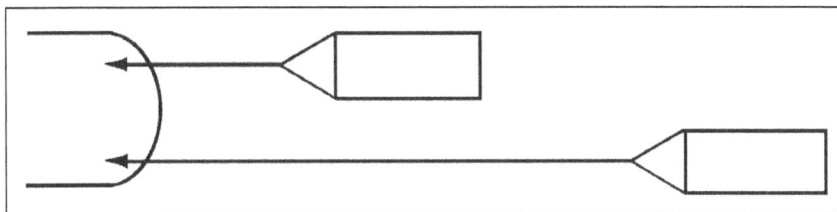

Figure 13-12. Honolulu Rig

CHRISTMAS TREE RIG

13-61. In this method, all of the barges tow from a single towing hawser (see Figure 13-13). This is done by the means of pendants or under riders shackled to flounder plates (sometimes called fish plates) inserted in the towing hawser.

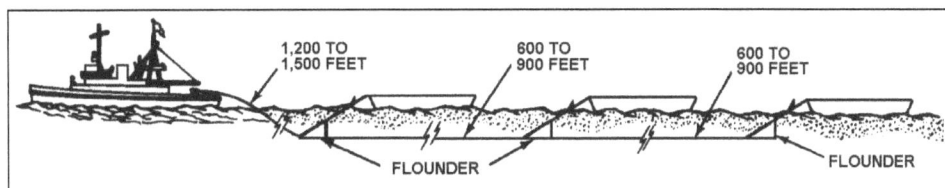

Figure 13-13. Christmas Tree Rig

MODIFIED CHRISTMAS TREE RIG

13-62. In the modified Christmas Tree rig, all of the tows are towed from a common flounder, but the last barge will tow as a separate unit (Figure 13-14).

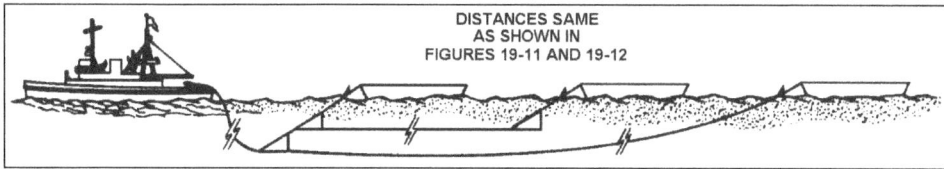

Figure 13-14. Modified Christmas Tree Rig

Note: Christmas Tree rigs are preferred for multiple tows. They are stronger and any one unit can be taken from the tow at anytime without disrupting the whole tow. The assistance of another tug is usually required to break up the Christmas Tree rig before entering port.

Chapter 14

Sea-Based Crew Gunnery Training

INTRODUCTION

14-1.   An operating environment characterized by persistent change mandates the use of realistic, mission-oriented training. Watercraft units must be prepared to operate in diverse environments and defend against a wide range of threats. Sea-based gunnery training helps prepare Soldiers and leaders for these challenges by providing an opportunity to improve task proficiency.

14-2.   Commanders and subordinate leaders need to plan training that reinforces performance standards for Soldiers and their respective units. Leaders should eliminate training distracters using a top down/bottom up approach to training. Sea-based training requires event discipline and incorporates realism into the training environment. Training with crew-served weapons demands that Soldiers effectively communicate and perform collective tasks.

14-3.   This chapter provides a framework to assist commanders in conducting sea-based crew gunnery training on Army watercraft. Guidelines are provided for the development of a sea-based gunnery training program that is designed to produce qualified gunners by training critical skills and facilitating teamwork, at the crew level, at sea. The chapter is split into three sections: fire commands, individual gunnery qualification, and collective gunnery training.

FIRE COMMANDS

14-4.   This section discusses the elements, terms, and types of fire commands. Standard fire command terminology and procedure helps mitigate risk by ensuring all gun crew members speak the same language and understand their respective roles and responsibilities. Fire commands are given to coordinate crew efforts and deliver effective fire on a target. Fire commands reduce confusion and deliver all required intelligence necessary to engage the target. Difference between effective communications can mean the difference between effective fire suppression and catastrophic loss of the vessel. A miscommunication in a shoot-no shoot situation can result in mission failure or the unauthorized engagement of a non-combatant target. Multiple firing positions and multiple targets increase the risk of an unauthorized engagement of a non-combatant target. Use of Sound Powered Phones or VHF radios reliability clearness require all elements in the shoot-no shoot situation be using the same commands for effective weapon engagement with little or no error in miscommunication. Fire commands are designed to increase the ability of the gun crew to instantly react to certain pre-determined situations. The vessel master issues commands that alert gun crew to react to a given situation in accordance with the rules of engagement (ROE) and escalation of force (EOF) restrictions.

ELEMENTS OF A FIRE COMMAND

14-5.   Table 14-1 lists the seven elements of a fire command. They are: alert, weapon/ammunition, description, direction, range, execution and termination. These elements provide the gun crew with the essential intelligence to engage the target.

Note: All elements of the fire command, except for the command of execution, may be given by the assistant gunner or gunner. The command of execution is given only by the vessel master.

Table 14.1 Elements of a Standard Fire Command

Element	Example	Remarks
Alert	"GUNNER"	Alerts the entire crew that someone in the crew will be firing an engagement using the mounted weapon.
Weapon	"CALIBER FIFTY"	Identifies the weapon to fire based on the threat.
Description	"SMALL CRAFT"	A clear and concise target description for the firer to identify.
Direction	"2 POINTS ON PORT BOW"	This is required to assist the gun crew in locating the target.
Range or Elevation	"600 METERS"	This is required to assist the gun crew in locating the target.
Execution	"FIRE"	The vessel master is the only crew member authorized to issue the command of execution. This cannot be delegated to the gunner.
Termination	"CEASE FIRE"	Any member of the crew can terminate an engagement

ALERT

14-6.   The alert element of the fire command alerts the gun crew to an impending engagement. It consists of the term: "GUNNER." This element is omitted if the gunner initiates the engagement.

WEAPON/AMMUNITION

14-7.   The weapon/ammunition element of the fire command tells the gun crew the type of weapon and ammunition to use for the engagement. Weapon/ammunition terms are Crew-served weapon (CSW) and Individual weapon.

DESCRIPTION

14-8.   The description element of the fire command tells the gunner what type of target to identify and engage. If there are multiple targets, the commander designates which target is to be engaged first. Example: "TWO SMALL CRAFT— LEAD CRAFT."

DIRECTION

14-9.   The direction element of the fire command tells the gun crew the general direction to the target. This element may be given using one or a combination of the following methods.

CLOCK METHOD

14-10.   The clock method is based on which way the vessel is facing, the bow of the vessel being twelve o'clock. Example: "ONE O'CLOCK."

POINT SYSTEM

14-11. The point uses the bow of the vessel as a starting point and names the target and gives the direction. Example: "SMALL CRAFT, 2 POINTS ON THE PORT BOW."

RANGE

14-12. The range element of the fire command tells the gun crew the range to the target. Gun crews should use this element if the gunner cannot determine the range to the target. Table 14-2 lists examples of range terms.

Table 14.2 Range Fire Commands

RANGE	FIRE COMMAND
1,000 meters.	"ONE THOUSAND."
1,200 meters.	"ONE TWO HUNDRED" OR "TWELVE HUNDRED."

EXECUTION

14-13. The execution element of the fire command gives the final authorization to engage the target. Only the vessel master can give the execution command. Examples: "FIRE" and "AT MY COMMAND."

TERMINATION OF ENGAGEMENT

14-14. Every engagement must be terminated. The vessel master announces "CEASE FIRE" to end an engagement. The vessel master has overall responsibility of the vessel and is responsible for terminating the engagement, but any member of the crew can give this command.

FIRE COMMAND TERMS

RESPONSE TERMS

14-15. The gunner can respond to the fire command using the word **IDENTIFIED** to confirm he/she has located the target(s). Saying IDENTIFIED by itself tells the vessel master that the gunner has confirmed the target as stated in the description. If appropriate, the gunner adds, FRIENDLY, NEUTRAL, or DOUBTFUL. The gunner can also say **CANNOT IDENTIFY.** The gunner uses this term to inform the vessel master that he/she cannot find the target. If the gunner cannot locate the target he/she says **CANNOT ENGAGE.** The gunner uses this term to inform the vessel master that he/she can identify the target, but for some reason, cannot engage it. If the gunner can engage the target he/she says **ON THE WAY or FIRING.** The gunner uses this term to inform all gun crew members that a weapon is being fired. Once the target is suppressed or destroyed the gunner says **TARGET DESTROYED or SUPPRESSED.** The gunner uses this term to inform the vessel master he/she has destroyed the target. The vessel master confirms destruction of the target.

REPEAT

14-16. If a gun crew member does not hear or understand an element of a fire command, he/she repeats the element in question. Example: Gunner: "RANGE." Watch: "FOUR HUNDRED."

CORRECTION

14-17. If the Watch makes a mistake, he/she announces "CORRECTION," corrects the mistaken element, then repeats all elements after the corrected element. Example: "GUNNER — SMALL CRAFT — ONE O'CLOCK — CORRECTION — SMALL CRAFT — THREE O'CLOCK."

TERMINATION

14-18. To terminate an engagement, the vessel master commands "CEASE FIRE." This stops the gunner from firing.

TYPES OF FIRE COMMANDS

INITIAL FIRE COMMANDS

14-19. Most engagements initiated by the vessel master begin with the initial fire command. When the vessel master decides to engage a target that is not obvious to the gunner, the vessel master must provide the gunner with the intelligence needed to engage the target effectively.

14-20. The vessel master must alert the gun crew and give the target description, direction, and execution commands (see Table 14-3). He/She should add the range element if he/she feels it is necessary to achieve a first round hit and if time permits for accurate range estimation. (The range element is announced unless a range finder is used during an engagement.)

Table 14.3 Sample CSW Commands

Element	Commander	Gunner
Alert	"GUNNER."	
Description	"SMALL CRAFT"	
Range	"200 METERS"	
Direction	"1 POINT ON PORT BOW."	
		"IDENTIFIED."
Execution	"FIRE."	
		"FIRING or ON THE WAY."
	"CEASE FIRE."	

REDUCED FIRE COMMAND

14-21. If the gunner identifies a threat, they can initiate a reduced fire command by giving an acquisition report, consisting of the target description and direction (see Table 14-4). Once the gunner gives the acquisition report, the vessel master must confirm the target and give the execution command before the gunner can engage the target.

Table 14.4 Reduced Fire Commands

Element	Commander	Gunner
Alert	(omitted)	
Weapon/Ammunition	(omitted)	
Description		"SMALL CRAFT."
Range		"200 METERS"
Direction		"ONE O'CLOCK."
Execution	"FIRE."	
		"FIRING or ON THE WAY."
	"CEASE FIRE."	

SUBSEQUENT FIRE COMMANDS

14-22. A subsequent fire command is used to make adjustments in direction and elevation, or change the rate of fire after an engagement is in progress. It can be given by the vessel master or gunner and includes the alert, correction, and execution commands. The alert portion of a subsequent fire command is nothing more than the vessel master or gunner announcing his/her sensing (the strike of the round in relation to the target). Corrections in direction are made first, then corrections in range. If the gunner can sense the strike of the round and make his own correction, this command may be unnecessary.

TRAINING AIDS, DEVICES, SIMULATORS AND SIMULATIONS (TADSS)

14-23. Prior to live fire training exercises, gunnery training should incorporate the use of TADSS (especially simulations and simulators) for the training of gunners, Vessel Crew and unit leaders. The Vessel Defense Simulator and Engagement Skills Trainer (EST) 2000 are discussed in the following paragraphs.

VESSEL DEFENSE SIMULATOR (VDS)

14-24. The Vessel Defense Simulator (VDS) system is intended to meet the joint training needs for crews operating and defending a vessel from armed threat at close range. The VDS has been specifically designed to provide a highly effective training and learning environment for defensive tactics, techniques, and procedures employing weapons in defense of vessels that are fitted with crew served and small arms. Training scenarios can be created for a wide range of threat situations, while in port or underway, which train crews to respond to real-time active hostile actions while operating in their normal mission activities.

14-25. The intended training purpose of the VDS is to enable total team training for vessel crews to include the command and leadership roles of officers and crew. The premise of the VDS is that it is not sufficient to simply have qualified gunners, it is essential that the entire team is trained and practiced in skills and knowledge for legal and effective management of weapons, personnel, ammunition, communications within the context of governing law and regulatory authority.

14-26. This training device consists of a vessel weapons simulator that replicates all of the key features and functions necessary to train essential skills required of a defensive tactical situation. Training is conducted, through assessment of hostile threat to neutralizing the threat and resuming the conduct of the vessel mission. This system description is organized to present the components of the simulator by functional group giving a description of the performance capability and equipment where appropriate for the VDS simulator.

14-27. The VDS can integrate with other VirtualShip® devices in the training facility to create a multiple ship training environment. The integrated training mode provides the means to integrate the VDS with co-located trainers. Interior ship sound-power phones are provided so that crew members may communicate between ship stations as they would aboard a vessel. All of the simulators that are integrated with VDS exchange network data to ensure that appropriate visual, motion, and aural cues are shared between all devices that are integrated.

ENGAGEMENT SKILLS TRAINER (EST) 2000

14-28. The EST 2000 replicates eleven weapons, including the rifle, carbine, pistol, grenade launcher, all machine guns, MK19, shotgun, and AT4. It has three modes of training: Marksmanship, Tactical collective, and Shoot/Don't shoot rules of engagement training. The EST 2000 supports realistic and comprehensive gated rifle marksmanship instruction, identifies Soldiers' needs by requiring them to satisfy gate requirements in order to progress, and facilitates any necessary remedial training prior to qualification. This unit/institutional, indoor, multipurpose, multilane, small arms, crew-served, and individual antitank training simulator is used to train and evaluate individual marksmanship training for initial entry Soldiers (BCT/OSUT). The EST 2000 provides unit sustainment training for active and reserve components in preparation for qualification on individual and crew served small arms live-fire weapons. The system provides unit collective tactical training for static dismounted infantry, scout, engineer, military police squads, and sustainment elements. EST 2000 simulates training events which lead to live-fire individual and crew weapon qualification which increases weapon, crew, fire team, and squad combat effectiveness. EST 2000 simulates squad collective, defensive, ambush, gunnery, and tactical tasks to train leaders of fire teams and squads in the command, control, and distribution of fires. EST 2000 will save ammunition, travel time, transportation costs, and other range support resources required to support functional gunnery training strategies and standards in weapon training.

SEA-BASED INDIVIDUAL GUNNERY QUALIFICATION

14-29. This section provides guidelines for sea-based qualification with Crew Served Weapons. Sea-based firing teaches the gunner techniques of fire that he will use in combat situations. During this training, the gunner is required to apply the fundamentals of gunnery learned in preparatory gunnery training and 10-meter firing. *Prior to conducting full-caliber live-fire sea-based gunnery training*, gunners should complete training in a simulation or simulator as required by the unit's weapons training program, and gunners must qualify on the crew-served weapon for their assigned station within the previous six months.

SEA-BASED QUALIFICATION

14-30. Sea-based qualification consists of three firing tables:

- Table 1-Vessel is "Dead in Water" to simulate at anchor or moored to a pier. Target is underway.
- Table 2-Vessel and target are underway.
- Table 3-Vessel and target are underway with increased range and speed of the target.

ENGAGEMENTS

14-31. There are three engagements per table:

- A bow target profile gives the gunner a picture as if the target is heading directly towards the vessel.
- A beam target profile gives the gunner a picture as if the target is running parallel to the vessel.
- A flank target profile gives the gunner a picture as if the target is approaching from an angle.

14-32. The gunner can fire up to five bursts (five to seven rounds per burst) at each engagement in Tables 1 and 2 and ten bursts (five to seven rounds per burst) at each engagement in Table 3. Instructors should encourage gunners to perform immediate action if a stoppage occurs during fire.

TABLE I	Vessel "Dead in Water" to simulate at anchor or moored to a pier. Target is underway.						
NGAGEMENT	STANDARD	AMMO	TARGET PROFILE	RANGE	TARGET SPEED	GO	NO GO
1	1 BURST HIT	30	BOW	1000-500	ANY		
2	1 BURST HIT	30	BEAM	500	5-15 knots (KTS)		
3	1 BURST HIT	30	FLANK	1000-500	5-15 KTS		

Table 1, Engagement 1

14-33. **Vessel "Dead in Water" to simulate at anchor or moored to a pier. Target is underway showing a bow profile.** Once the gunner and assistant gunner are in position, the vessel master or mate instructs the team to prepare to fire by the command ACTION.

- At the command ACTION, the machine gun crew places the machine gun into action.
- The following fire command is given. The gunner and assistant gunner repeat each element of the fire command as it is given. Below is a sample:
  - FIRE MISSION (The gunner loads.)
  - 2 POINTS ON PORT BOW (The gunner focuses on the target or target area.)
  - TARGET ONE THOUSAND (The gunner adjusts sights and acquires the sight picture.)

- COMMENCE FIRING (The gunner fires when ready after this command from the vessel master.)
- The assistant gunner loads the 30-round belt of ammunition and the gunner fires a burst of five to seven rounds at the 1000-meter, bow profile target.
- If the gunner misses:
    - The gunner uses the tracer to sense the location of the beaten zone and adjusts his point of aim.
    - The gunner fires another five- to seven-round burst.

Note: Engagement 1 ends at the 500 meter range and the score is recorded.

Table 1, Engagement 2

14-34. **Vessel "Dead in Water" to simulate at anchor or moored to a pier. Target is underway at 5 to 15 KTS showing a beam profile.** Once the gunner and assistant gunner are in position, the vessel master or mate instructs the team to prepare to fire by the command ACTION.

- At the command ACTION, the machine gun crew places the machine gun into action.
- The following fire command is given. The gunner and assistant gunner repeat each element of the fire command as it is given. Below is a sample:
    - FIRE MISSION (The gunner loads.)
    - ON PORT BEAM (The gunner focuses on the target or target area.)
    - TARGET FIVE HUNDRED (The gunner adjusts sights and acquires the sight picture.)
    - COMMENCE FIRING (The gunner fires when ready after this command from the vessel master.)
- The assistant gunner loads the 30-round belt of ammunition and the gunner fires a burst of five to seven rounds at the 500-meter, beam profile target.
- If the gunner misses the target:
    - The gunner uses the tracer to senses the location of the beaten zone and adjusts his point of aim.
    - The gunner fires another five- to seven-round burst as required.

Note: Engagement 2 ends when the rounds are expended and then the score is recorded.

Table 1, Engagement 3

14-35. **Vessel "Dead in Water" to simulate at anchor or moored to a pier. Target is underway at 5 to 15 KTS showing a flank profile.** Once the gunner and assistant gunner are in position, the vessel master or mate instructs the team to prepare to fire by the command ACTION.

- At the command ACTION, the machine gun crew places the machine gun into action.
- The following fire command is given. The gunner and assistant gunner repeat each element of the fire command as it is given. Below is a sample:
    - FIRE MISSION (The gunner loads.)
    - 1 POINT ABAFT PORT BEAM (The gunner focuses on the target or target area.)
    - TARGET ONE THOUSAND (The gunner adjusts sights and acquires the sight picture.)
    - COMMENCE FIRING (The gunner fires when ready after this command from the Vessel master.)
- The assistant gunner loads the 30-round belt of ammunition and the gunner fires a burst of five to seven rounds at the 1000-meter, flank profile target.
- If the gunner misses the target:
    - The gunner uses the tracer to senses the location of the beaten zone and adjusts his point of aim.

    ■   The gunner fires another five- to seven-round burst as required.

Note: Engagement 3 ends at the 500 meter range and the score is recorded.

TABLE II	Vessel is underway and target is underway.						
ENGAGEMENT	STANDARD	AMMO	TARGET PROFILE	RANGE	TARGET SPEED	GO	NO GO
1	1 BURST HIT	30	BOW	1000-500	ANY		
2	1 BURST HIT	30	BEAM	500	5 KTS		
3	1 BURST HIT	30	FLANK	1000-500	5 KTS		

Table 2, Engagement 1

14-36. **Vessel is underway and the target is underway showing a bow profile .** Once the gunner and assistant gunner are in position, the vessel master or mate instructs the team to prepare to fire by the command ACTION.

- At the command ACTION, the machine gun crew places the machine gun into action.
- The following fire command is given. The gunner and assistant gunner repeat each element of the fire command as it is given. Below is a sample:
  - ■ FIRE MISSION (The gunner loads.)
  - ■ 2 POINTS ON PORT BOW (The gunner focuses on the target or target area.)
  - ■ TARGET ONE THOUSAND (The gunner adjusts sights and acquires the sight picture.)
  - ■ COMMENCE FIRING (The gunner fires when ready after this command from the vessel master.)
- The assistant gunner loads the 30-round belt of ammunition and the gunner fires a burst of five to seven rounds at the 1000-meter, bow profile target.
- If the gunner misses:
  - ■ The gunner uses the tracer to senses the location of the beaten zone and adjusts his point of aim.
  - ■ The gunner fires another five- to seven-round burst.

Note: Engagement 1 ends at the 500 meter range and the score is recorded.

Table 2, Engagement 2

14-37. **Vessel is underway and the target is underway showing a beam profile, speed not more than 5 KTS.** Once the gunner and assistant gunner are in position, the vessel master or mate instructs the team to prepare to fire by the command ACTION.

- At the command ACTION, the machine gun crew places the machine gun into action.
- The following fire command is given. The gunner and assistant gunner repeat each element of the fire command as it is given. Below is a sample:
  - ■ FIRE MISSION (The gunner loads.)
  - ■ ON PORT BEAM (The gunner focuses on the target or target area.)
  - ■ TARGET FIVE HUNDRED (The gunner adjusts sights and acquires the sight picture.)
  - ■ COMMENCE FIRING (The gunner fires when ready after this command from the vessel master.)
- The assistant gunner loads the 30-round belt of ammunition and the gunner fires a burst of five to seven rounds at the 500-meter, beam profile target.
- If the gunner misses the target:
  - ■ The gunner uses the tracer to senses the location of the beaten zone and adjusts his point of aim.

■   The gunner fires another five- to seven-round burst as required.

Note: Engagement 2 ends when the rounds are expended and then the score is recorded.

Table 2, Engagement 3

14-38. Vessel is underway and the target is underway showing a flank profile, speed not more than 5 KTS. Once the gunner and assistant gunner are in position, the vessel master or mate instructs the team to prepare to fire by the command ACTION.

- At the command ACTION, the machine gun crew places the machine gun into action.
- The following fire command is given. The gunner and assistant gunner repeat each element of the fire command as it is given. Below is a sample:
  - FIRE MISSION (The gunner loads.)
  - 1 POINT ABAFT PORT BEAM (The gunner focuses on the target or target area.)
  - TARGET ONE THOUSAND (The gunner adjusts sights and acquires the sight picture.)
  - COMMENCE FIRING (The gunner fires when ready after this command from the vessel master.)
- The assistant gunner loads the 30-round belt of ammunition and the gunner fires a burst of five to seven rounds at the 1000-meter, flank profile target.
- If the gunner misses the target:
  - The gunner uses the tracer to senses the location of the beaten zone and adjusts his point of aim.
  - The gunner fires another five- to seven-round burst as required.

Note: Engagement 3 ends at the 500 meter range and the score is recorded.

TABLE III	Vessel is underway and target is underway.							
ENGAGEMENT	STANDARD	AMMO	TARGET PROFILE	RANGE	TARGET SPEED	GO	NO G
1	1 BURST HIT	60	BOW	1500-1000	ANY		
2	1 BURST HIT	60	BEAM	1000	15 KTS		
3	1 BURST HIT	60	FLANK	1500-1000	15 KTS		

Table 3, Engagement 1

14-39. Vessel is underway and the target is underway showing a bow profile. Once the gunner and assistant gunner are in position, the vessel master or mate instructs the team to prepare to fire by the command ACTION.

- At the command ACTION, the machine gun crew places the machine gun into action.
- The following fire command is given. The gunner and assistant gunner repeat each element of the fire command as it is given. Below is a sample:
  - FIRE MISSION (The gunner loads.)
  - 2 POINTS ON PORT BOW (The gunner focuses on the target or target area.)
  - TARGET ONE THOUSAND (The gunner adjusts sights and acquires the sight picture.)
  - COMMENCE FIRING (The gunner fires when ready after this command from the vessel master.)
- The assistant gunner loads the 30-round belt of ammunition and the gunner fires a burst of five to seven rounds at the 1500-meter, bow profile target.
- If the gunner misses:
  - The gunner uses the tracer to senses the location of the beaten zone and adjusts his point of aim.

- The gunner fires another five- to seven-round burst.

Note: Engagement 1 ends at the 1000 meter range and the score is recorded.

Table 3, Engagement 2

14-40. **Vessel is underway and the target is underway showing a beam profile, speed 5- 15 KTS.** Once the gunner and assistant gunner are in position, the vessel master or mate instructs the team to prepare to fire by the command ACTION.

- At the command ACTION, the machine gun crew places the machine gun into action.
- The following fire command is given. The gunner and assistant gunner repeat each element of the fire command as it is given. Below is a sample:
  - FIRE MISSION (The gunner loads.)
  - ON PORT BEAM (The gunner focuses on the target or target area.)
  - TARGET FIVE HUNDRED (The gunner adjusts sights and acquires the sight picture.)
  - COMMENCE FIRING (The gunner fires when ready after this command from the vessel master.)
- The assistant gunner loads the 30-round belt of ammunition and the gunner fires a burst of five to seven rounds at the 1000-meter, beam profile target.
- If the gunner misses the target:
  - The gunner uses the tracer to senses the location of the beaten zone and adjusts his point of aim.
  - The gunner fires another five- to seven-round burst as required.

Note: Engagement 2 ends when the rounds are expended and then the score is recorded.

Table 3, Engagement 3

14-41. **Vessel is underway and the target is underway showing a flank profile, speed 5- 15 KTS.** Once the gunner and assistant gunner are in position, the vessel master or mate instructs the team to prepare to fire by the command ACTION.

- At the command ACTION, the machine gun crew places the machine gun into action.
- The following fire command is given. The gunner and assistant gunner repeat each element of the fire command as it is given. Below is a sample:
  - FIRE MISSION (The gunner loads.)
  - 1 POINT ABAFT PORT BEAM (The gunner focuses on the target or target area.)
  - TARGET ONE THOUSAND (The gunner adjusts sights and acquires the sight picture.)
  - COMMENCE FIRING (The gunner fires when ready after this command from the vessel master.)
- The assistant gunner loads the 30-round belt of ammunition and the gunner fires a burst of five to seven rounds at the 1500-meter, flank profile target.
- If the gunner misses the target:
  - The gunner uses the tracer to senses the location of the beaten zone and adjusts his point of aim.
  - The gunner fires another five- to seven-round burst as required.

Note: Engagement 3 ends at the 1000 meter range and the score is recorded.

STOPPAGE.

14-42. If a stoppage occurs, the gunner must apply immediate action. If the stoppage is reduced, he continues to fire the course.

- If a stoppage occurs that cannot be reduced by immediate action, the gunner raises his hand and awaits assistance.
- Once the stoppage is reduced, the gunner completes firing beginning with the next task.
- If a stoppage is caused by an error on the part of the gunner, additional time is not permitted. The gunner receives the score he earned before the stoppage occurred.
- If it is necessary to replace the machine gun, the gunner must zero the new weapon. The gunner can fire the exercise again.
- Gunners who cannot fire a task or cannot complete firing in the time allowed (because of malfunctions) can finish the exercise in an alibi run after all other gunners' complete firing. They fire only those tasks they failed to engage because of the malfunction.

SEA-BASED QUALIFICATION SCORECARD

LAST NAME	FIRST NAME	MI	RANK
HOLNICK	RAY	G	SGT

DATE (YYYY/MM/DD)	UNIT
2013 / 05 / 02	97TH TRANS CO.

TABLE I	Vessel "Dead in Water" to simulate at anchor or moored to a pier. Target is underway.						
ENGAGEMENT	STANDARD	AMMO	TARGET PROFILE	RANGE	TARGET SPEED	GO	NO GO
1	1 BURST HIT	30	BOW	1000-500	ANY	✓	
2	1 BURST HIT	30	BEAM	500	5-15 KTS	✓	
3	1 BURST HIT	30	FLANK	1000-500	5-15 KTS	✓	

TABLE II	Vessel is underway and target is underway.						
ENGAGEMENT	STANDARD	AMMO	TARGET PROFILE	RANGE	TARGET SPEED	GO	NO GO
1	1 BURST HIT	30	BOW	1000-500	ANY	✓	
2	1 BURST HIT	30	BEAM	500	5 KTS	✓	
3	1 BURST HIT	30	FLANK	1000-500	5 KTS	✓	

TABLE III	Vessel is underway and target is underway.						
ENGAGEMENT	STANDARD	AMMO	TARGET PROFILE	RANGE	TARGET SPEED	GO	NO GO
1	1 BURST HIT	60	BOW	1500-1000	ANY	✓	
2	1 BURST HIT	60	BEAM	1000	15 KTS	✓	
3	1 BURST HIT	60	FLANK	1500-1000	15 KTS	✓	

GUNNERS SIGNATURE

Ray Holnick

GRADER'S PRINTED OR TYPED NAME	GRADER'S SIGNATURE
SFC GERARD TUCKER	*Gerard Tucker*

Format: Sample of Sea-Based Qualification Scorecard

SEA-BASED COLLECTIVE GUNNERY TRAINING

14-43. Collective training allows units to practice higher-level tasks in training environments that mimic the joint operating environment. This section outlines a training strategy for collective gunnery that may be used as part of a Live-Fire Exercise (LFX) or multi-echelon training event.

14-44. This Collective Training strategy applies to all weapons available on Army watercraft. In this document, mounted crew-served weapons include the M2, MK19, and M249. If additional weapon systems are introduced onto vessels, then they may also be incorporated into collective training. The commander has the flexibility to select which weapon system(s) to use for gunnery training based on mission requirements, availability of equipment, and other considerations.

14-45. By integrating support elements, aviation and indirect-fire support with training, commanders can improve the tactical value of gunnery training. Commanders should emphasize mission-specific Rules of Engagement (ROE) as necessary. Training should be conducted in a variety of weather and sea state conditions as determined by the unit commander or vessel master. Unit leadership should conduct risk assessments prior to all levels of training. Safety is paramount, particularly when conducting live fire exercises.

CONCEPT

14-46. Collective gunnery training measures a unit's proficiency in executing collective tasks. Leaders are trained and evaluated on the following skills:

VESSEL MASTER:
- Distribution and control of fires.
- Conducting collective tasks, both pure and combined arms, in accordance with the applicable Combined Arms Training Strategies (CATS).

SQUAD/SECTION LEADERS:
- Distribution and control of fires for fire teams within their sections.
- Conducting squad/section-level collective tasks in accordance with the applicable CATS.

PRINCIPLES OF FIRE CONTROL

14-47. Effective fire control requires a unit to rapidly acquire the enemy and mass the effects of fires to achieve decisive results in the close fight. When planning and executing direct fires, the commander and subordinate leaders must know how to apply several fundamental principles. The purpose of these principles of direct fire is not to restrict the actions of subordinates. Applied correctly, these principles help the team accomplish its primary goal in any direct fire engagement–to both acquire first and shoot first giving subordinates the freedom to act quickly upon acquisition of the enemy. This discussion focuses on the following principles:
- Mass the effects of fire.
- Destroy the greatest threat first.
- Avoid target overkill.
- Employ the best weapon for the target.

MASS THE EFFECTS OF FIRE

14-48. The team must mass its fires to achieve decisive results. Massing entails focusing fires at critical points and distributing the effects. Random application of fires is unlikely to have a decisive effect. For example, concentrating the team's fires at a single target may ensure its destruction or suppression.

ASSESS AND DESTROY THE MOST DANGEROUS THREAT FIRST

14-49. The order in which the team engages enemy forces is in direct relation to the danger they present. The threat posed by the enemy depends on the enemy's weapons, range, and positioning. Presented with multiple targets, a unit will, in almost all situations, initially concentrate fires to destroy the greatest threat, and then distribute fires over the remainder of the enemy force.

AVOID TARGET OVERKILL

14-50. Use only the amount of fire required to achieve necessary effects. Target overkill wastes ammunition and ties up weapons that are better employed acquiring and engaging other targets. The idea of having every weapon engage a different target, however, must be tempered by the requirement to destroy the greatest threats first.

EMPLOY THE BEST WEAPON FOR THE TARGET

14-51. Using the appropriate weapon for the target increases the probability of rapid enemy destruction or suppression and saves ammunition. The team has many weapons with which to engage the enemy. Target type, range, exposure, weapons and ammunition availability, and desired target effects are key factors in determining the weapon and ammunition that should be employed. Additionally, leaders should consider individual crew capabilities when deciding on the employment of weapons. The commander organizes and arrays his forces based on the enemy, and desired effects of fires.

FIRE CONTROL MEASURES

14-52. Fire control measures are the means by which the commander or subordinate leaders control fires. Application of these concepts, procedures, and techniques assist the unit in acquiring the enemy, focusing fires on the enemy, distributing the effects of the fires, and preventing fratricide. At the same time, no single measure is sufficient to effectively control fires. At the company team level, fire control measures will be effective only if the entire unit has a common understanding of what they mean and how to employ them. The following discussion focuses on the various fire control measures employed by the company team.

SECTOR OF FIRE

14-53. A sector of fire is a defined area that must be covered by direct fire. Leaders assign sectors of fire to subordinate elements, crew-served weapons, and individual Soldiers to ensure coverage of an area of responsibility. In assigning sectors of fire, commanders and subordinate leaders consider the number and types of weapons available.

DIRECTION OF FIRE

14-54. A direction of fire is an orientation or point used to assign responsibility for a particular area on the battlefield that must be covered by direct fire. Leaders designate directions of fire for the purpose of acquisition or engagement by subordinate elements, crew-served weapons, or individual Soldiers.

ENGAGEMENT CRITERIA

14-55. Engagement criteria are a specific set of conditions that dictate initiation of fires. Engagement criteria, specify the circumstances in which subordinate elements are to engage. The circumstances can be based on a friendly or enemy event.

RULES OF ENGAGEMENT

14-56. ROEs specify the circumstances and limitations under which forces may engage; they include definitions of combatant and noncombatant elements and prescribe the treatment of noncombatants. Factors influencing ROE are national command policy, the mission and commander's intent, the operational environment, and the law of war. ROE always recognize a Soldier's right of self-defense; at the same time, they clearly define circumstances in which he/she may fire.

ENGAGEMENT TECHNIQUES

14-57. Engagement techniques are effects-oriented fire distribution measures.

- **Point Fire** - Point fire entails concentrating the effects of a unit's fire against a specific, identified target. When leaders direct point fire, all of the unit's weapons engage the target, firing until it is destroyed or the required time of suppression has expired. Employing converging fires from dispersed positions makes point fire more effective because the target is engaged from multiple directions. The unit may initiate an engagement using point fire against the most dangerous threat, and then revert to area fire against other less threatening point targets.
- **Area Fire** - Area fire involves distributing the effects of a unit's fire over an area in which enemy positions are numerous or are not obvious.
- **Simultaneous Fire** - Units employ simultaneous fire, to rapidly mass the effects of their fires or to gain fire superiority.
- **Alternating Fire** - In alternating fire, pairs of elements continuously engage the same point or area target one at a time.
- **Sequential Fire** - Sequential fire entails the subordinate elements of a unit engaging the same point or area target one after another in an arranged sequence. Sequential fire can also help to prevent the waste of ammunition, and sequential fire permits elements that have already fired to pass on information they have learned from the engagement.

COLLECTIVE GUNNERY

14-58. Collective Gunnery provides a useful framework for evaluating unit performance and leadership at the crew, vessel, and company level. DA PAM 350-38, Standards in Training Commission (STRAC), provides ammunition resourcing information for each weapon system to conduct collective training.

14-59. Watercraft units may be employed in unique formations across the battlefield, so commanders should plan training based on mission requirements and unit-specific characteristics. Vessels may operate in concert as a tactical convoy. Vessels may also be assigned to unique areas of the battlefield and operate individually. This training strategy allows the commander (or vessel master) to select the number and formation of vessels involved in training.

14-60. It is ultimately the commander's responsibility to determine the nature and frequency of the engagements. In order to provide Soldiers with complex, realistic training, a commander may decide to conduct multiple engagements at the same time. The commander may also choose which weapon systems to use during engagement, or he may delegate this responsibility accordingly.

14-61. The following exercises test the application of fire from two or more guns against sea based targets. They test collective individual skills, fire control, leader skills, adjustment of fire, methods of target engagement, and the control of two or more weapon systems. **Ammunition** - As authorized for Collective Training, per each weapon system table, in DA PAM 350-38 Chapter 9.

Note: Prior to 1 of the 3 exercises, the Vessel Master will engage the target with nonlethal M203 rounds to enable the M203 Grenadier to be properly certified on the employment techniques of BA39 Flash/Bang grenade with one reload as well as providing the Vessel Master training on the EOF decision making process.

EXERCISE 1

14-62. Vessel "Dead in Water" to simulate at anchor or moored.

- **ACTION** - Engage target with multiple crew served weapons.

- **CONDITIONS** - Vessel "Dead in Water" to simulate at anchor or moored and a remotely operated sea-based target underway at any speed during the hours of daylight or under degraded conditions.
- **STANDARDS** - Gun crews must cover 50 percent of the target with effective fire while performing the following:
  - The leader lays the guns on his target.
  - The leader issues the fire command for engaging the target.
  - Gunners engage.
  - Observers ensure that fires remain on the target and adjust appropriately.

EXERCISE 2

14-63. Vessel is underway and target is underway at 5 KTS.

- **ACTION** - Engage target with multiple crew served weapons.
- **CONDITIONS** - Vessel is underway and remotely operated sea-based target is underway at a speed of 5 KTS during the hours of daylight or under degraded conditions.
- **STANDARDS** - Gun crews must cover 50 percent of the target with effective fire while performing the following:
  - The leader lays the guns on his target.
  - The leader issues the fire command for engaging the target.
  - Gunners engage.
  - Observers ensure that fires remain on the target and adjust appropriately.

EXERCISE 3

14-64. Vessel is underway and target is underway at 15 KTS.

- **ACTION** - Engage target with multiple crew served weapons.
- **CONDITIONS** - Vessel is underway and remotely operated sea-based target is underway at a speed of 15 KTS during the hours of daylight or under degraded conditions.
- **STANDARDS** - Gun crews must cover 50 percent of the target with effective fire while performing the following:
  - The leader lays the guns on his target.
  - The leader issues the fire command for engaging the target.
  - Gunners engage.
  - Observers ensure that fires remain on the target and adjust appropriately.

EVALUATION

COMMANDER'S ASSESSMENT

14-65. The standards of execution should be derived from unit collective tasks and battle drills. Collective tasks provide a specific set of criteria for commanders to use when determining overall unit proficiency. The commander can use training and evaluation outlines (T&EOs) to track performance steps and measures. This data provides a basis for designating a task as either trained (T), needs practice (P), or untrained (U).

14-66. It may be difficult for a single person to evaluate sea-based maneuver and fires, so it is important to consider the physical location of the senior evaluator. The commander may also solicit reports from experienced subordinates in order to improve his assessment.

GUNNERY SCORES

14-67. Unit evaluation should also include a measurement of target accuracy. Range characteristics, such as the type of targets available, will affect this standard. In some cases, range equipment may provide a feedback mechanism for firing. In other cases, a commander may have to make an assessment of target destruction by simple observation. In general, gunnery scores should exceed fifty-percent target destruction. Commanders may opt to increase the gunnery standard as they see fit.

Chapter 15

Rigging (Seamanship)

INTRODUCTION

15-1.   This chapter describes the different types of blocks used in shipboard rigging and their nomenclature and maintenance. It also discusses the requirements for inspecting standing rigging. Also included are the formulas that are used to compute the safe working load and breaking strain for fiber, synthetic, and wire rope, hooks, shackles, and turnbuckles.

DESCRIPTION OF BLOCKS

15-2.   A block consists of one or more pulleys or sheaves fitted in a wood or metal frame. Each block has one or more straps of steel or rope that strengthen the block and, in most cases, support the sheave pin. By inserting a hook or shackle in the strap, the block itself may be suspended or a load applied to the block. If the block has a becket to which the fall is spliced, the becket is also secured to the strap. A block with a rope led over the sheave is convenient in applying power by changing the direction of the pull. Used in conjunction with rope and another block, it becomes a tackle and increases the power applied on the hauling part (described later in this chapter).

DETERMINE THE SIZE OF BLOCK TO USE WITH FIBER LINE

15-3.   The size of a block is found by measuring the length of the cheek of that block. The constant is 3 and the circumference of the line is the line size. The circumference of the line to be used will determine the size of the block needed. Blocks for fiber lines come in the following sizes: 4, 5, 6, 7, 8, 10, 12, and 14 inches.

Formula: line size (LS) x circumference (C) = size of block (SB)

Example

Determine the size of block to use for a 3 1/2-inch fiber line.

Line size to be used is 3 1/2 inches.
3 1/2 inches (LS) times a constant of 3
3 1/2 x 3 = 10 1/2 inches

15-4.   The closest sizes are 10-inch and 12-inch blocks. Go to the next larger size to select the 12-inch block. Blocks are designed for use with a certain size of rope. Therefore, they should never be used with rope of a larger size. Rope bent over a small sheave will be distorted, and any great strain applied will damage it and may even result in the rope wearing on the frame.

DETERMINE THE SIZE OF BLOCK TO USE WITH WIRE ROPE

15-5.   It is impossible to give an absolute minimum size for wire rope sheaves because of the factors involved. However, experience has shown that the diameter of a sheave should be at least 20 times the diameter of the wire rope. An exception to this is a 6 x 37 wire and other flexible wire for which smaller sheaves can be used because of their greater flexibility. The construction of the wire rope has a great deal to do with determining the minimum diameter of sheaves to be used (Figure 15-1). The stiffer the wire rope, the larger the sheave diameter required.

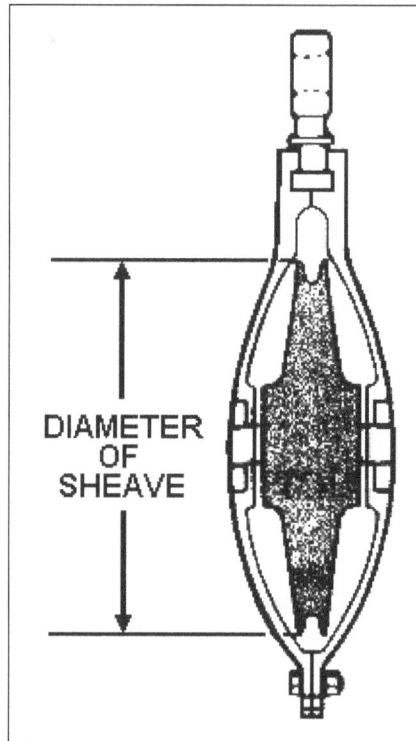

Figure 15-1. Wire Rope Block

COMMON CARGO BLOCK

15-6.   The three types of cargo blocks most frequently seen on ships are the diamond, oval, and roller bearing. Each of these blocks is described below.

- **Diamond Block.** A single-sheave diamond block is shown in Figure 15-2, but there may be many more, depending on the use of the block. Sheaves of this type of block are usually bushed with a high-grade bronze alloy, and the pins are equipped with grease fittings. Sheave bushings should be lubricated with hard graphite grease (such as Federal Specification VV-G-671, grade 0).

- **Oval Block.** Oval blocks are built to the same specifications as diamond blocks except that the cheeks are oval instead of diamond shaped. The most common use of these two blocks is for topping lifts of cargo booms.

- **Roller Bearing Block.** Head, heel, and many of the fairlead blocks are of the roller bearing type. These blocks have cast steel cheeks and sheaves. The sheaves are equipped with roller bearing assemblies. The pin is provided with a grease fitting. Roller bearing blocks are used where high-speed operation is essential. See Figure 15-2.

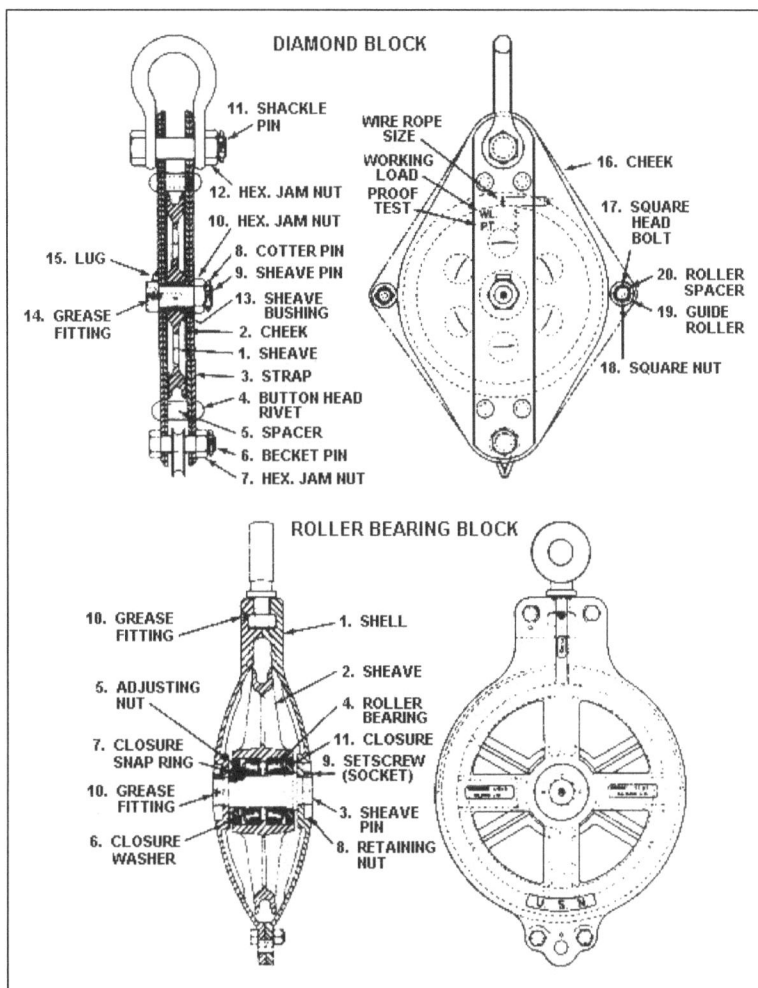

Figure 15-2. Common Cargo Blocks

NAMING A BLOCK

15-7.   Regardless of type, a cargo block is usually named for its location in the cargo rig. The block at the head of the boom through which the whip runs is called the head block. The block at the foot, which fairleads the wire to the winch, is the heel block. A small single-sheave block in the middle of most booms is called the slack wire block because it prevents slack in a whip from hanging down in a bight. Blocks in the topping lift are upper and lower topping lift blocks. A fairlead block, called a check block, is permanently fixed by welding or bolting one cheek to a bulkhead, davit, and so on. Another fairlead block is a snatch block, which is cut at the swallow (the hole the line reeves through), hinged on one side, and fitted with a hasp on the other. This permits the block to be opened and clamped on a line rather than reeving the end of the line through. Tail blocks are single blocks usually used alone with a whip or as a runner.

Note: When ordering a block, five things must be specified: wood or metal, size, type rig (with or without becket), and number of sheaves.

TYPES OF RIGS

15-8.   Blocks may be single, double, treble, and so on. That is, they may be fitted with one, two, three, or more sheaves, respectively. When used in a tackle, one of the blocks must be fitted with a becket to which one end of the line is spliced. When the hook, shackle, and swivel are fitted on the blocks they are called rigs. Figure 15-3 shows various types of rigs and fittings.

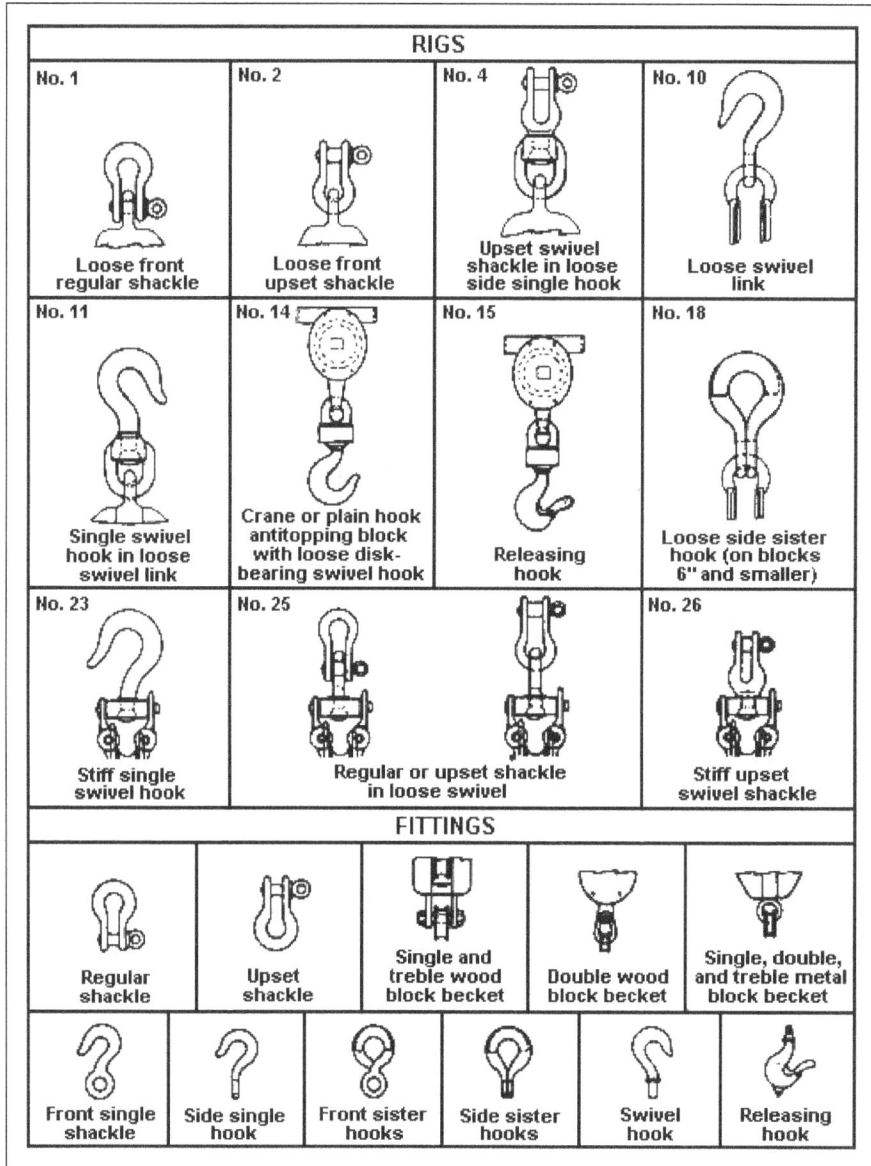

RIGS

No. 1	No. 2	No. 4	No. 10
Loose front regular shackle	Loose front upset shackle	Upset swivel shackle in loose side single hook	Loose swivel link
No. 11	**No. 14**	**No. 15**	**No. 18**
Single swivel hook in loose swivel link	Crane or plain hook antitopping block with loose disk-bearing swivel hook	Releasing hook	Loose side sister hook (on blocks 6" and smaller)
No. 23	**No. 25**		**No. 26**
Stiff single swivel hook	Regular or upset shackle in loose swivel		Stiff upset swivel shackle

FITTINGS

Regular shackle	Upset shackle	Single and treble wood block becket	Double wood block becket	Single, double, and treble metal block becket
Front single shackle	Side single hook	Front sister hooks	Side sister hooks	Swivel hook	Releasing hook

Figure 15-3. Various Rigs and Fittings

COMBINATIONS OF BLOCKS AND TACKLES

15-9.   Tackles are designated in two ways. One is the number of sheaves in the blocks that are used to make the tackle, such as single whip, gun tackle, or twofold purchase. The other designation is according to the purpose for which the tackle is used, such as yard tackles, stay tackles, or fore-and-aft tackles. Only the most commonly used combinations found aboard ship are shown in Figure 15-4 and described as follows.

- **Single Whip.** Consists of one single-sheave block fixed to a support with a rope passing over the sheave.
- **Runner.** Consists of a single block, but the block is free to move. One end of the rope is secured to the support with the weight attached to the block.
- **Gun Tackle.** Consists of two single blocks. It takes its name from the use made of it in hauling muzzle-loading guns back into battery after the guns are fired and reloaded.
- **Luff Tackle (Jigger).** Consists of a double and a single block.
- **Twofold Purchase.** Consists of two double blocks.

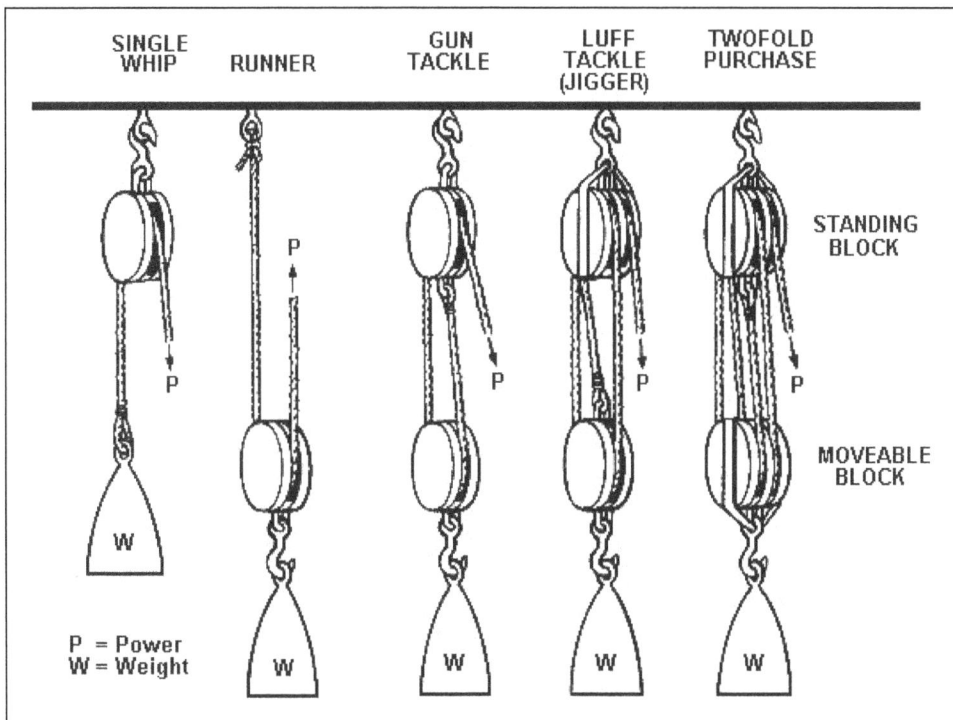

Figure 15-4. Blocks and Tackles

REEVING BLOCKS AND TACKLES

15-10. The preferred method of reeving multiple sheave blocks is referred to as the "right-angle method of reeving". With this method, one block (usually the head block) rests on the edge of its plates and cheeks, and the other block rests on its cheek. The sheaves are at right angles to each other (Figure 15-5). The advantages of using the right-angle method of reeving are that it reduces the chances of the rope chafing or of the blocks turning.

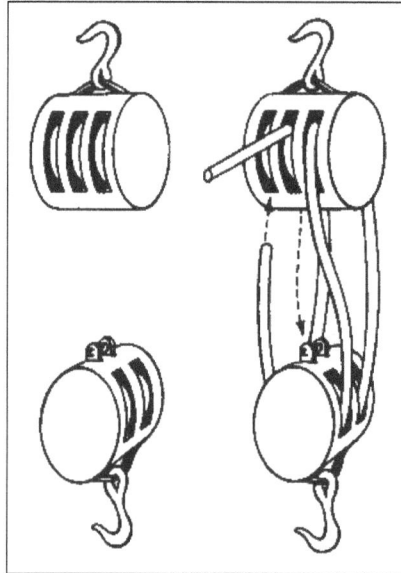

Figure 15-5. Blocks at Right Angles

- **Reeving a Double Luff Tackle.** A double luff tackle consists of a triple sheave and a double sheave block. The right-angle method of reeving is shown in Figure 15-6.

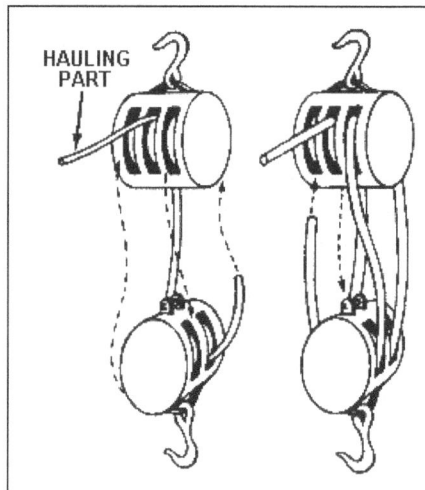

Figure 15-6. Reeving a Double

- **Reeving a Threefold Purchase.** The same method used to reeve a threefold purchase is used in reeving the double luff tackle (see Figure 15-7). After the line has been reeved through the last sheave, the final step is to make an eye-splice around the thimble and then bolt it into the becket.

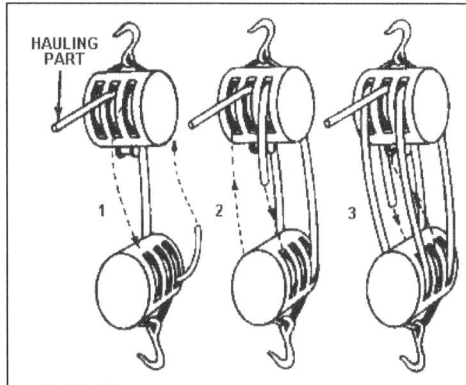

Figure 15-7. Reeving a Threefold Luff Tackle Purchase

DETERMINING THE MECHANICAL ADVANTAGES OF TACKLES

15-11. The mechanical advantage of a simple tackle is determined by counting the number of parts of the moving lines at the moveable block. The moveable block is the block that is attached to the weight to be moved (see Figure 15-8). Friction is not considered in the following example: If a load of 10 pounds requires 10 pounds to lift it, the mechanical advantage is 1. If a load of 40 pounds requires only 10 pounds of power to lift it, then the mechanical advantage is 4 to 1, or 4 units of weight lifted for each unit of power applied.

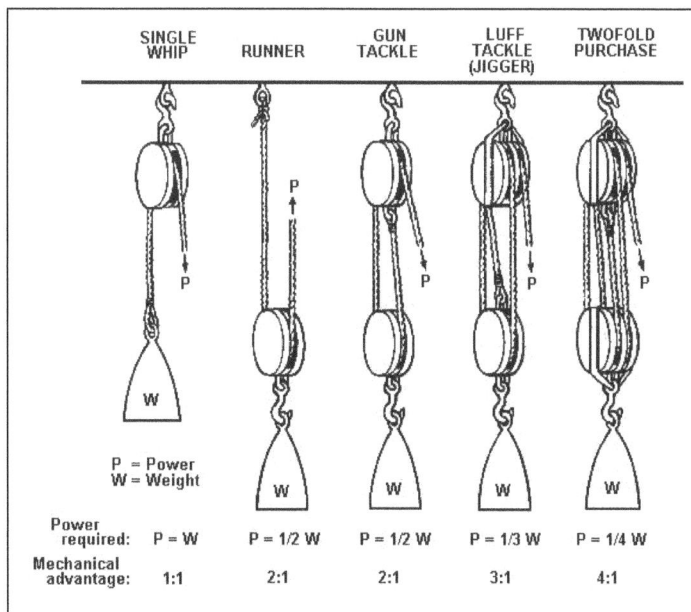

Figure 15-8. Mechanical Advantages of Tackles

COMPUTING FRICTION

15-12. A certain amount of the force applied to a tackle is lost through friction. Friction in a tackle is the rubbing of ropes against each other or against the frame or shell of a block, the passing of the ropes over the sheaves, and the rubbing of the pin against the sheaves. This loss in efficiency of the block and tackle must be added to the weight being lifted when determining the power required to lift a given load. Roughly 10 percent of the load must be added to the load for every sheave in the tackle. For example, what would be the loss of efficiency due to friction when picking up 500 pounds and using a twofold purchase?

Weight of load is 500 pounds.

10 percent of the weight of the load is 50 pounds.

With a twofold purchase, there are four sheaves.

4 (sheaves) x 50 pounds (10 percent of weight) = 200 pounds loss in efficiency due to friction.

COMPUTING BREAKING STRENGTH AND SAFE WORKING LOAD

15-13. When working with line, it is essential that you do not overload it because doing so is dangerous and costly. An overloaded line may part and injure someone in the vicinity. Even if it does not part, its useful life is shortened every time it is overloaded. For these reasons, you need to know a line's breaking strength and safe working load. The manufacturer's data gives the BS of a line, but to learn the line's SWL, you must apply an SF. An SF is a number by which the BS is divided to find the range in which it is safe and economical to operate the rope. Table 15-1 shows, even under the best of conditions, that the allowance for safety is considerable.

Table 15-1. Safety Factor of Line

Line	Working Conditions		
	Best	Average	Poor
Manila	5	10	15
Nylon Polyester	3	4	6
Polypropylene Polyethylene	5	6	8

USEFUL FORMULAS FOR LINES

15-14. When the manufacturer states the size and BS of its lines, use these figures for determining the strength of line. If this information is not available, then use the rule of thumb to compute the SWL and the BS. The following rules of thumb (see Table 15-2) give only approximate results. However, the error will be on the side of safety because of the constants used in the formula.

Table 15-2. Rule of Thumb for Computing the Safe Working Load and Breaking Strength

Type of Line	Constant
Sisal	160
Manila	200
Three-strand nylon	500
2-in-1 braided nylon	600

15-15. The formula for SWL in pounds is: Line circumference squared times the constant for line = SWL

Examples

3-inch sisal:
3 x 3 x 160=
9 x 160 = 1,440 pounds SWL

3-inch manila:
3 x 3 x 200 =
9 x 200 = 1,800 pounds SWL

3-inch, three-strand nylon:
3 x 3 x 500=
9 x 500 = 4,500 pounds SWL

3-inch, 2-in-1 braided nylon:
3 x 3 x 600 =
9 x 600 = 5,400 pounds SWL

15-16. An SF of 5 is generally used in marine operations. Multiply this by the SWL to find the BS of a fiber line. This is the amount of weight in pounds required to part the line. If you are given the BS of a line, divide it by the safety factor 5 to find the SWL.

Note: The safety factor of 5 is valid when using new line or line that is in good condition. As line ages and wears through use, the safety factor drops. Old line may have a safety factor of 3.

SWL AND BS FOR WIRE ROPE

15-17. Useful formulas for determining the SWL of several grades of wire rope have constants not to be confused with safety factors. For example, the formula for the SWL in STONs (2,000 pounds) for extra improved plow steel wire rope is—

Diameter squared times 10 = SWL in STONs

To find the SWL of 1-inch, 6 x 19, extra improved plow steel wire rope: 1 x 1 x 10 = 10 STONs

15-18. A figure relatively constant in marine operations, especially for new wire rope, is the SF of 5. It is used with the SWL to find the breaking strength or strain:

SWL x 5 = BS
10 x 5 = 50 STONs

15-19. The formulas for improved plow steel, plow steel, and mild plow steel (6 x 19 wire rope) are as follows:

Improved plow steel and plow steel:
Diameter squared x 7 = SWL in STONs
SWL x SF = BS in STONs

Mild plow steel:
Diameter squared x 6 = SWL in STONs
SWL x SF = BS in STONs

COMPUTING THE BREAKING STRENGTH OF A BLOCK AND TACKLE

15-20. Breaking strength determines the ultimate strength of the block and tackle. When computing the breaking strength of a block and tackle think of this as the load that your line should be expected to handle on a regular basis. Computing the correct breaking strength will safeguard expensive equipment and also protect the lives of personnel.

DETERMINING BREAKING STRESS

15-21. Perform the following steps to determine breaking stress.
- Step 1. Determine the friction for the block and tackle.
- Step 2. Determine the total weight to be lifted.
- Step 3. Determine the strain on the hauling part of the block and tackle.
- Step 4. Apply the breaking stress formula to compute the breaking stress of the block and tackle.

Note: The SF for the hauling part is always 5. The formula—safety factor (SF) times strain on hauling part (SHP) = breaking strength (BS) for the block and tackle.

- Step 5. Compare the breaking stress to the figures shown in the line strength table (see Table 15-3). The SWL of the line used should be greater than the computed BS for the block and tackle.

Example

1. Determine the breaking strain for a twofold block and tackle that is going to be used to lift a 500-pound weight.

2. Determine the minimum size manila line that has an SWL capable of making the lift.

Procedure. Friction (F) is computed at 10 percent per sheave.

Step 1. Determine the friction. For a block and tackle, 10 percent times the number of sheaves equals the percent of friction. Using a twofold purchase, there are four sheaves, giving a loss of efficiency of 40 percent.

Step 2. Determine the total weight (TW) to be lifted. The original weight (W) to be lifted is 500 pounds. There is a 40 percent loss of efficiency that must be added to that weight to be lifted (40 percent x 500 = 200 pounds). The formula for total weight is—

W + F = TW

500 + 200 = 700 pounds total weight to be lifted.

Step 3. Determine the SHP. The mechanical advantage (MA) for a twofold purchase is 4. The formula is—

$$TW \div MA = SHP$$

$$700 \div 4 = 175 \text{ pounds SHP}$$

Step 4. Compare the SHP to the line strength shown in Table 15-3. Select an SWL that exceeds the computed SHP for the block and tackle. You would use 1 1/2-inch manila line, which has an SWL of 450 pounds for making the lift.

Note: The information in Table 15-3 is computed in pounds for new line. For line that has been used, these figures will decrease. Old line may have only 60 percent of strength shown in pounds for a given size of line.

Table 15-3. Line Strength Table

Size in Inches	Manila		Three-Strand Nylon		2-in-1 Braided Nylon	
	SWL	BS	SWL	BS	SWL	BS
1	200	1,000	500	2,500	600	3,000
1 1/2	450	2,250	1,125	5,625	1,350	6,750
2	800	4,000	2,000	10,000	2,400	12,000
2 1/2	1,250	6,250	3,125	15,625	3,750	18,750
3	1,800	9,000	4,500	22,500	5,400	27,000
3 1/2	2,450	12,250	6,125	30,625	7,350	36,750
4	3,200	16,000	8,000	40,000	9,600	48,000
4 1/2	4,050	20,250	10,125	50,625	12,150	60,750
5	5,000	25,000	12,500	62,500	15,000	75,000
5 1/2	6,050	30,250	15,125	75,625	18,150	90,750
6	7,200	36,000	18,000	90,000	21,600	108,000
6 1/2	8,450	42,250	21,125	105,625	25,350	126,750
7	9,800	49,000	24,500	122,500	29,400	147,000
7 1/2	11,250	56,250	28,125	140,625	33,750	168,750
8	12,800	64,000	32,000	160,000	38,400	192,000
8 1/2	14,450	72,250	36,125	180,625	43,350	216,750

COMPUTING SAFE WORKING LOAD FOR HOOKS, SHACKLES, AND TURNBUCKLES

15-22. Calculated or predicted design loads are compared to a baseline strength in computing the safety factor for hooks, shackles, and turnbuckles. All hooks, shackles, and turnbuckles will be tested before being used.

15-23. Compute the SWL of a Hook. The diameter of a hook is measured where the inside of the hook starts its arc. The constant for a hook is 2/3.

Formula: Diameter squared times the constant (2/3) = SWL of hook in STONs

  Step I. Measure diameter of the hook to be used.

  Step 2. Use the constant of 2/3.

  Step 3. Apply the formula to determine the SWL of the hook in STONs.

Example

Determine the SWL of a 3-inch hook.

(3 x 3 = 9)(C = 2/3)

9 x 2/3 = 6 STONs SWL

15-24. Compute the SWL of a Shackle. Measure the diameter of the shackle at its side. The constant for shackles is 3.

Formula: Diameter squared times the constant (3) = SWL in STONs

  Step I. Measure the diameter at the side of the shackle.

  Step 2. Use the constant of 3.

  Step 3. Apply the formula to determine the SWL in STONs for a shackle.

Example

Determine the SWL of a shackle that has a diameter of 2 inches.

(2 x 2 = 4)(C = 3)

4 x 3 = 12 STONs SWL

15-25. Compute the SWL of a Turnbuckle. To determine the SWL for turnbuckles, measure the diameter of the threaded rod (Figure 15-9) and check the SWL in Table 15-4.

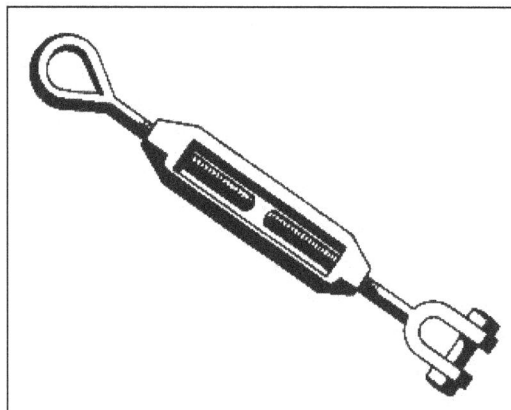

Figure 15-9. Threaded Rod on Turnbuckle

Table 15-4. Turnbuckle Rod SWL Table

Values in STONs (2,000 Pounds)	
Rod Diameter (in Inches)	SWL (in STONs)
1/2	.9
5/8	1.5
7/8	2.2
1	3.1
1 1/8	5.1
1 1/4	6.6

MAINTENANCE AND OVERHAUL OF BLOCKS

15-26. Blocks, like other equipment exposed to the elements, will become useless if they do not receive proper maintenance. The bearing and bushing will wear if they are not properly lubricated. The shells and accessories will deteriorate if they are not properly preserved. Maintenance for the fiber rope and the wire rope blocks is discussed as follows.

FIBER ROPE BLOCKS

15-27. These types of blocks should be disassembled periodically and inspected and lubricated. A mixture of white lead and tallow, or graphite and grease, should be used. To disassemble a block, remove the becket bolt and becket, pry off the keeper, and drive out the pin. To loosen the strap in the frame, tap the bottom with a hammer. Then if you cannot pull it out by hand, insert a marlinespike in the U of the strap and drive it out by tapping on the marlinespike with the hammer. Figure 15-10 shows a disassembled block.

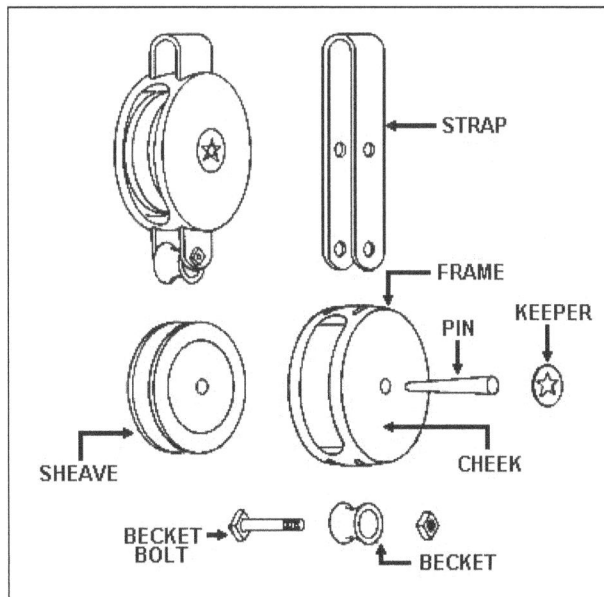

Figure 15-10. A Disassembled Block

15-28. Inspect the frame of the block for any cracks or splits and for any signs of the sheave wearing on the frame. If there are any worn spots on the inside of the frame, check the pin to see if it is bent. Check the hooks or shackles for any sign of distortion. A bent pin or a distorted hook or shackle is no longer safe. Dropping a wooden block can split its frame. Never paint a wooden block because a coat of paint could hide a split. Instead, use clear shellac or varnish or several coats of linseed oil. Metal in constant use is subject to fatigue. Frequently and carefully inspect blocks in running rigging for any signs of distortion or wear. Immediately replace any doubtful block and, if the cost warrants, send it to a shipyard for testing. Inspect and replace any suspected wooden blocks. Many parts for blocks are available separately--for example, rigs for wire rope blocks. Before replacing an entire block, consult the supply officer to see if you can get a replacement for any part that is defective.

WIRE ROPE BLOCKS

15-29. These types of blocks used in cargo handling rigs and others in continuous use should be disassembled frequently and inspected for wear. However those used only occasionally seldom need to be disassembled if they are kept well lubricated. Two types of wire rope blocks are the diamond and oval blocks and the roller bearing block. Refer back to Figure 15-2, to do the following:
- **Diamond and Oval Blocks.** To remove the sheave from a diamond or oval block, take out the cotter pin (8) and remove the hexagon nut (10) from the sheave pin (9). Drive out the sheave pin. For a diamond block it is necessary to loosen all bolts holding the cheeks together and to remove one before the sheave will slide out. With an oval block it is necessary only to loosen the bolts.
- **Roller Bearing Blocks.** To disassemble a roller bearing block, loosen the setscrews (9) and remove the retaining nuts (8). Take out the bolts holding the shell together and remove the shell. Remove the closure snap rings (7), adjusting nut (5), closure washer (6), and closure (11). Now remove the pin, then the bearings from the sheave.

STANDING RIGGING

15-30. Standing rigging, usually of 6 x 19-inch galvanized, high-grade plow steel wire rope is used to support the masts. The fore and aft supports are called stays and the supports running athwart ship are shrouds. Stays and shrouds are set up at the lower end with turnbuckles. Vibration often causes turnbuckles to back off. To prevent this, keepers are installed on most turnbuckles in standing rigging. The effectiveness of shrouds and stays is reduced considerably if they are allowed to become slack. Inspect standing rigging periodically and tightened if necessary. Use the following procedure when considerable adjustments are required.
- Slacken all stays and shrouds so that there are no unbalanced forces applied to the mast.
- Take up the slack as uniformly as possible until sag is substantially eliminated from all stays and shrouds, and turnbuckles are hand tight. Measure the distances between the ends of the turnbuckle bolts.
- Tighten each turnbuckle so that it is shortened by a distance equal to 1 inch for each 60 feet of stay length.

15-31. Insulators should present clean surfaces. They should not be painted, tarred, varnished, or coated in any way. All electrical grounds on standing rigging should be inspected periodically for excessive deterioration at points of contact between different metals.

INSPECTIONS OF RIGGING

15-32. A weekly inspection of all booms and their rigging and associated fittings is conducted by the mate and boatswain. Whenever a boom is to be used for hoisting or lowering a load equal to its rated capacity, as shown on the heel of the boom, the chief mate should be notified. He/She will make a thorough inspection of the boom and its associated fittings and rigging before the lift is made. Whenever signs of deterioration are found, defective components should be replaced or renewed as soon as possible. If the inspection indicates a dangerous condition or weakness of any component, this should be reported without delay, and the boom in question should not be operated until it is repaired or replaced.

GROUNDING MASTS

15-33. Unless otherwise directed, mast shrouds should be grounded at the deck to prevent accumulation of static charges. One method of grounding shrouds is shown in Figure 15-11. Most electrical insulation and grounds on metallic standing rigging should be inspected periodically for deterioration at points of contact between dissimilar metals. When deterioration is evidenced, the connections should be thoroughly cleaned and replaced as required.

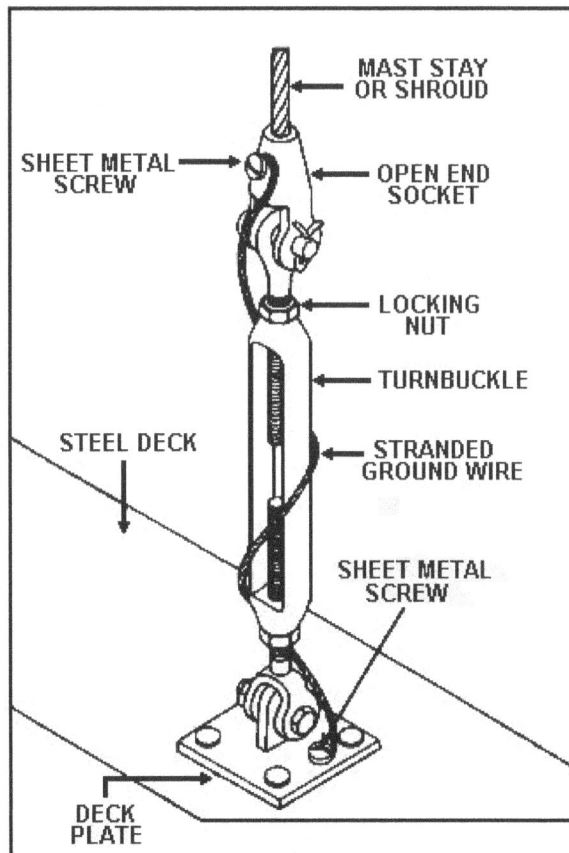

Figure 15-11. Grounding a Shroud

This page intentionally left blank.

Chapter 16

Ground Tackle

INTRODUCTION

16-1.   Ground tackle consists of all the equipment used in anchoring. This includes the anchors, anchor cables or chain, connecting devices, and the anchor windlass. This chapter discusses these items and their nomenclature, maintenance, and use.

ANCHORS

16-2.   An anchor works like a pickaxe. When the pick is driven into the ground, it takes a great deal of force to pull it loose with a straight pull on the handle. However, by lifting the handle, leverage is obtained which breaks it free. In the same way, the anchor holds because the anchor chain or cable causes the pull on the anchor to be in line with its shank. When it is desired to break the anchor free, the chain is taken in and this lifts the shank of the anchor and gives the leverage needed to loosen the anchor's hold. The primary function of an anchor is to hold the ship against the current and wind. On landing craft, stern anchors are also used to prevent broaching on the beach and to assist in retracting from the beach.

NOMENCLATURE

16-3.   The following describes the different parts of an anchor (see also Figure 16-1).
- **Ring (Shackle).** Device used to shackle the anchor chain to the shank of the anchor. The ring is secured to the top of the shank with a riveted pin.
- **Shank.** The long center part of the anchor running between the ring and the crown.
- **Crown.** The rounded lower section of the anchor to which the shank is secured. The shank is fitted to the crown with a pivot or ball-and-socket joint that allows a movement from 30 degrees to 45 degrees either way.
- **Arms.** The parts that extend from each side of the crown.
- **Throat.** The inner curved part of an arm where it joins the shank.
- **Fluke or palm.** The broad shield part of the anchor that extends upward from the arms.
- **Blade.** That part of the arm extending outward below the fluke.
- **Bill or pea.** Tip of the palm or fluke.

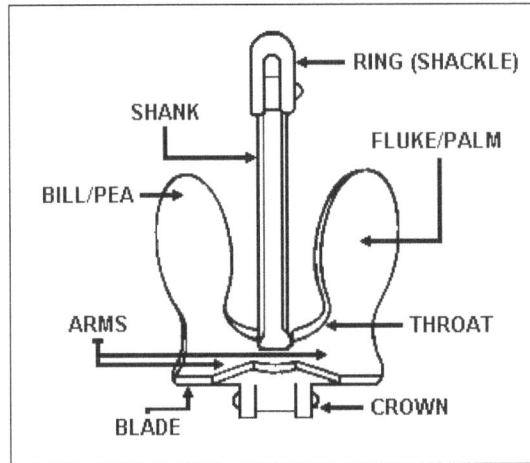

Figure 16-1. Nomenclature of an Anchor

TYPES

16-4.   Three types of anchors used aboard Army vessels are the stockless, the lightweight, and the mushroom (see Figure 16-2).

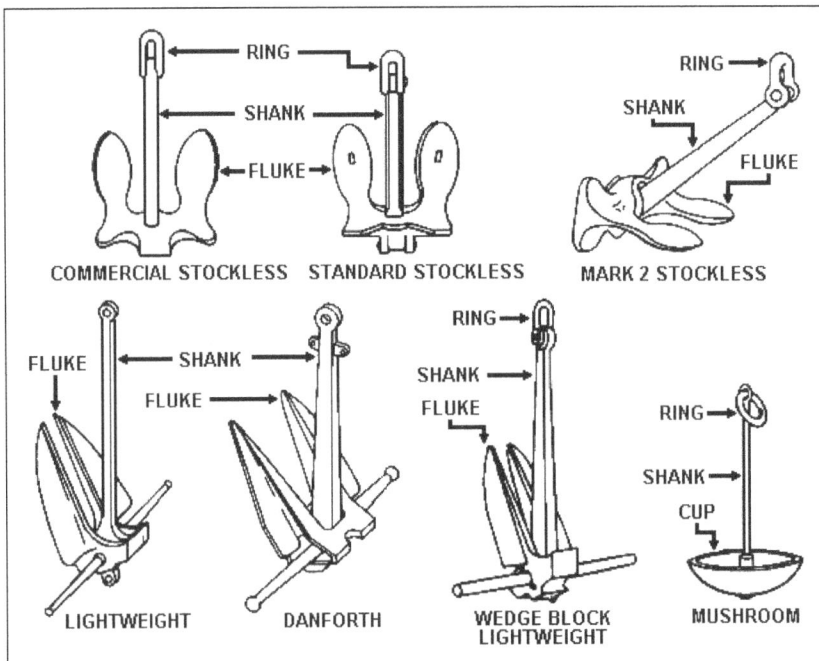

Figure 16-2. Types of Anchors

Note: All vessels, 380 tons and over, must carry a spare bow anchor. Seagoing tugs must carry a kedge anchor.

ANCHOR CHAIN

16-5.   Modern anchor chain is made of die-lock chain with studs. The size of the link is designated by its diameter, called wire diameter.   The Federal Supply Catalog lists standard sizes from ¾ inch to 4 3/4 inches. Wire diameter is measured at the end and a little above the centerline of the link. The length of a standard link is 6 times its diameter and the width is 3.6 times its diameter. All links are studded; that is, a solid piece is forged in the center of the link. Studs prevent the chain from kinking and the links from pounding on adjacent links. They also further strengthen the chain up to 15 percent.

CHAIN NOMENCLATURE

16-6.   A chain is made up of many parts besides links. A variety of equipment is required to use and maintain the chain.

- **Standard Shots.** The lengths of chain that are connected to make up the ship's anchor chains are called shots. A standard shot is 15 fathoms (90 feet) long.

- **Detachable Links.** Shots of anchor chain are joined by a detachable link. The detachable link (see Figure 16-3) consists of a C-shape link with two coupling plates which form one side and stud of the link. A taper pin holds the parts together and is locked in place at the large end by a lead plug. Detachable link parts are not interchangeable. Therefore, matching numbers are stamped on the C-link and on each coupling plate to ensure identification and proper assembly. You will save time and trouble when trying to match these parts if you disassemble only one link at a time and clean, slush, and reassemble it before disassembling another. When reassembling a detachable link, make sure the taper pin is seated securely. This is done by driving it in with a punch and hammer before inserting the lead plug over the large end of the pin. Detachable link toolbox sets contain tools, including spare taper pins and lock plugs, for assembling and disassembling links and detachable end links.

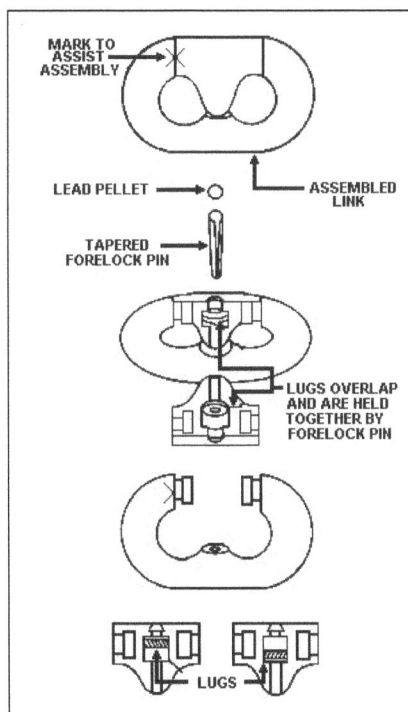

Figure 16-3. Detachable Link

● **Bending Shackles.** Bending shackles are used to attach the anchor to the chain.

Note: The slush, a preservative and lubricant, is a mixture of 40 percent white lead and 60 percent tallow by volume. If the white lead/tallow mixture is not available, grease (MIL-G-23549A) may be substituted.

● **Chain Swivels.** Furnished as part of the outboard swivel shot, chain swivels reduce kinking or twisting of the anchor chain.
● **Outboard Swivel Shots.** Standard outboard swivel shots consist of detachable links, swivel, end link, and bending shackle. They are used on most vessels to attach the anchor chain to the anchor. These shots vary in length up to approximately 6 1/2 fathoms and are also termed bending shots. The taper pin in the detachable link, located in the outboard swivel shot, is additionally secured with a wire-locking clip.

MAKEUP OF AN ANCHOR CABLE

16-7. An anchor cable is an assembly of a number of individual units properly secured together (see Figure 16-4). These units are connected to the anchor by means of a swivel piece made up of shackles, swivels, and special links.

Note: Each shot of chain is joined together with a detachable link.

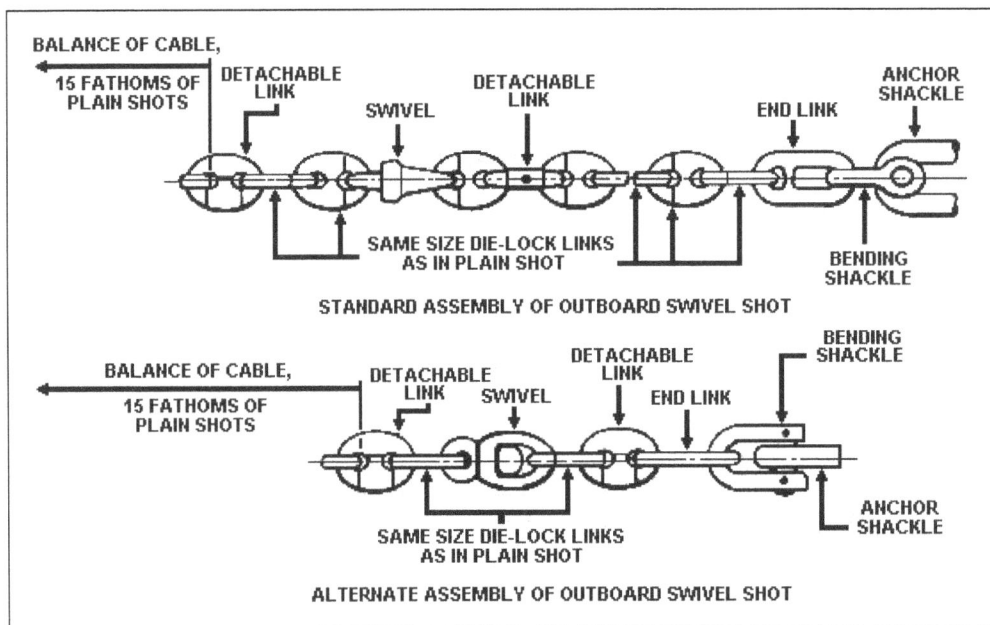

Figure 16-4. Connecting Anchor to Anchor Cable

MARKING THE ANCHOR CHAIN

16-8.   For the safety of every ship, the ship's officers and the boatswain must know at all times the scope or how much anchor chain has been paid out. To make this information quickly available, a system of chain markings is used. Figure 16-5 shows the standard system for marking an anchor chain.

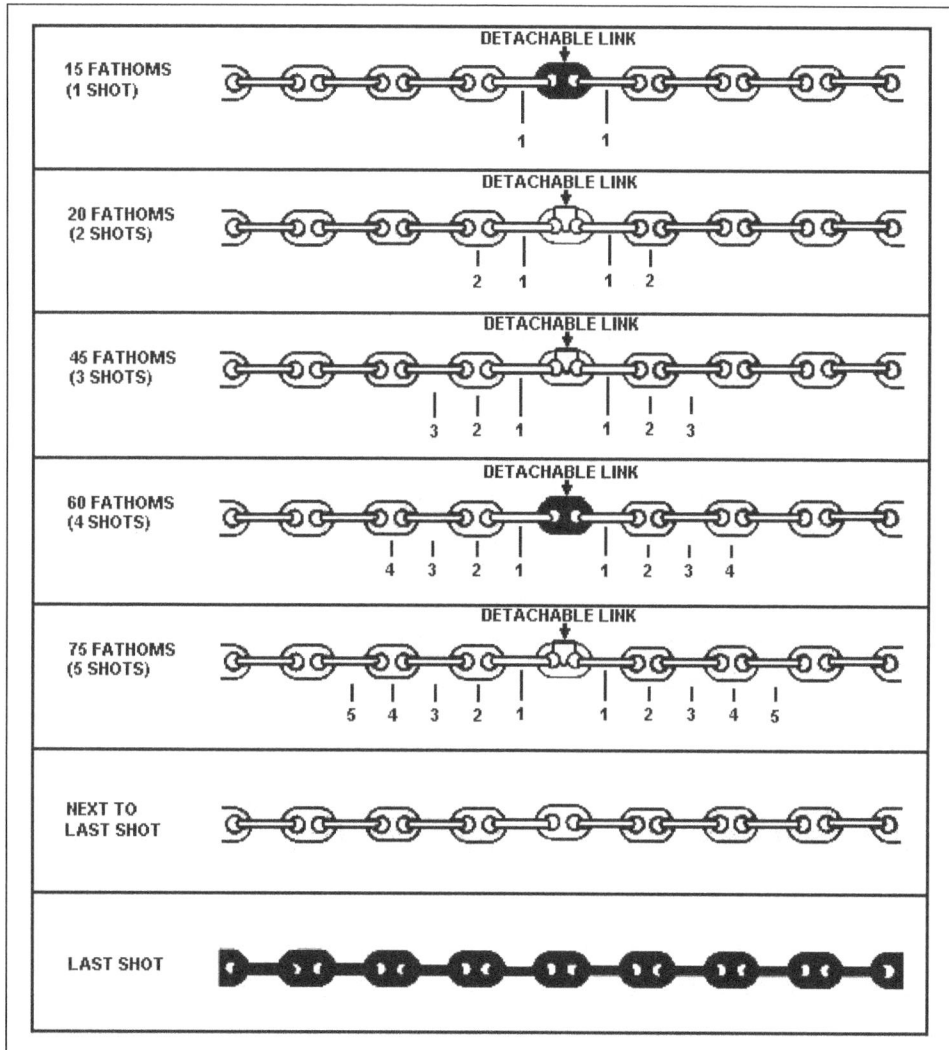

Figure 16-5. Standard Anchor Chain Markings

COLOR MARKINGS

16-9.   The tools required for color marking an anchor chain are wire brush, paint brush, rags, and paint (red, white, blue, and yellow enamel paint).

- **15 fathoms (1 shot).** The detachable link is painted red, and one link on each side is painted white.
- **30 fathoms (2 shots).** The detachable link is painted white, and two links on each side are painted white.
- **45 fathoms (3 shots).** The detachable link is painted blue, and three links on each side are painted white.
- **60 fathoms (4 shots).** The detachable link is painted red, and four links on each side are painted white.
- **75 fathoms (5 shots).** The detachable link is painted white, and five links on each side are painted white.

16-10.   Paint each link in the next to last shot yellow. The yellow alerts you that you are running out of chain. Paint each link in the last shot red.

Note 1: 1 fathom = 6 feet. There are 15 fathoms (90 feet) in a shot of anchor chain.

Note 2: This method is used through the entire marking procedure alternating red, white, and blue for detachable links as appropriate.

WIRE MARKINGS

16-11.   In addition to color markings, wire markings may also be used. The purpose of the wire marking is to let you count the shots by feel during blackout conditions or if the markings on the chain are worn off or rusted over.

- **lst shot.** One turn of wire on the first stud from each side of the detachable link.
- **2nd shot.** Two turns of wire on the second stud from each side of the detachable link.
- **3nd shot.** Three turns of wire on the third stud from each side of the detachable link.
- **4th shot.** Four turns of wire on the fourth stud on each side of the detachable link.
- **5th shot.** Five turns of wire on the fifth stud on each side of the detachable link.
- **6th shot.** Six turns of wire on the sixth stud on each side of the detachable link.

THE ANCHOR WINDLASS

16-12.   The anchor windlass is installed on vessels primarily for handling and securing the anchor and anchor chain. Windlasses are provided with capstans or catheads, which are used for handling mooring lines when docking and undocking.

16-13.   There are two general types of windlasses installed aboard Army harbor craft. These are the horizontal shaft windlass (Figure 16-6) and the vertical shaft windlass (Figure 16-7).

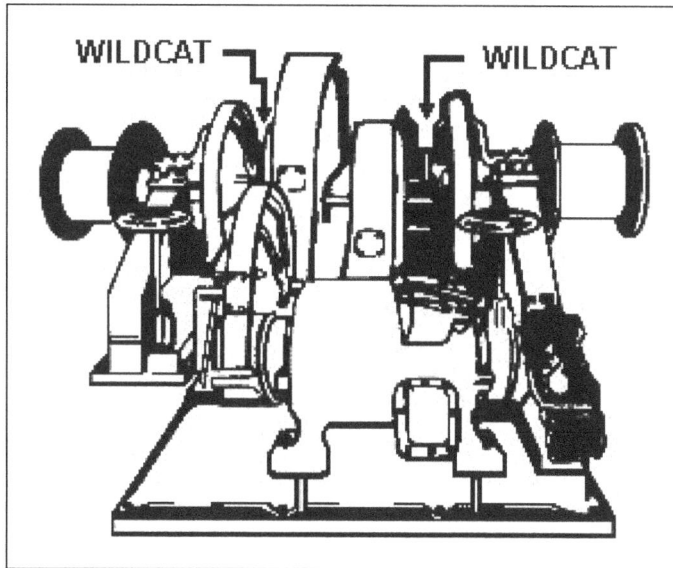

Figure 16-6. Horizontal Shaft Windlass

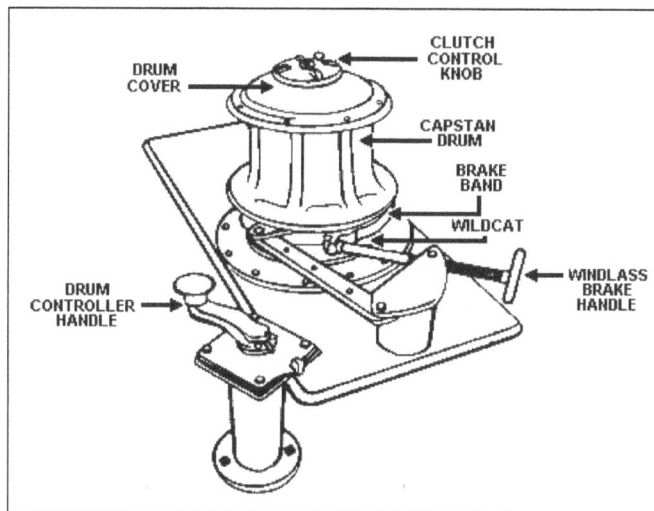

Figure 16-7. Vertical Shaft Windlass

HORIZONTAL SHAFT

16-14. This type of windlass is usually a self-contained unit with the windlass and windlass motor mounted on the same bedplate. It handles both the port and starboard anchors and is found aboard large vessels. Figure 16-8 shows the side view of this windlass.

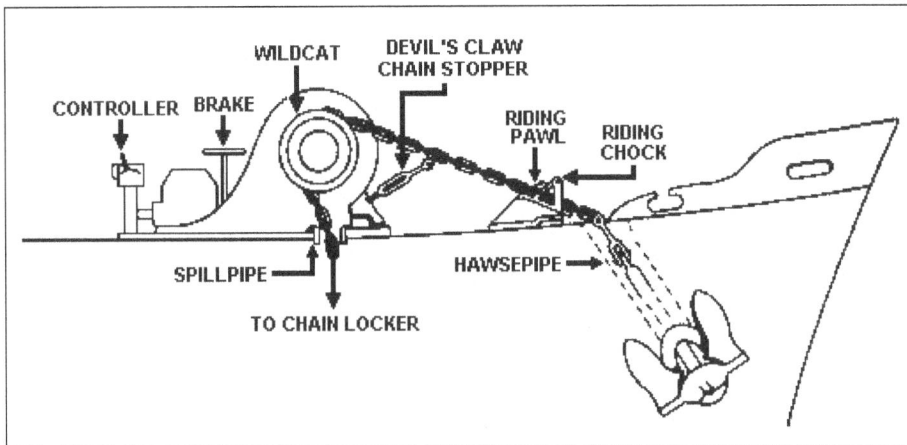

Figure 16-8. Side View of Horizontal Shaft Anchor Windlass

VERTICAL SHAFT

16-15. This type of windlass is found on tugs and barges. With the vertical shaft windlass, the power source is located below the deck with only the wildcat and capstan showing above the deck. The controller for the windlass is also above deck. This type of windlass can handle only one anchor.

TERMINOLOGY

16-16. Although there is a difference in construction and appearance between the horizontal and the vertical shaft windlass, they do share a common terminology. Definitions of parts of equipment used in anchoring, starting at the anchor and working aft, are as follows:

- **Hawsepipe.** Openings in the forward part of the ship where the shank of the anchor is stowed.
- **Buckler plate.** A heavy steel plate that is "dogged down" by butterfly nuts when the vessel is at sea. The buckler plate covers the hawsepipe opening on deck and prevents water from rushing up the hawsepipe and spilling on deck.
- **Riding chock.** A metal fairlead for the anchor chain. It prevents the chain from fowling on deck and also holds the riding pawl.
- **Riding pawl.** A safety stopper that works like a ratchet on the links of the chain. It is lifted up to the "open" position when the anchor chain is run out. When heaving the chain in, the pawl is "closed" or dropped in the after side of the riding chock. The pawl bounces over the incoming chain. However, if an emergency occurs, such as the wildcat jumping out of gear, the pawl will catch on a link of the anchor chain and hold the chain and keep it from running out.
- **Chain stopper.** A turnbuckle inserted in a short section of chain with a pelican hook or a devil's claw attached to one end and a shackle on the other end. The stopper chain is screwed at the base of the windlass. In operation, the devil's claws are used when the vessel is setting out to sea. The claws are put on a link of the chain and the turnbuckle is set up, acting as a permanent stopper. On some ships, a pelican hook is used.
- **Wildcat.** A sprocket wheel in the windlass with indentations for the links of the anchor chain. The wildcat, when engaged, either hauls in or pays out the anchor chain. When disengaged, the wildcat turns freely and the only control of the anchor chain is the friction brake.

- **Friction brake.** A band which bears on a flywheel. By tightening up on the band by means of the brake handle, the wildcat can be controlled.
- **Locking ring.** A device, with pigeon holes, into which a bar is placed to lock the wildcat to the hoisting gear of the engine. The locking ring is usually turned forward to disengage the wildcat and turned aft to engage it. On the capstan the wildcat is engaged or disengaged by turning the capstan barrel cover.

LETTING GO THE ANCHOR—GENERAL PROCEDURES

16-17. Certain procedures are required when preparing to let go the anchor. For this discussion, assume that the anchors are secured for sea with the spill pipes cemented in. This is a practice of good seamanship for ships operating at sea, where there are many days between ports or when heavy weather is expected.

WARNINGS

1. Only crewmembers on the anchor detail will be permitted on the bow.

2. Crewmembers will not stand between the capstan and the hawsepipe when letting go the anchor.

16-18. Use the following procedures prior to entering port or when planning to use the anchor.
- Make sure the devil's claw assembly is taut.
- Engage the wildcat and release the brake.
- "Walk out" enough chains to break out the cement plugs in the spill pipes and free the chains.
- "Walk" the chain back to the original position.
- Clean area around the chain and anchor.
- Release the devil's claw or hooks.
- Put the riding pawls in the OPEN position.
- Make sure that the anchor is not frozen or jammed in the hawsepipe. The best way to do this is to "walk out" the anchor until it is clear of the hawsepipe.
- After "walking out" the anchor, set the brake tight and disengage the wildcat.
- The anchor is free.

Note: On vessels having two anchors, get both anchors ready.

OPERATING THE CAPSTAN ANCHOR WINDLASS

16-19. The Markey type VEV-16 anchor windlass (see Figure 16-9) with the single vertical capstan barrel is the type described here. Although there are other types of vertical capstan barrel anchor windlasses in the system, their method of operation will, in most cases, be similar. The capstan barrel is keyed to the main shaft and is in continuous motion while the motor is running. The wildcat is driven off the capstan by two axial keys that may be engaged or disengaged by turning the capstan barrel cover (clutch) (see Figure 16-10). An inner drum, to which the capstan barrel cover is bolted, provides two axial cams that engage the two keys and moves them up or down as the barrel cover is turned. Indicator plates show key engagement and spring ball locks are provided to hold the shifter mechanism in either position.

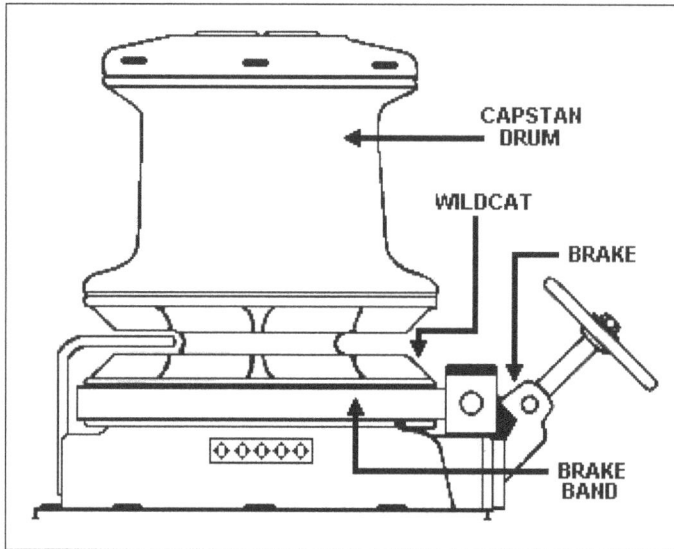

Figure 16-9. Capstan/Wildcat and Brake

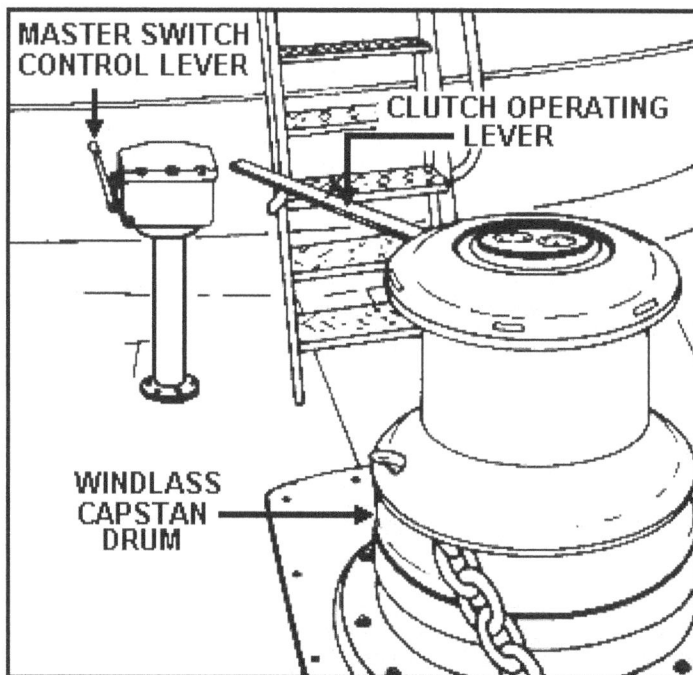

Figure 16-10. Turning the Drum Brake

DROPPING THE ANCHOR WITHOUT POWER

16-20. Use the following steps when dropping the anchor without power:

WARNING

Safety goggles must be worn when dropping the anchor without power. There will be a great deal of rust, sparks, dirt, and debris flying about as the chain runs out.

- Remove and stow the buckler plates.
- Make sure that the brake is on by fully turning the hand wheel counterclockwise.
- Disengage the axial keys by turning the capstan barrel cover.
- Remove the chain stopper and open the riding pawl.
- Let go the anchor by releasing the brake (turning the hand wheel clockwise). The anchor and chain will run freely when the brake is released. Use the brake to control the running speed of the chain.

WARNING

DO NOT allow the chain to run too slowly. The brake, when slipping continuously, will develop excessive heat and may burn out. However, the chain must not run so fast that it will jump out of the wildcat.

Note: Once the anchor has hit the bottom and the strain is taken off the anchor chain, there should be a natural slowing down in the rate that the chain pays out.

- To stop lowering the anchor, turn the brake hand wheel counterclockwise. This applies the brake.
- To secure, close the riding pawl and replace the chain stopper.

Note 1. As soon as the anchor hits bottom, during daylight hours, raise the anchor ball; during darkness turn on the anchor lights and shut off the navigation lights.

Note 2. During daylight, after the ship is anchored, the union jack should be hoisted and flown from the jack staff. The national ensign should be shifted from the gaff to the flag staff at the stern.

RAISING THE ANCHOR

16-21. Use the following procedures to raise the anchor.

- Turn the stop switch lever of the controller to the ON position and check for power.

WARNING

Whenever the controller is left unattended, the stop switch must be put in the OFF position.

- Check to see that the brake is on.
- Turn the capstan barrel cover to engage the two axial keys. This will put the wildcat in motion when the controller is operated.
- Put the controller handle in the hoist position and take a strain on the anchor chain, then stop.
- Drop the riding pawl.
- Release the chain stopper.
- Open the brake.
- Put the controller handle in the hoist position and raise the anchor. You will usually feel a "surge" or release of strain on the anchor chain when the anchor breaks free of the bottom.

Note: As the anchor chain is coming in, crew members should be stationed at the hawsepipe with a fire hose. The chain should be thoroughly washed down and freed of all mud, silt, and debris.

- House the anchor.
- Put on the brake.
- Replace the chain stopper.
- Disengage the axial keys by turning the capstan barrel cover.

Note 1. As soon as the anchor breaks free of the bottom, during daylight hours, drop the anchor ball. During darkness, switch off the anchor lights and turn on normal navigation lights.

Note 2. Once the "anchor ball" is lowered, the ship is underway. The union jack is lowered and the national ensign is transferred to the gaff.

OPERATING THE HORIZONTAL ANCHOR WINDLASS

16-22. Anchor windlasses and their accompanying equipment vary in size and shape depending on the type and size ship on which they are used. However, the procedure for releasing the anchor remains the same.

LETTING GO THE ANCHOR

16-23. To let go the anchor do the following:
- Inspect the anchor windlass, anchor chain, hawsepipe, and anchor to see that they are free and clear for letting go the anchor.
- Make sure that the brake is set tight (Figure 16-11).

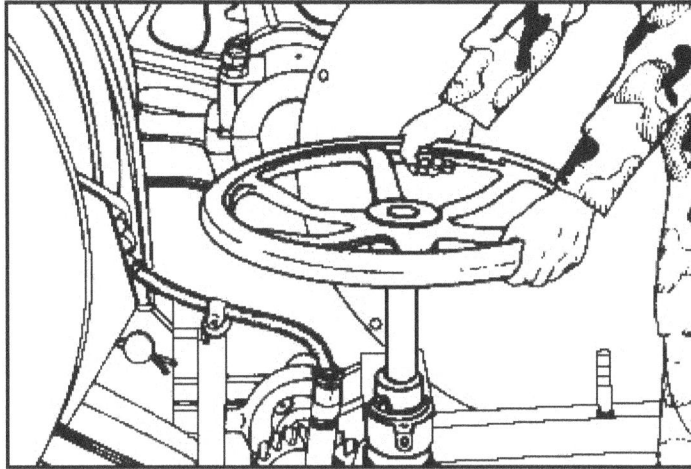

Figure 16-11. Setting Up the Brake

- Disengage the wildcat (Figure 16-12).

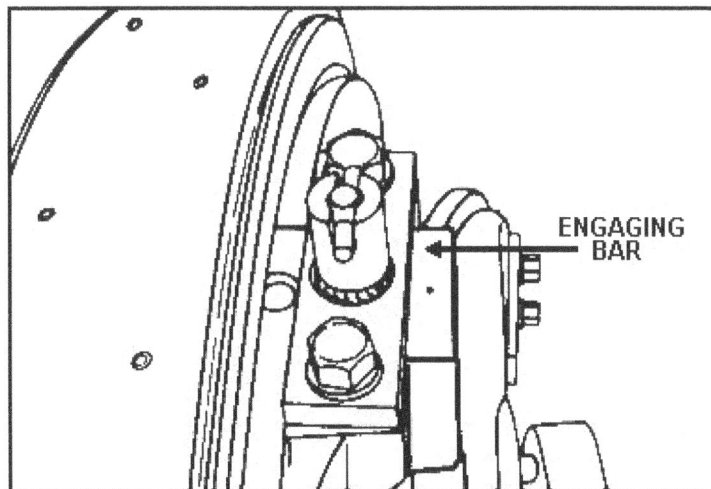

ENGAGING
BAR

Figure 16-12. Wildcat Engaged

- Lift the locking ring key on the locking ring (Figure 16-13).

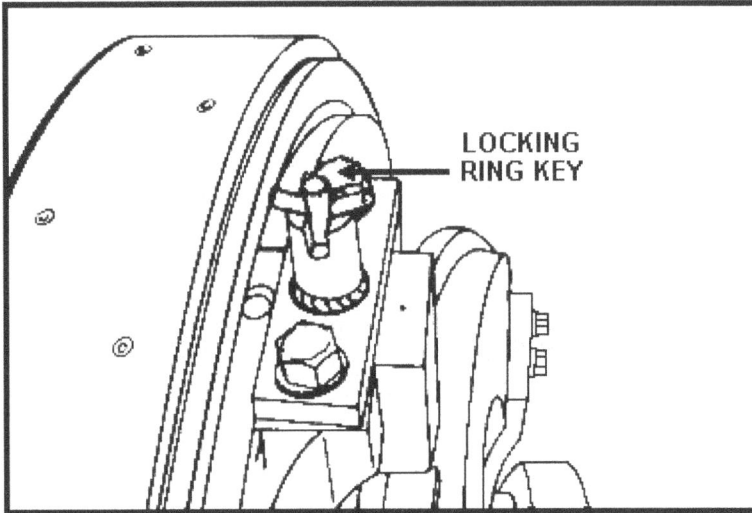

Figure 16-13. Lifting Locking Ring

- Insert the anchor bar into the pigeonhole on the locking ring (Figure 16-14). The locking ring is then turned forward to disengage the wildcat.

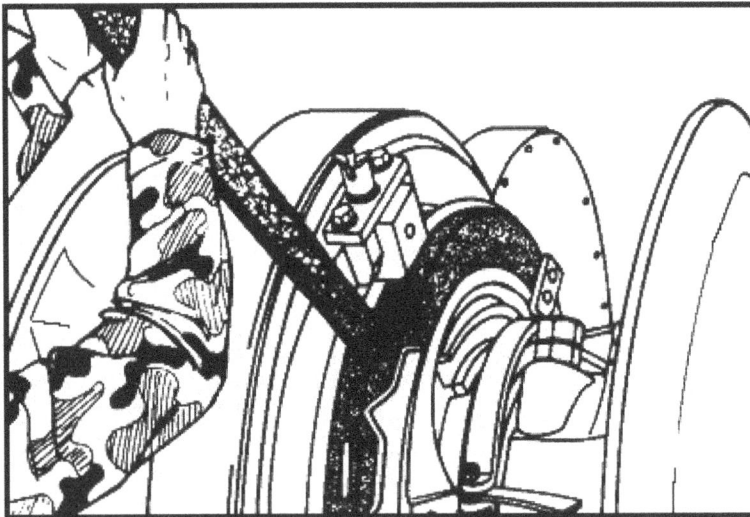

Figure 16-14. Inserting Anchor Bar

● Wildcat is disengaged (Figure 16-15).

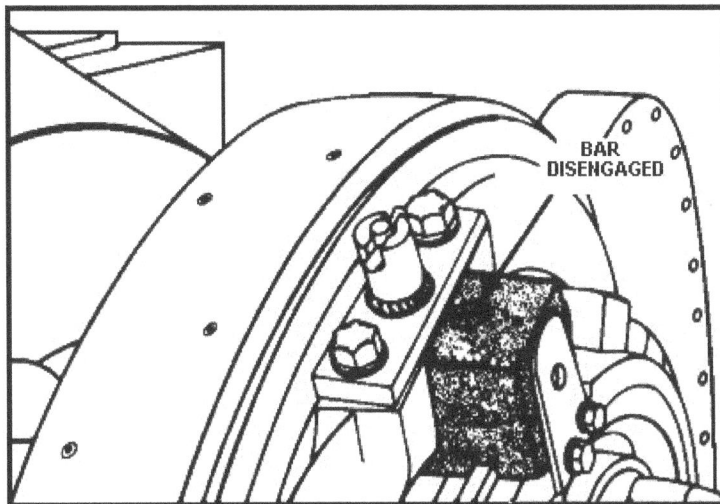

Figure 16-15. Wildcat Disengaged

● Remove the chain stopper. The turnbuckle is secured at the base of the windlass. The devil's claw is put on a link of the anchor chain, and the turnbuckle is set up, acting as a permanent stopper.

● Slack off anchor chain stopper (Figure 16-16).

Figure 16-16. Slacking Off Chain Stopper

● Take off the devil's claw (Figure 16-17).

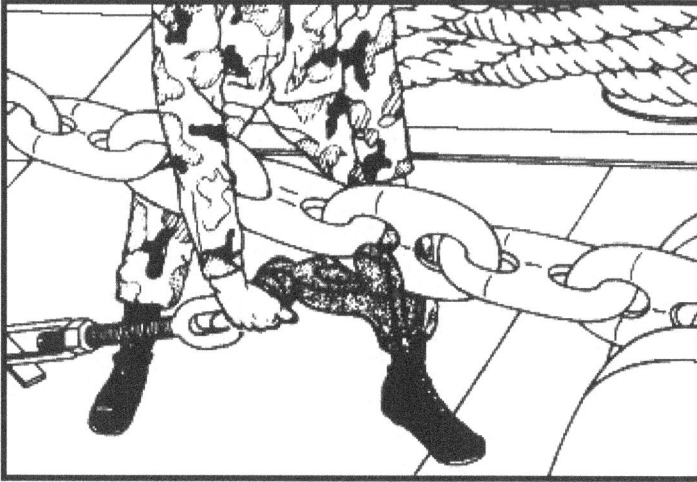

Figure 16-17. Removing the Devil's Claw

● Lift open the riding pawl (Figure 16-18).

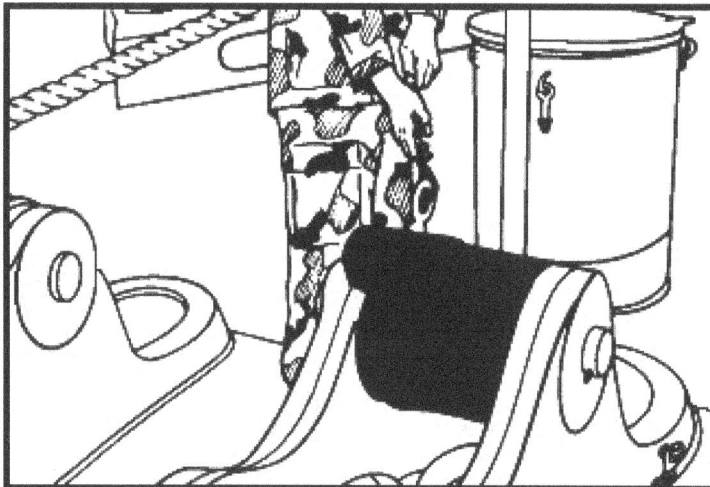

Figure 16-18. Opening Riding Pawl

● On command from the bridge, let go the anchor by releasing the brake. Wear your safety goggles and keep your head turned to the side to protect your eyes from flying rust, sparks, and dirt from the anchor chain. Usually one can sense when the anchor hits the bottom--there is a noticeable slackening in speed of the chain paying out.

● Once the chain has hit bottom and slowed up in paying out, tighten the brake to where you can control the paying out of the chain.

Note: During daylight hours, as soon as the anchor hits the bottom, raise the anchor ball. Next, raise the union jack and shift the national ensign from the gaff to the flag staff. During hours of darkness or restricted visibility, as soon as the anchor hits the bottom, turn on the anchor lights and shut off the navigation lights.

● Secure the brake. Use the anchor bar or a "valve wrench" to set up on the brake (Figure 16-19).

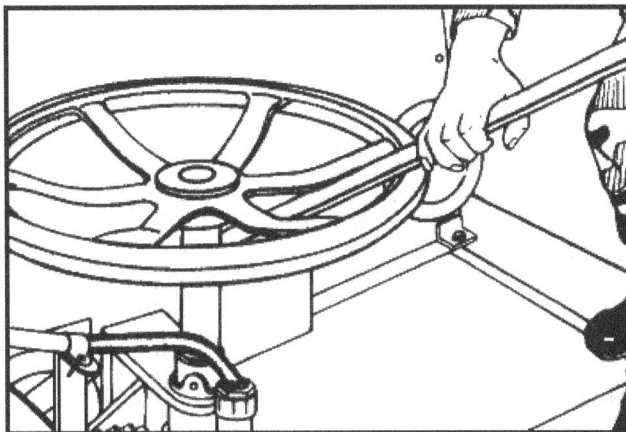

Figure 16-19. Setting Brake

● Replace the stopper. After the brake has been set up, then hook up the devil's claw and secure the stopper (Figure 16- 20). This will aid in holding the anchor and take some of the strain from the brake.

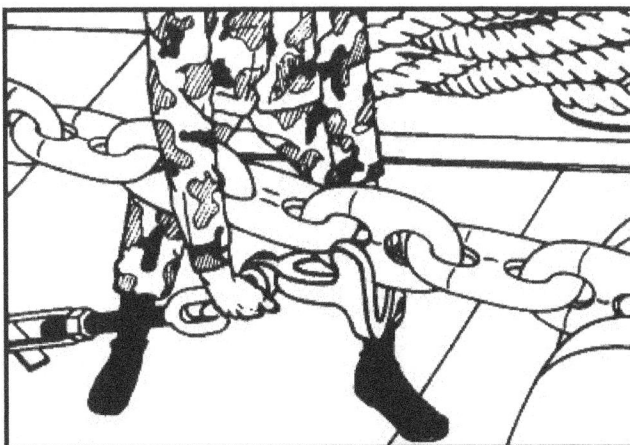

Figure 16-20. Securing Devil's Claw

- Close the riding pawl (Figure 16-21).

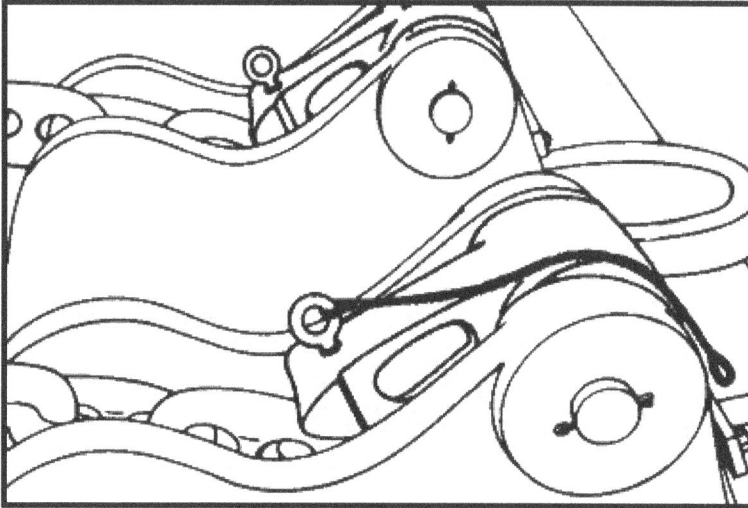

Figure 16-21. Closing Riding Pawl

RAISING THE ANCHOR

16-24. Use the following procedures when raising the anchor.

- Inspect the anchor windlass, chain, hawsepipe, and anchor to see that they are free and clear.
- Turn on the power switch and then push the control handle forward to the lower position (Figure 16-22). Put it in this position just long enough to hear the power turn on, and then bring the control handle back to the stop position.

Figure 16-22. Checking Controller

- Make sure that the brake is secured.
- Engage the wildcat.
- Lift locking ring key on locking ring (Figure 16-23).

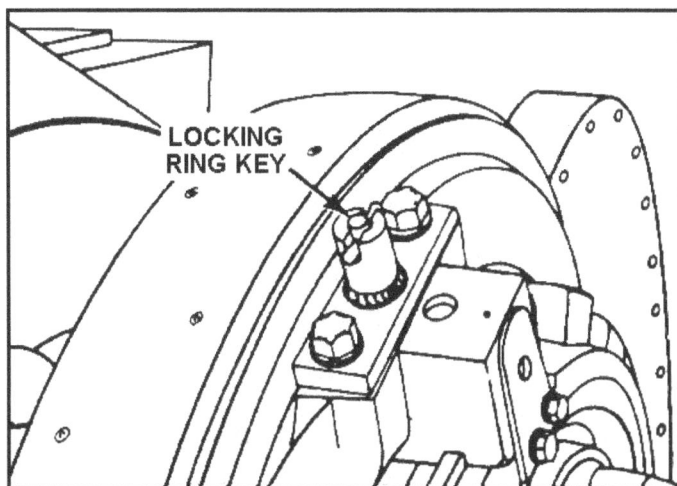

Figure 16-23. Lift Locking Ring Key

- Using the anchor bar, turn the locking ring to engage the wildcat (Figure 16-24). Figure 16-25 shows the wildcat engaged.

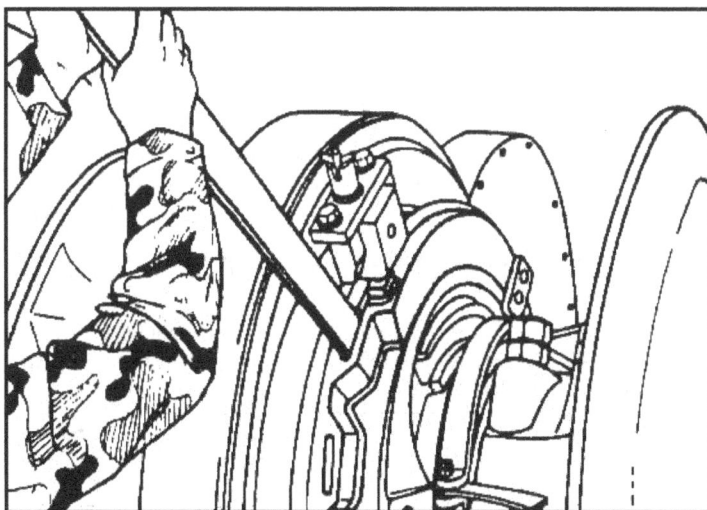

Figure 16-24. Engaging Wildcat

Figure 16-25. Wildcat Engaged

- Remove the anchor chain stopper.
- Turn on the power and, using the windlass control, pull back on the handle. Take just enough strain to where you hear the engaging bar hit the spoke of the windlass, and then stop.
- Release the brake. Slack off on the brake until it is free.
- Raise anchor on anchor windlass control.
- House the anchor.
- Put on the brake.
- Replace the chain stopper.
- Disengage the windlass.

Note 1: If the locking ring does not turn freely, go back to the control lever, push it forward and move locking ring forward about 1 inch. This will take the pressure off the engaging bar.

Note 2: Crew members should be stationed at the hawsepipe with a fire hose. The chain should be thoroughly washed down and freed of all mud, silt, and debris as it is hauled in.

Note 3: When the anchor breaks free of the bottom, during daylight hours, lower the anchor ball and the union jack; shift the national ensign back to the gaff. During hours of darkness, shut off anchor lights and turn on navigation lights.

SEQUENCE OF WEIGHING ANCHOR

16-25. Figure 16-26 shows the seven stages of weighing anchor.

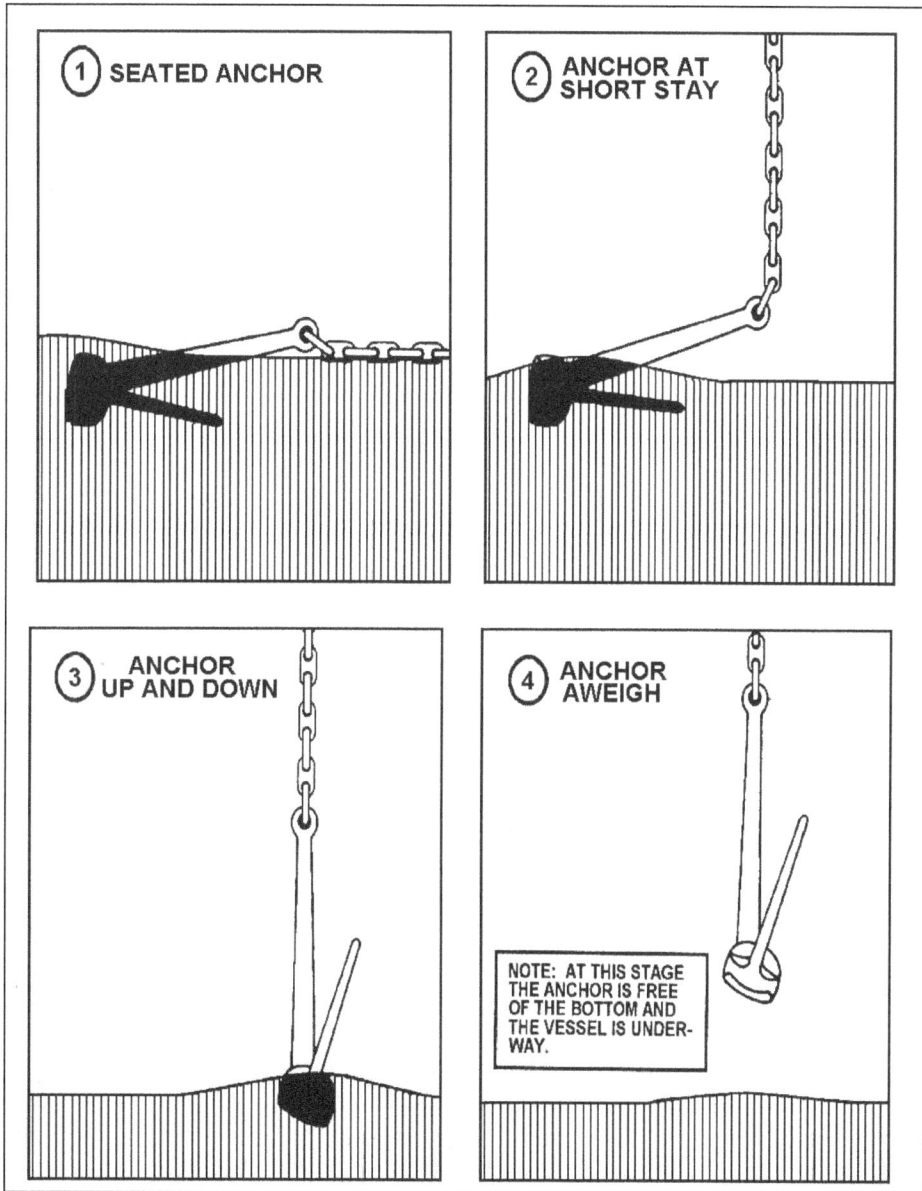

Figure 16-26. Sequence of Weighing Anchor

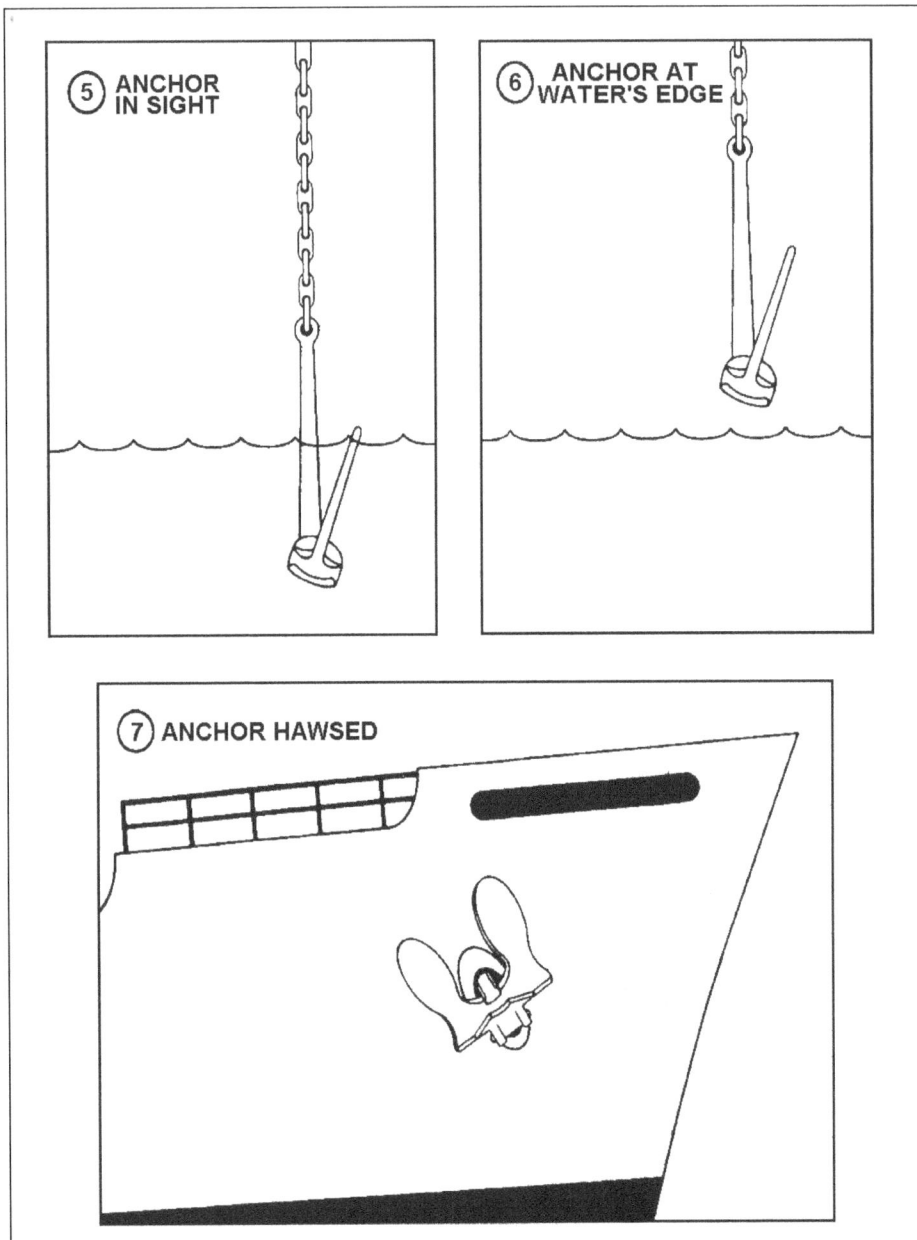

Figure 16-26. Sequence of Weighing Anchor (continued)

MAINTENANCE

16-26. The external maintenance of the anchor windlass and the periodic maintenance and checks of the anchor chain are the responsibility of the deck department.

MAINTENANCE OF THE WINDLASS

16-27. Maintenance and adjustment of equipment should be continued during periods when it is not in use to prevent deterioration and to provide dependable operation. Inspect windlass weekly and operate as necessary to ensure that equipment is in proper condition. Each wildcat is equipped with an externally contracting brake flat band operated by a hand wheel. This brake can be used to hold the anchor and chain and to control the rate of descent. Inspect this brake regularly for wear, maladjustment, and defective parts. Consult the applicable windlass technical manual for detailed instructions for maintenance and adjustment of the brake. Failure of the wildcat brake can result in loss of the anchor and chain. Lubrication instructions are provided in the applicable TM lubrication chart. These instructions should be followed as to grades of lubricant, frequency of application, and points of application.

16-28. If the windlass is not used frequently, it should be lubricated before each operation in accordance with the applicable TM. Rotation of the windlass by power during lubrication will distribute the lubricant evenly. The locking mechanism can be disengaged and the chain held by engaging the wildcat brake. After using the windlass, lubricate the equipment to prevent rusting and freezing of adjacent parts and to protect finished surfaces from corrosion. Check the mounting frame to ensure that nuts and hold-down bolts are tight. Chip, scrape, preserve, and paint the frame, catheads, brake bands, and external parts of the brake band. Keep them free of rust.

MAINTENANCE OF THE ANCHOR CHAIN

16-29. Only minor maintenance can be performed on anchor chains. High-strength, welded chain and appendages can only be overhauled and heat-treated by shipyards meeting the requirements of the DOD.

Maintaining Chain Identification Marks

16-30. Each shot of anchor chain usually bears a serial number that is stamped, cut, or cast on the inner side of the end links at the time of manufacture. In the case of cast steel chain, this number is preceded by the letters C.S. If an end link is lost or removed from a shot, this identification number should be cut or stamped on the side of the new end link of the altered shot. The studs of forged-iron and forged-steel, fire-welded links have the wire diameter of the links imposed on the reverse side, with the opposite side indicated in raised letters. Cast steel and some types of high-strength, welded steel chain have these markings on the studs of alternate links only.

Restrictions as to Use of Chain Appendages

16-31. During makeup or repair, anchor chain appendages should be restricted to the purposes for which they are intended. The intended uses are obvious, but particular attention should be given to the uses of the detachable link.

Periodic Maintenance

16-32. Semiannually, all anchor chains of sizes up to and including 1 1/2 inches should be arranged on deck and examined throughout their length. If necessary, they should be scaled and cleaned of rust and foreign matter. Detachable links should be disassembled, examined for excessive wear or corrosion, and replaced as necessary. When the stock of detachable links is exhausted, new high-strength detachable links will replace the standard detachable links in sizes from 3/4 inch to 1 3/4 inches inclusive. These new links will have proof loads equal to the breaking load of the standard detachable links. Before reassembly, coat the new links with white lead. The detachable link, located in the outboard swivel shot, is fitted with a corrosion-resisting steel locking wire, which serves to hold the taper pin in position. Disassembly of this link requires the removal and probable destruction of the locking wire. A replacement wire of the same type should be carefully examined, put in order, and, if needed, coated with red-lead primer, Military

Specification MIL-P-17545; zinc-chromate primer, Federal Specification TT-P-645; or Military Specification MIL-P-8585. This should be followed by one coat of black enamel, Military Specification MIL-P-15146.

16-33. When facilities permit, the chain links should be preheated prior to both the primer and final coat of painting. A temperature of 250 degrees F (121 degrees C) is recommended, but a lower temperature of 150 degrees F (66 degrees C) will decrease the drying time. In cold weather, apply some heat to counteract the natural thickening of paint. This can be accomplished by using an immersion-type electric heater or a steam coil. When left standing for a considerable period, the turpentine substitute can evaporate to such an extent that it will cause thickening of the paint. The addition of solvent will remedy such a condition.

Note: Vessels which receive anchor chains that have been coated with either red-lead primer or zinc-chromate primer and black enamel or black-asphalt varnish should have this coating left intact and covered with one coat of black enamel, Military Specification MIL-P-15146.

Replacement of Worn Chain

16-34. Replace any part of the chain that has corroded or worn so that the mean diameter is reduced to 90 percent of its normal diameter. However, replace only if the diameters of the remaining links allow continued use. If it appears uneconomical to replace worn parts, the chain should be surveyed. If replacements are made, the new links, shackles, or parts should be heat-treated, proof-tested, and, in the case of wrought iron, heat-treated again. In each case a complete report should be made containing the following information:

- Material composition of the chain.
- Shot number.
- Length of each shot.
- Nature of work actually performed on the chain.
- Date of such work.
- Cost.

16-35. This report should reference the file number of the correspondence authorizing the work involved. This report should also include disposition of the chain after the heat treatment.

Chapter 17

Damage Control

INTRODUCTION

17-1.   Damage control is based on the premise that the safety and life of a ship depends on watertight integrity. This chapter describes some emergency procedures that can be used in the event the ship's hull has been punctured and watertight integrity has been lost. The procedures described are emergency measures taken by the damage control team to maintain watertight integrity of the ship in the event of accident, collision, or grounding.

DAMAGE CONTROL PROGRAM

17-2.   If a ship's hull is punctured, watertight integrity is lost. If enough water is allowed to enter the hull and is uncontrolled, the ship will sink. There is no such thing as a "little leak". Any size leak is a cause for alarm. Through damage control, this "leak" may be either stopped or reduced to a point where the ship's pumps can control any excess water.

DAMAGE CONTROL TEAM

17-3.   Along with other emergency duties (fire and lifeboat), certain crew members are also assigned to an emergency squad or damage control team. This team may consist of the chief officer, an engineer, bosun, and two or more seamen and enginemen. There should be sufficient skills among the team members to perform the tasks required in an emergency.

DAMAGE CONTROL KITS

17-4.   Army vessels are authorized to carry specified damage control kits. These damage control kit items should be set aside and used only for their intended purpose.

PURPOSE OF THE DAMAGE CONTROL TEAM

17-5.   In the event of fire, collision, grounding, or hostilities, one of the damage control team's missions is to assist in maintaining the watertight integrity of the ship. Many ships have been lost because no real effort was made to save them. When plugging leaks, the ultimate aim is to stop the leak permanently. The amount of water entering a vessel through a hole varies directly with the area of the hole and with the square root of its depth. Realistically, if you can reduce the flow of water by more than 50 percent, it is a job well done. Also, the ship's pumps should be able to handle whatever water is left. The values in Figure 17-1 show how important it is to put some kind of plug into any hole right away. Damage control also consists of either shoring up decks that are weakened or strengthening bulkheads between flooded compartments. Although all damage control work is temporary, it must be strong enough to allow the ship to make it back to port safely.

SHORING

17-6.   The term "shoring" involves two phases:

- Stopping or reducing the inflow of water.
- Bracing or shoring up the damaged or weakened members of the ship's structure by transferring and spreading the pressures to other portions of the ship.

17-7.   Shoring also includes the processes of patching and plugging.

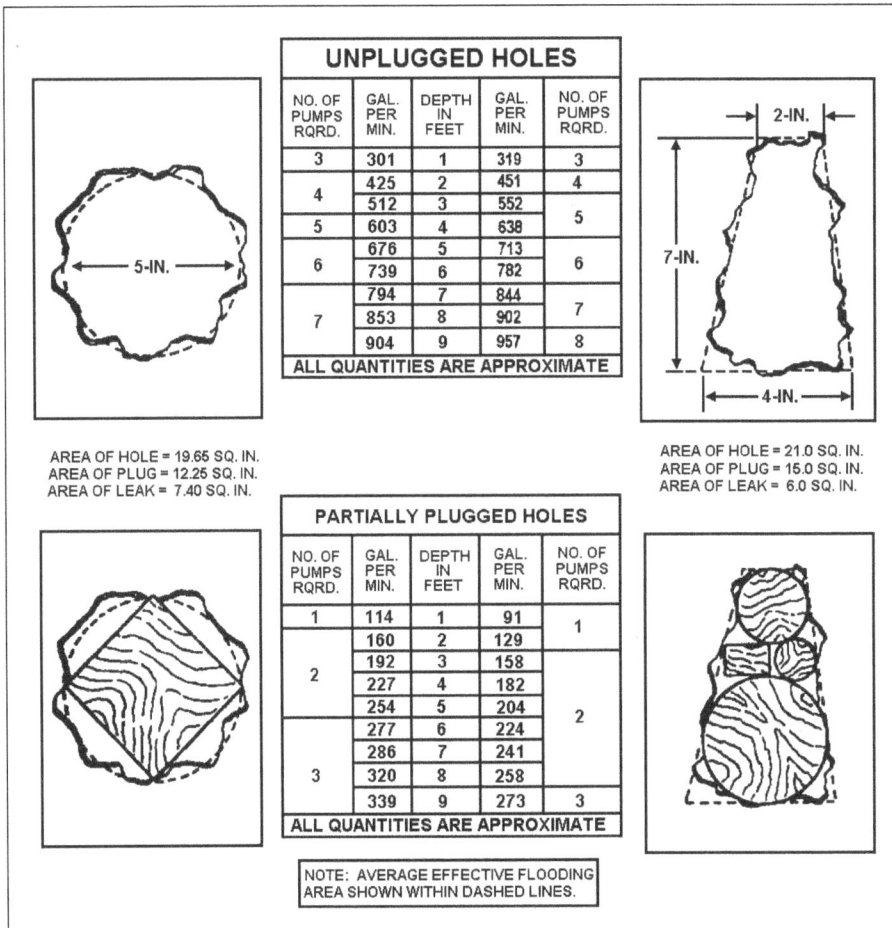

UNPLUGGED HOLES

NO. OF PUMPS RQRD.	GAL. PER MIN.	DEPTH IN FEET	GAL. PER MIN.	NO. OF PUMPS RQRD.
3	301	1	319	3
4	425	2	451	4
	512	3	552	5
5	603	4	638	
	676	5	713	6
6	739	6	782	
	794	7	844	7
7	853	8	902	
	904	9	957	8
ALL QUANTITIES ARE APPROXIMATE				

AREA OF HOLE = 19.65 SQ. IN.
AREA OF PLUG = 12.25 SQ. IN.
AREA OF LEAK = 7.40 SQ. IN.

AREA OF HOLE = 21.0 SQ. IN.
AREA OF PLUG = 15.0 SQ. IN.
AREA OF LEAK = 6.0 SQ. IN.

PARTIALLY PLUGGED HOLES

NO. OF PUMPS RQRD.	GAL. PER MIN.	DEPTH IN FEET	GAL. PER MIN.	NO. OF PUMPS RQRD.
1	114	1	91	1
	160	2	129	
2	192	3	158	
	227	4	182	
	254	5	204	2
	277	6	224	
3	286	7	241	
	320	8	258	
	339	9	273	3
ALL QUANTITIES ARE APPROXIMATE				

NOTE: AVERAGE EFFECTIVE FLOODING
AREA SHOWN WITHIN DASHED LINES.

Figure 17-1. Flooding Effect Comparison: Unplugged Holes Versus Partially Plugged Holes

CLEARING THE DECKS

17-8.   The first step in effective shoring is to clear the decks. Damage serious enough to produce a hole in the hull usually leaves wreckage scattered about the area. The damaged area should be cleared quickly to permit the damage control team to do a quick, adequate, and safe job. Most loose wreckage can be removed by hand. At other times cutting and breaking are required. This requires the use of mauls, sledges, axes, heavy cold chisels, pinch bars, power drills, power chisels, and air hammers. If fire accompanies the damage, burning bedding, stores, and supplies must be removed. A devil's claw (a homemade long-handled rake device) made of steel is handy for this purpose. Shoring tools such as saws, 2-foot squares, hammers, and hatchets are stowed in the ship's damage control locker. Additional equipment may be devised, limited only by the ingenuity of the ship's crew.

SPEED

17-9.   Step two in shoring is speed. Seconds count, especially if there is a hole below the waterline. Each member of the damage control team must be able to think fast and improvise shoring with whatever material is available. More than once; items such as life jackets, mattresses, pillows, and mess tables have proven to be satisfactory temporary shoring material.

PREPARATION

17-10. The third step is preparation. Only through regular drills can skills be developed that will enable each man to do a fast, effective shoring job under adverse conditions. A thorough training program should be established to train all crewmembers. Damage control lockers must be clean and orderly. All tools should be placed in secure mountings, yet be readily detachable.

SHORING PRINCIPLES

17-11. Observe the following basic principles when shoring damaged or weakened members of a ship's structure.

- Spread the pressure. Make full use of strength members by anchoring shores against beams, stringers, frames, stiffeners, and stanchions (see Figure 17-2). Place the legs of shoring against strong-backs at angles from 45 degrees to 90 degrees (see Figure 17- 3).

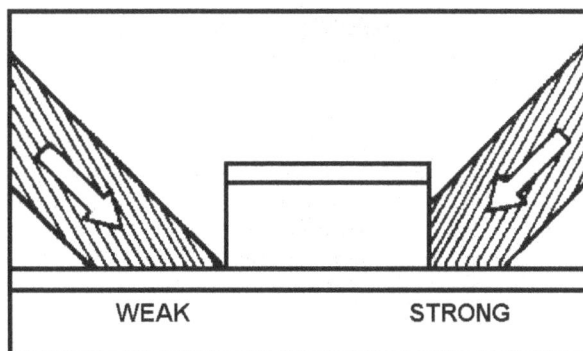

Figure 17-2. Anchoring Fit

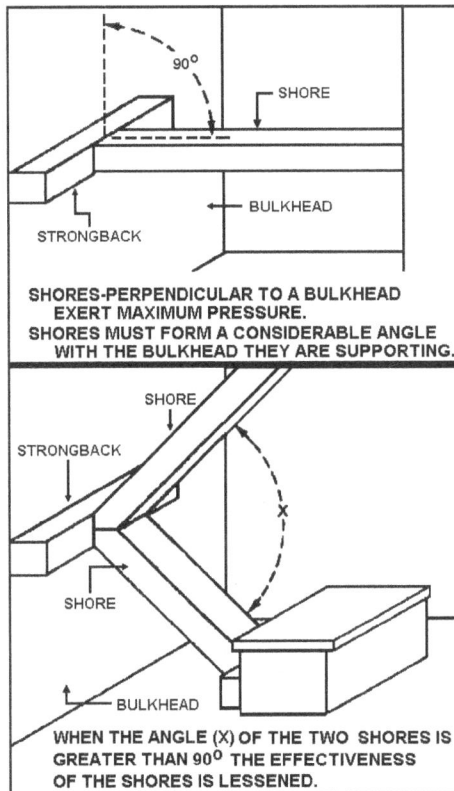

Figure 17-3. Correct Shoring Angles

- Plan shoring to hold the bulkhead as it is (see Figure 17-4). Do not attempt to force a warped, sprung, or bulged bulkhead back into place.
- Secure all shoring. Use nails and cleats to ensure that shoring will not work out of place.
- Inspect shoring periodically. The motion of the ship often can produce new stresses that will cause even carefully placed shoring to work free. Inspect all shoring regularly, particularly when the ship is underway.

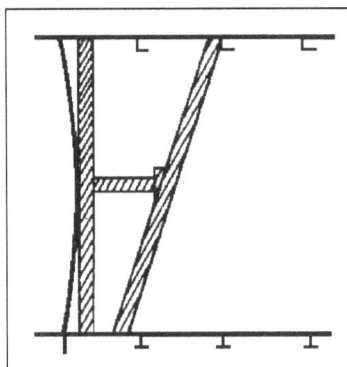

Figure 17-4. Shoring for Bulging Plate

BRACING

17-12. In addition to breaks and cracks in the hull, severe damage to a ship can impose stresses on bulkheads adjacent to the damaged area. The internal bulkheads of a ship are not designed to withstand a great amount of internal water pressure and must be immediately braced. It is unlikely that any two bracing jobs will ever be handled in the same manner, even among the same class ships. The location and extent of damage present individual problems to test the common sense and good judgment of the shoring party. Each case is different. The following fundamentals serve as a starting point. Brace weakened or damaged bulkheads against decks, overhead beams, stanchions, and hatch coamings. It is important to allow a three-point distribution of pressure. At the same time it is equally important to avoid damage to flanges, stiffeners, and deck beams. Place shores, so the pressure they receive produces direct compression. However, never place a shore deliberately below the point of compression so that bowing results. It is better to install several shores at close intervals because a bowed shore is dangerous to personnel and ship safety. When relatively long shores support heavy pressures, there is an even greater tendency of the shores to bow. Figure 17-5 shows shoring against horizontal pressure.

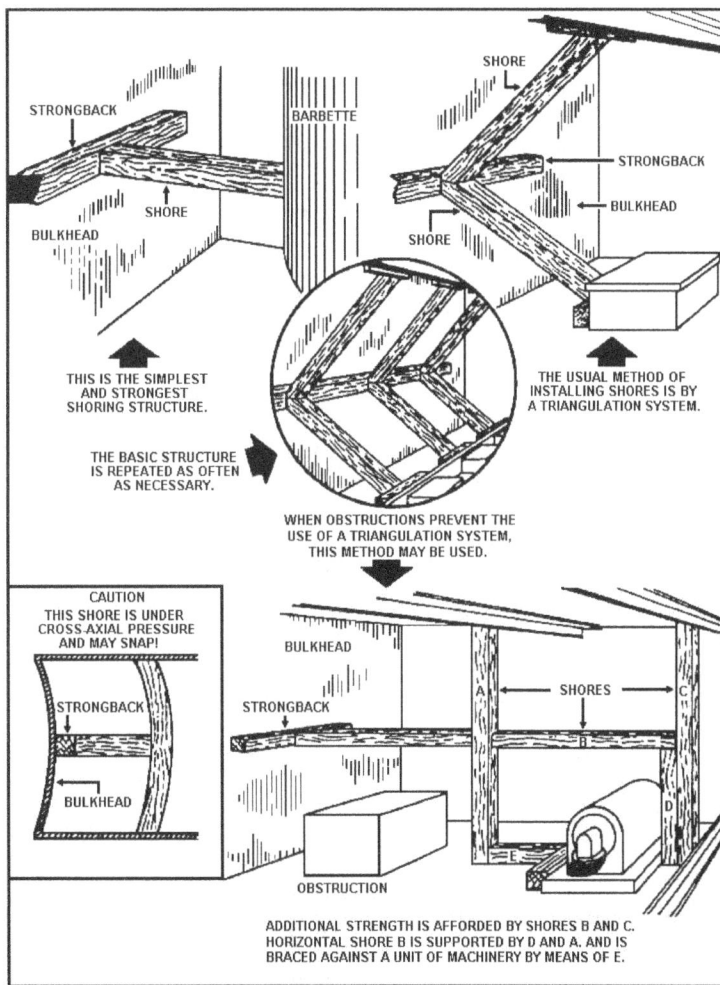

Figure 17-5. Shoring Against Horizontal Pressure

> **CAUTION**
>
> 1. Never use a shore that is longer than 30 times its minimum thickness. Therefore, the maximum length of a standard 4-inch by 4-inch shore must not exceed 10 feet (120 inches).
>
> 2. Sometimes, it is impossible to foresee where new stresses will cause bowing. If a timber begins to bow, the pressure should be relieved immediately to prevent snapping. If there is danger of a shore jumping out as the ship works, the shore should be held in position with nails and cleats.

17-13. Secure the butt ends of shores against only undamaged members of the ship's structure such as hatches, stanchions, machinery foundations, frames, and coamings. Strongbacks may help distribute pressure on a bulkhead or deck. Each strongback must be supported with a number of shores placed to exert pressures perpendicular to the bulkhead. Use wedges to anchor and tighten shores in place. They should be driven in uniformly from both sides so the end of the shore will not be forced out of position (Figure 17-6). When the butt of a shore is anchored against a joint having protruding rivet heads, the shore is anchored with a shole (a plank or plate with pockets chiseled out for the rivet heads). This prevents the end of the shore from splitting. As the shoring job progresses, it must be checked carefully to ensure that all wedges are exerting the same amount of force on the member being shored in place. The desired force should be obtained by using as few wedges as possible.

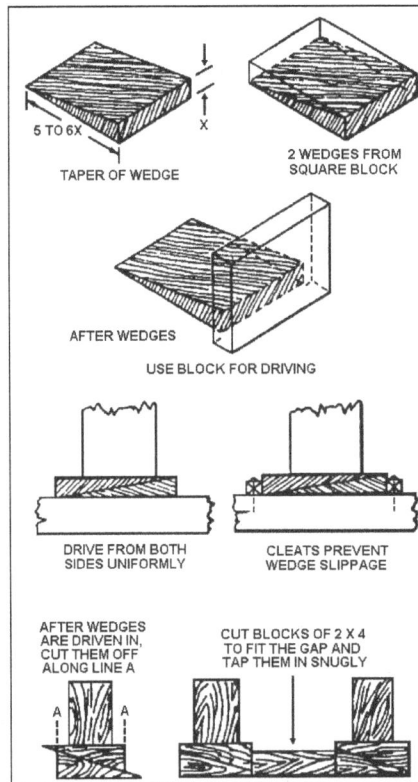

Figure 17-6. Driving Wedges

17-14. About half the shoring job is getting the right size shore. Practice in using the measuring batten will help in an actual emergency. The ends of the batten should be fitted firmly into the recesses selected for anchoring the timbers, and the exact measurement from each leg should be transferred carefully to the shoring. The most rapid and accurate way to measure shores for cutting is by using an adjustable shoring batten similar to the one shown in Figure 17-7. To use the shoring batten, extend it to the required length and lock it with the thumbscrews on the length-locking device. Then measure the angles of cut by adjusting the hinged metal pieces at the ends of the batten and lock the angle-locking device in place. Lay the batten along the shore. Mark and cut the timber to the proper length and angle. Shores should be cut 1/2 inch shorter than the measured length to allow space for wedges.

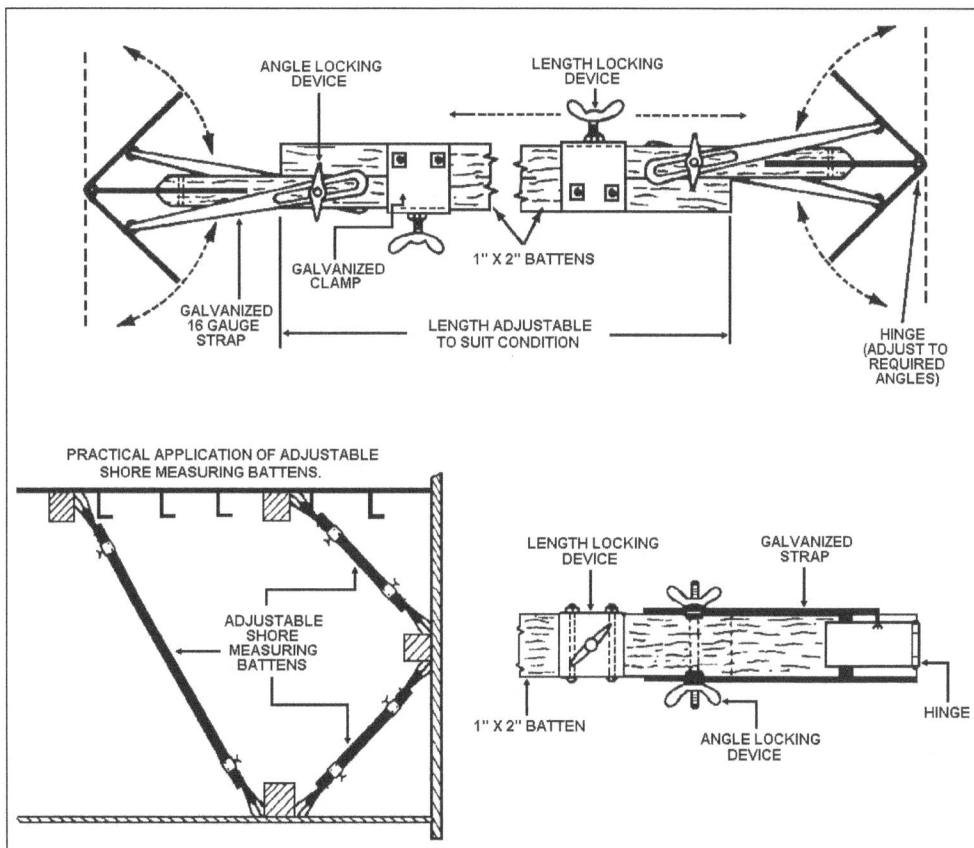

Figure 17-7. Adjustable Shore Measuring Batten

17-15. Shoring may be confined to one compartment of the ship. However, if the pressure on a deck or bulkhead is so great that the next deck or bulkhead anchorage cannot safely withstand the pressure, place the shoring in adjacent areas to distribute pressures (Figure 17-8). The work of bracing often can be expedited by having chainfalls, blocks and tackles, and jacks available for use in moving heavy weights back into their original position. Dry sand can be sprinkled on oily decks as a safeguard against slipping.

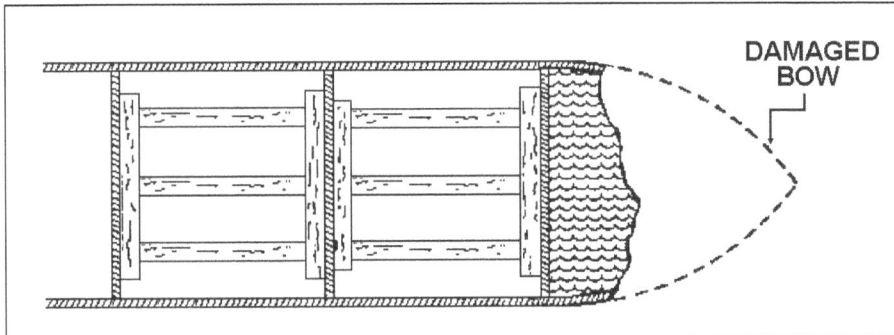

Figure 17-8. Shoring Spread to Adjacent Compartments

17-16. Most shoring is used to support bulkheads endangered by structural damage or weakness caused by concussion or by the pressure of floodwater. Shoring up a flooded compartment requires that particular attention be given to the heavier pressures existing at the lower sections of the bulkheads. These pressures increase with the height of water in the compartment. The area of greatest pressure can also move from one area to another due to pitching and rolling of the ship. When a hatch or door is used to support shoring, the entire hatch cover or door should be shored over. Never place shoring ends and wedges directly against such openings, as they are the weakest parts of the bulkhead or deck. The pressure should be spread over the hatch or door and onto supporting structures. Closely allied with shoring are the other basic damage control operations. Emergency lighting and lines for submersible pumps must be rigged. Ruptured fire mains and other liquid carrying lines must be isolated or repaired. The entire operation must be attacked with determination and with an open mind to cope with conditions that never seem to parallel those in a reference book.

USE OF CARPENTER'S SQUARE IN SHORING

17-17. If a shoring batten is not available, measure the shores for length by using a folding rule or a steel tape and a carpenter's square (see Figure 17-9). The step-by-step procedure for measuring shores is as follows:

- Measure the distance A from the center of the strongback to the deck. Then measure the distance B from the edge of the anchorage to the bulkhead, subtracting the thickness of the strongback.
- Lay off the measurements A and B on a carpenter's square, using the ratio 1 inch to 1 foot. Rule measurement is taken to the nearest 1/16 inch. To maintain the ratio of 1 inch to 1 foot, use the information in Table 17-1.
- Measure the diagonal distance between A and B. Figure 17-9, shows this distance as 7 7/8 inches. Because of the 1 inch to 1 foot ratio, the distance in feet would be 7 7/8 feet or 7 feet 10 1/2 inches.
- Subtract a 1/2 inch since shore should be cut 1/2 inch shorter than the measured distance. The final length of the shore should be 7 feet 10 inches.

17-18. The carpenter's square may also be used to measure the angles of cut and to mark the shore for cutting (see Figure 17-10).

Table 17-1. Ratio Conversion Table

Actual Rule Measurement	Measurement on Carpenter's Square
3/4 inch	1/16 inch
1 1/2 inches	1/8 inch
2 1/4 inches	3/16 inch
3 inches	1/4 inch
3 3/4 inches	5/16 inch
4 1/2 inches	3/8 inch
5 1/4 inches	7/16 inch
6 inches	1/2 inch
6 3/4 inches	9/16 inch
7 1/2 inches	5/8 inch
8 1/4 inches	11/16 inch
9 inches	3/4 inch
9 3/4 inches	13/16 inch
10 1/2 inches	7/8 inch
11 1/4 inches	15/16 inch
12 inches	1 inch

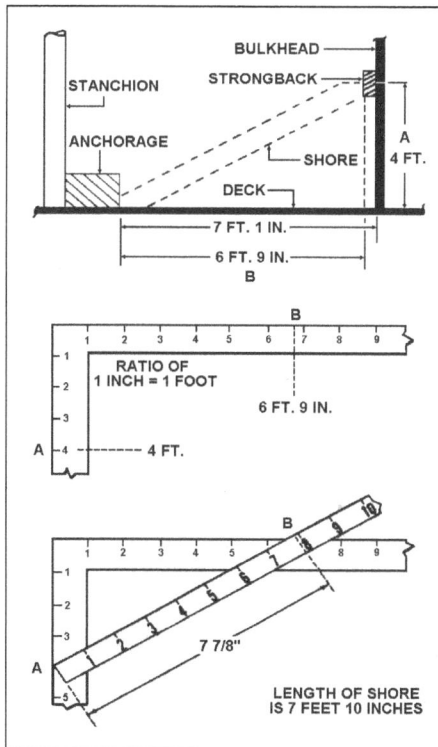

Figure 17-9. Measuring Length of Shore

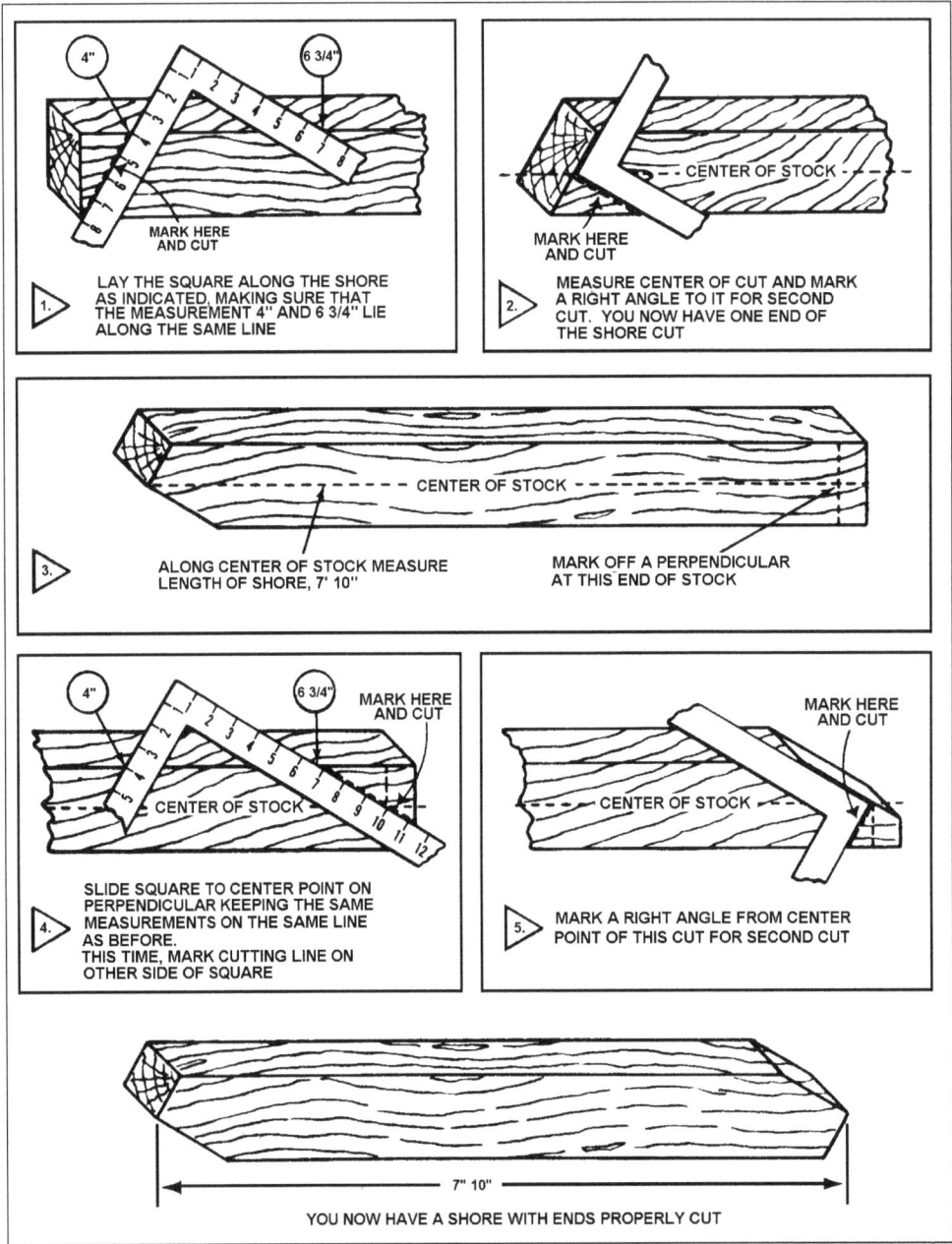

Figure 17-10. Cutting the Angles of a Shore

PLUGGING

17-19. Plugging is a technique used for filling small holes with a suitable material to stop the flow of water until permanent repairs can be made. Holes up to 6 inches in diameter can usually be plugged by driving in wooden plugs or wedges.

WOODEN PLUGS

17-20. Plugs made of bare, soft wood perform best because they soak up water, swell, and hold firmly in place.

- Painted wood does not swell, and should be used only in emergencies.
- Square-end plugs hold better than conical plugs.
- Additional sealing properties can be obtained by wrapping the plugs with cloth.
- Use "oakum" (a black sticky fibrous material made from old hemp) if carried aboard ship. Coat the plugs with oakum before putting them in the hole.

LEAD PLUGS

17-21. Plugs cut from sheet lead are effective in stopping leaks when a plate has pulled loose from its rivets. Often small leaks can be stopped by driving in lead slugs, strips, or plugs.

CRACKS

17-22. Cracks are dangerous because they may enlarge and spread. If time permits, drill a small hole at each end of the crack (see Figure 17-11). This will prevent it from cracking any further.

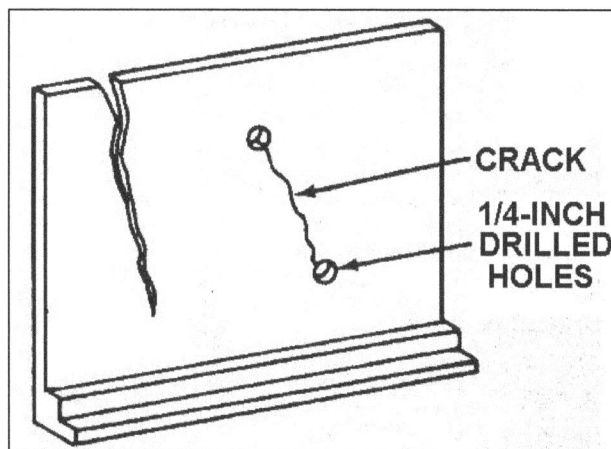

Figure 17-11. Cracks with Holes

17-23. The drilled holes should be plugged with either machine or wood screws. A flat piece of rubber or canvas backed up with a board should be laid across the crack and held in place with shoring (see Figure 17-12). This type of patch should be inspected frequently as it tends to shift and slip as the ship works.

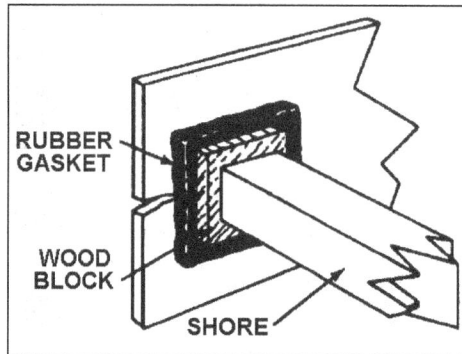

Figure 17-12. Shoring over Drilled Crack

PATCHING

17-24. Patching is used to cover larger holes with sections of improvised or prefabricated material. This TC only describes the procedures for applying a soft patch because in damage control you are only interested in stopping or controlling the inflow of water. The soft patch is for temporary repair only.

PILLOWS, MATTRESSES, AND BLANKETS

17-25. Pillows, mattresses, and blankets can be rolled up and shoved into holes. They can be rolled around a wooden plug or a timber to increase their size and to provide rigidity. Such plugs cannot be relied upon, as they have a tendency to be torn out of the holes by action of the sea. This is an expedient to retard the flow of water entering the vessel until a more suitable patch can be applied. Figure 17-13, shows the use of mattresses installed inside and outside the hull as a patch. Placing mattresses inside will reduce the possibility of the patch being knocked away by the sea. If innerspring mattresses are used, at least two thicknesses of blanket should be used as a facing. Over a period of time, feather pillows are not as effective as folded blankets for patches. Feathers in the pillow get wet and tend to lump at one end.

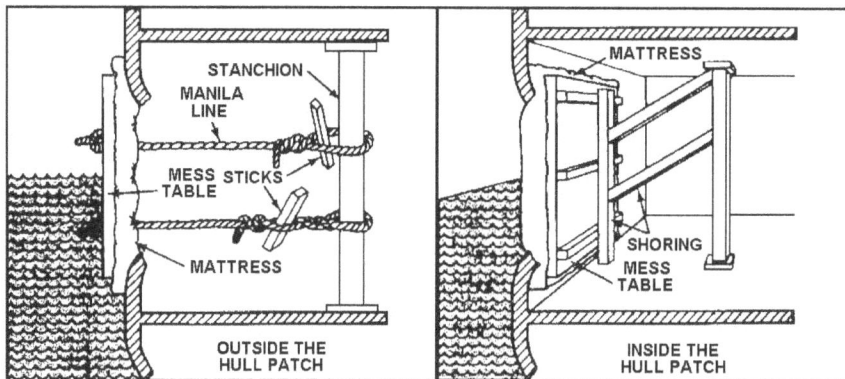

Figure 17-13. Example of Hull Patching Using a Mattress

HINGED PLATE PATCH

17-26. A variation of the plate patch is called a hinged plate patch (see Figure 17-14). This is a circular plate, 18 inches or less in diameter, cut in two, and so hinged that it can be folded and pushed through a hole from inside the vessel. The plate should be fitted with a gasket, such as a pillow, and also a line for securing to the vessel. Using water diving equipment, this patch can be applied over a submerged hole without going outside the vessel. This patch is for use over relatively small holes, as it has no vertical support to hold it in place.

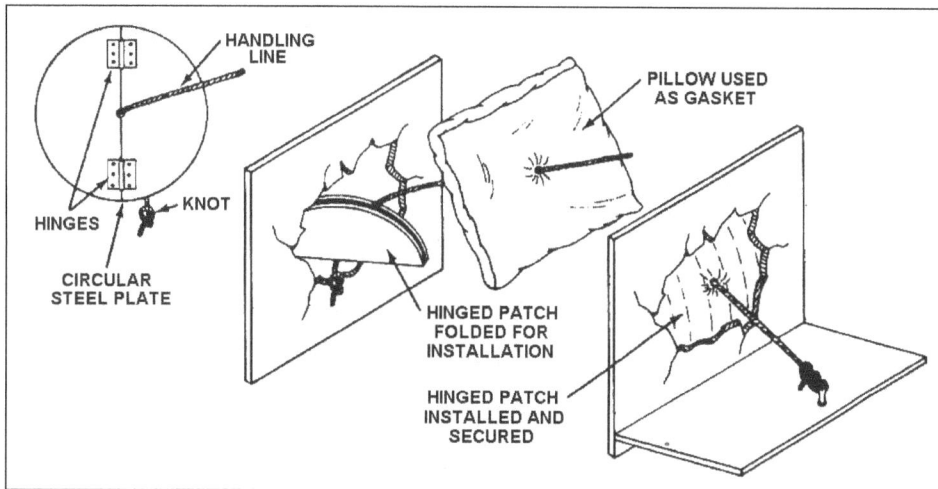

Figure 17-14. Hinged Plate Patch

BUCKET PATCH

17-27. An ordinary galvanized bucket can be used in a variety of ways to stop leaks. It can be pushed into a hole, bottom first, to form a metal plug, or it can be stuffed with rags and put over a hole. It can be held in place by shoring or by using a hook bolt.

USE OF A HOOK BOLT FOR SECURING PATCH

17-28. A hook bolt is a long bolt having the head end shaped so that the bolt can be hooked to plating through which it has been inserted. The common types are the T, J, and L (see Figure 17-15). The long shanks are threaded and provided with nuts and washers. Steel or wooden strongbacks are used with them. The bolt has no regular head. The head end of the bolt is inserted through a hole and the bolt rotated or adjusted until it cannot be pulled back through the hole. A pad or gasket, backed by a plank or strongback, is then slid over the bolt and the patch secured in place by taking up on the nut. It is generally necessary to use these bolts in pairs. Figure 17-16 shows an installed patch using two J-type hook bolts. Hook bolts can be used in combination with various patches such as the folding plate and the bucket. Figure 17-17, shows how to patch a hole using the folding T-type hook bolt.

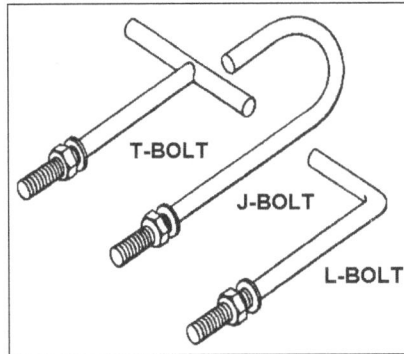

Figure 17-15. Types of Hook Bolts

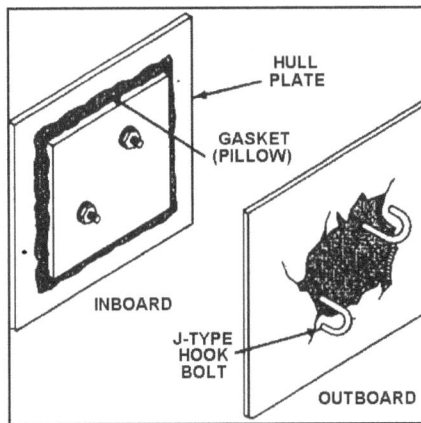

Figure 17-16. Patching Using Hook Bolts

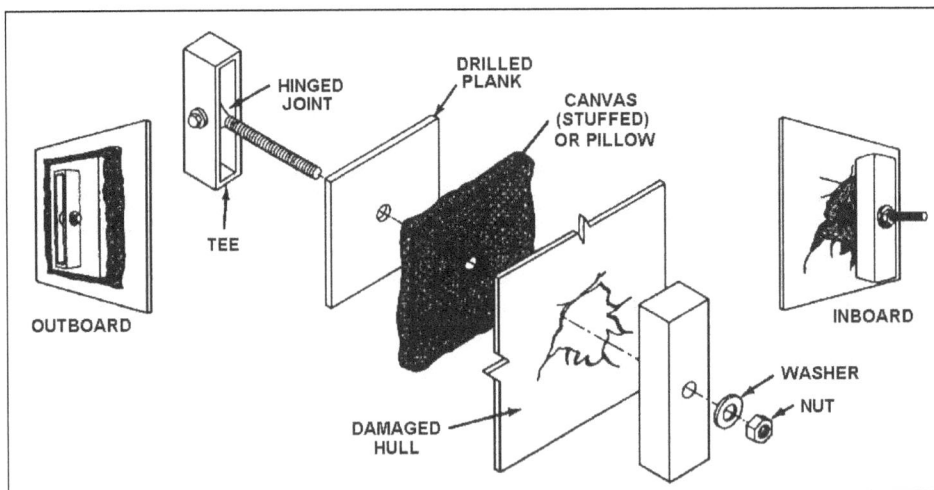

Figure 17-17. Patching Using the Folding T-Type Hook Bolt

PIPE REPAIR

17-29. Piping system leaks usually accompanies any large hole in the hull. Soft patches can seal holes and cracks in low-pressure lines and water lines. Install a soft patch on a pipe as follows (see 17-18):

- Opening is plugged with soft wood plugs or wedges (the flow of water must not be retarded by driving an excessive amount of wood into the pipe).
- Trim plugs and wedges flush with the outside of the pipe.
- Wrap rubber sheeting over the damaged area and back it with light sheet metal held in place with bindings of wire or marline.

17-30. Stop minor pipe leaks with a jubilee patch (an adjustable strap with a flange on each edge). These can be made up as needed. The patch is shaped by bending sheet metal around a cylinder and turning out the flanges and then clamped in place (see Figure 17-19). The flanges may have to be reinforced as pressure increases (Figure 17-20).

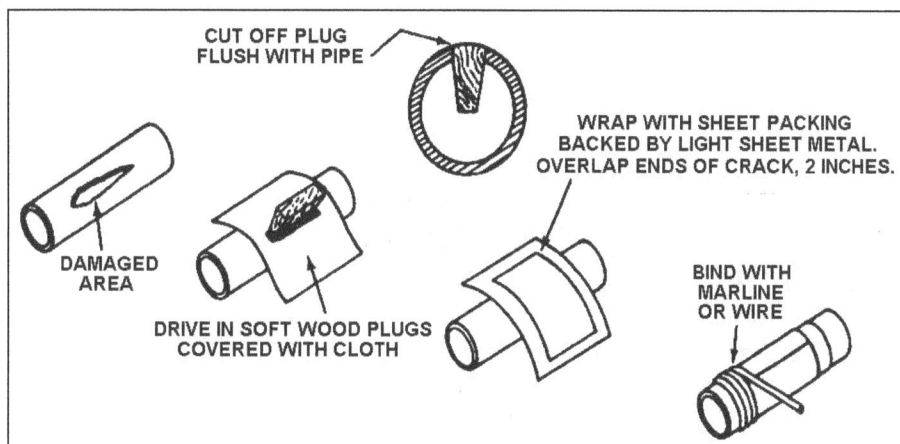

Figure 17-18. Installing a Soft Patch on a Pipe

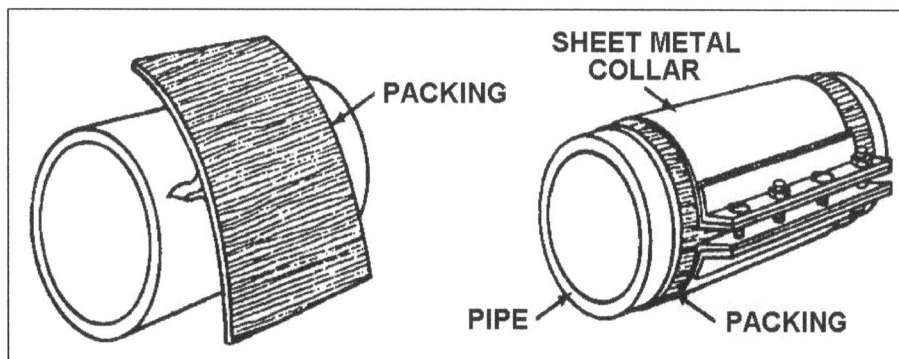

Figure 17-19. Jubilee Pipe Patches

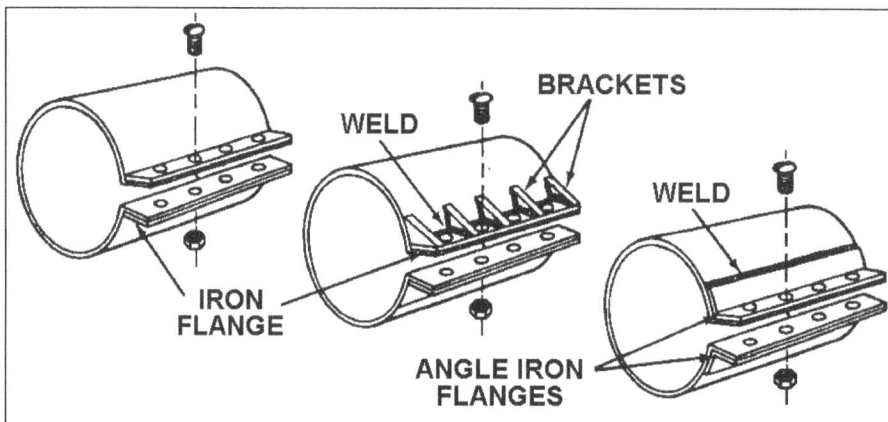

Figure 17-20. Three Types of Reinforced Clamps

EMERGENCY DAMAGE CONTROL METALLIC PIPE REPAIR KIT

17-31. Most water, fuel, and gas lines can be repaired and restored to the system within 30 minutes if the contents of the emergency damage control metallic pipe repair kit are applied properly. In addition to repair or patching of piping, certain materials, which may be used to patch small cracks and ruptures in flat metal surfaces, are included in the kit. Materials in the kit may be obtained separately through appropriate supply channels whenever a need arises to replace them. You do not need to obtain another complete new kit. A complete kit contains the following materials:

- Four cans epoxy, resin, 400 grams each.
- Four cans liquid hardener, 100 grams each.
- Four cans paste resin, 300 grams each.
- Four cans paste hardener, 75 grams each.
- One piece woven roving cloth, 24" x 10 ".
- One piece void cover, 8" x 36".
- One piece polyvinyl chloride (PVC) film, 36" x 72".
- One chalk line, 1/8 pound.
- Four pairs of gloves.
- Two eye shields.
- Four wooden spatulas.
- One sheet of emery cloth, 9" x 11".
- One instruction manual.

DESCRIPTION OF MATERIALS

17-32. The following describes the basic materials found in the kit. The discussion of factors related to plastics is given to help you gain a better understanding of the kit and its use.

- **Void Cover.** The void cover is a resin-treated glass cloth that can be cut and formed to cover the damaged area. It is sufficiently rigid to give support to the patch.
- **Woven Roving Cloth.** The woven roving cloth is made of a short-staple, glass fiber woven into a thick, fluffy cloth. During the application of a plastic patch this cloth is coated with the resin-hardener mixture and either wrapped around or placed over the damaged area. The glass cloth provides the main strength of the patch and also provides a means of applying the resin-hardener mixture.
- **Film (PVC).** The plastic film, referred to as PVC, is a thin transparent polyvinyl chloride material. It is used as a separating film for the flat patch to prevent the patch from sticking to the

backup plate or other supports. In the pipe patch, it is used to cover the entire patch and retain the activated resin around the patch. Kraft wrapping paper may be used as a substitute if necessary.

- **Resins and Hardeners.** The liquid and paste resins are of the epoxy type. The liquid and paste hardeners are chemical compounds used to harden the resins. The resins and the hardeners are packaged in premeasured amounts. For proper mixture and better results, mix the complete contents of the hardener in the smaller can with the complete contents of the resin in the larger can.

CAUTION

DO NOT mix hardener with resin until all preparations have been completed. DO NOT intermix liquid resin and paste hardener or paste resin and liquid hardener.

17-33. When the resins and the hardeners are mixed together, a chemical reaction occurs which causes the mixture to harden (liquid mixture, approximately 12 minutes; paste mixture, approximately 17 minutes). This reaction is exothermic, which means that heat is given off. For approximately 12 to 17 minutes the temperature increases gradually until it reaches 120 degrees F to 135 degrees F. At this point a sudden sharp rise in temperature is known as kick over. It is at this temperature that the resin-hardener mixture begins to solidify and change color from gray to light brown. The peak temperature (350 degrees F) can be observed through the external change of the patch. The resin-hardener mixture begins to cool slowly due to the poor thermal conductivity of the materials. After kick over, the mixture continues to harden and increase in strength. This process is referred to as curing. Approximately 30 minutes after kick over (the sharp rise in temperature), the patch is strong and hard and cool enough to use. Pressure should not be restored to the system until the patch has cured. The patch is considered sufficiently cured when the bare hand can be placed on it without discomfort from heat.

17-34. Several factors contribute to the control of kick over. The most important factor is the temperature. Both the initial temperature of the activated resin mixture and the temperature of the atmosphere, affect the kick over time. However, of these two temperatures, the initial temperature of the activated resin has the greater effect. When the temperature of the resin and the hardener prior to mixing is increased, the kick over time decreases. Conversely, when the temperature of the resin and hardener prior to mixing decreases, the kick over time increases. Knowledge of controlling kick over is necessary since it corresponds to application of working time. This means that when the initial temperature of the mixture is 73 degrees F, the patching material must be placed over the rupture within 12 minutes. Once the resin and the hardener are mixed together, the chemical reaction cannot be stopped. Therefore, the patch should be completely applied before kick over occurs.

17-35. Figure 17-21 shows the relationship of the kick over time of the resin temperature. If you know the resin temperature at the time of mixing, you will be able to determine the amount of time available to apply the patch before kick over occurs. You can see that if the resin temperature is 80 degrees (point A), the kick over will occur in less time than if the resin temperature were 60 degrees (point B). The difference in resin temperatures represents an application working time of 9 minutes versus 18 minutes.

Note: If the initial resin temperature exceeds 80 degrees F the temperature should be reduced by artificial means to 73 degrees F prior to mixing. This lowering of the temperature allows for additional application working time.

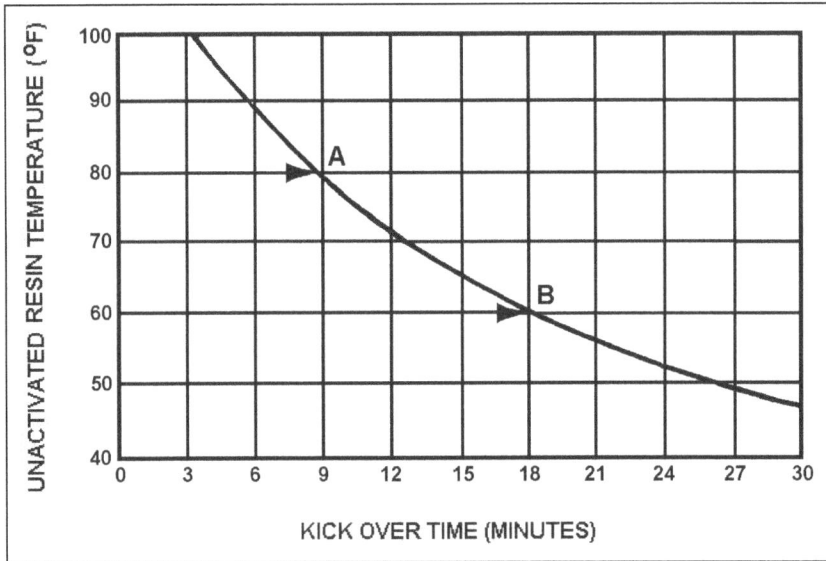

Figure 17-21. Resin Temperature Versus Kick Over Time Graph

MATERIALS REQUIRED FOR PIPE PATCH

17-36. Table 17-2 shows approximate quantities of materials required for pipe patches. The top figure in the boxes shows quantities in the amount of resin and hardener mixture in grams. The second group of figures, immediately below is the dimensions in inches of woven roving cloth.

ADVANTAGES OF THE PLASTIC PATCH

17-37. From the damage control viewpoint, the main advantages of the plastic patch are versatility, simplicity, effectiveness, speed of application, and durability. The plastic patch can be successfully applied to a variety of damaged surfaces (such as smooth edges or jagged protruding edges). Since the plastic has excellent adhesive qualities it can be readily applied to steel, cast iron, copper, copper-nickel, brass, bronze, and galvanized metals. The plastic materials and the plastic patch may be readily prepared and applied. By following the instructions outlined in the instruction manual included in the kit, a person with little or no experience can readily prepare the materials and apply a plastic patch. Applying a plastic patch is comparable to applying a battle dressing used in first aid. If the materials are properly prepared and the application procedures are duly followed, the plastic patch will be 100 percent effective. If leakage occurs through a plastic patch, it is likely that proper preparation and application procedures have not been followed. The speed of application will vary somewhat with the size and type of rupture and also with local working conditions. When proper preparation procedures are followed, an inexperienced crew, who has had a minimum amount of training, can apply a simple patch to a 4-inch pipe in 10 minutes or less. The type and size of the structure to which the patch is applied do not materially affect the time involved in patching. However, some types of damage may require more initial preparation.

Table 17-2. Table of Approximate Quantities of Materials Required for Pipe Patches

Length of rupture (inches)	1	1 ½	2	2 ½	3	4	5	6
1 / 2	200 8 x 12	290 8 x 18	400 8 x 25	500 8 x 31	500 8 x 37	770 8 x 48	960 8 x 60	1150 8 x 72
1	220 9 x 12	325 9 x 18	450 9 x 25	550 9 x 31	675 9 x 37	865 9 x 48	1080 9 x 60	1300 9 x 72
2	240 10 x 12	360 10 x 18	500 10 x 25	620 10 x 31	740 10 x 37	960 10 x 48	1200 10 x 60	1440 10 x 72
3	260 11 x 12	400 11 x 18	550 11 x 25	680 11 x 31	800 11 x 37	1050 11 x 48	1320 11 x 60	1585 11 x 72
4	290 12 x 12	435 12 x 18	600 12 x 25	750 12 x 31	890 12 x 37	1150 12 x 48	1440 12 x 60	1725 12 x 72
5	312 13 x 12	470 13 x 18	650 13 x 25	800 13 x 31	960 13 x 37	1250 13 x 48	1560 13 x 60	1875 13 x 72
6	336 14 x 12	500 14 x 18	700 14 x 25	875 14 x 31	1040 14 x 37	1350 14 x 48	1680 14 x 60	2020 14 x 72
7	360 15 x 12	540 15 x 18	750 15 x 25	925 15 x 31	1110 15 x 37	1450 15 x 48	1800 15 x 60	3160 15 x 72
8	385 16 x 12	575 16 x 18	800 16 x 25	1000 16 x 31	1185 16 x 37	1550 16 x 48	1920 16 x 60	2300 16 x 72
9	408 17 x 12	612 17 x 18	850 17 x 25	1054 17 x 31	1258 17 x 37	1632 17 x 48	2040 17 x 60	2448 17 x 72

APPLICATION OF PLASTIC PATCH

17-38. When applying a plastic patch, you will see that as the individual patch materials are applied, the patch becomes progressively wider. Figure 17-22 shows the relative positions of the patch materials to one another.

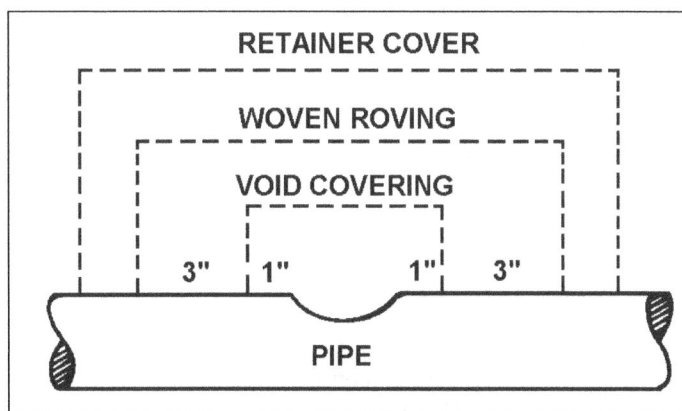

Figure 17-22. Relative Positions of Patch Materials

17-39. The buildup in the patch length during application must be considered initially when determining the length of the patch to be applied. Where suitable, allow the patch to extend at least 4 to 5 inches on either side of the rupture. In addition to the size of the rupture, the width of the patch may also depend on the location of the rupture in the pipe system. For example, an elbow rupture may require a wider patch than the same size rupture would require in a straight section of pipe. Certain specific preparations must be made prior to the actual application of the plastic patch. These are as follows:

- Secure or isolate the rupture area in the piping system.
- Remove all lagging.
- Clean the area around the rupture and remove all grease, oil, dirt, paint, and other foreign matter. If grease or oil is present, use an approved solvent such as ethyl chloroform. If this solvent is not available, scrape and wipe the surface until it is clean. When a clean surface is obtained, the surface may be further abraded for better adhesion. An abrasive cloth is furnished with the kit.
- Make sure that the entire pipe surface is dry.
- Where practical, simplify the rupture by bending or removing irregular projections. This may be accomplished by cutting or burning.

CAUTION

It is of the utmost importance that no explosive conditions exist prior to using spark-producing tools or burning equipment.

17-40. Determine the materials that will be required, such as the amount of woven roving cloth and the amount of resins and hardeners. For example, a 2-inch rupture in a 2-inch diameter pipe will require 500 grams of activated resin and a piece of woven roving cloth that is at least 25 inches long. Cut the woven roving cloth wide enough to extend at least 3 to 4 inches on either side of the rupture.

SIMPLE PIPE PATCH

17-41. The following are step-by-step procedures for applying the simple pipe patch.

- Put on eye shields and gloves. Then open the liquid resin can and the liquid hardener can.
- Add hardener to the resin and mix thoroughly for approximately 2 minutes or until a uniform gray color is observed. (Note that the entire contents of the liquid hardener in the smaller can, is the correct proportion for mixing with the entire contents of the larger can of liquid resin.)
- Coat both sides of the void cover with the resin-hardener mixture and tie the void cover over the rupture with chalk line as shown in Figure 17-23, step A.
- Lay the woven roving cloth on a clean flat surface. Starting at one end of the cloth, pour on resin-hardener mixture and spread evenly over the entire surface of the cloth using the spatula provided in the kit. Only one side of the woven roving cloth needs to be impregnated. Be sure that the edges are well impregnated with the resin-hardener mixture.
- Center the woven roving cloth over the void cover with the impregnated side toward the pipe. Wrap it around the pipe not less than three turns and preferably not more than four turns (see Figure 17-23, step B).
- Wrap the PVC film around the entire patch making at least two complete turns. Tie the PVC film with the chalk line, starting from the center of the patch and working toward one end, making 1/2-inch spacing between spirals (see Figure 17-23, step C). Tie this end securely but do not sever the line.
- Make one spiral back to the center of the patch, then working to the opposite end, form the center of the patch. Make 1 1/2-inch spacing between spirals and again secure the line. After 30 to 40 minutes the patch should be sufficiently cured to restore the pipe to service.

17-42. Remember that for best results the temperature of the liquid resin and the liquid hardener should be approximately 70 degrees F before mixing. The patch will cure in approximately one hour from the initial mixing time. After an hour, pressure may be restored to the piping system. In emergencies, if the temperature of the resins and the hardeners is below 50 degrees F, applying external heat with hot-air heaters may accelerate kick over. However, the external heat must be applied gradually because excessive application of heat will cause the plastic patch to be extremely porous.

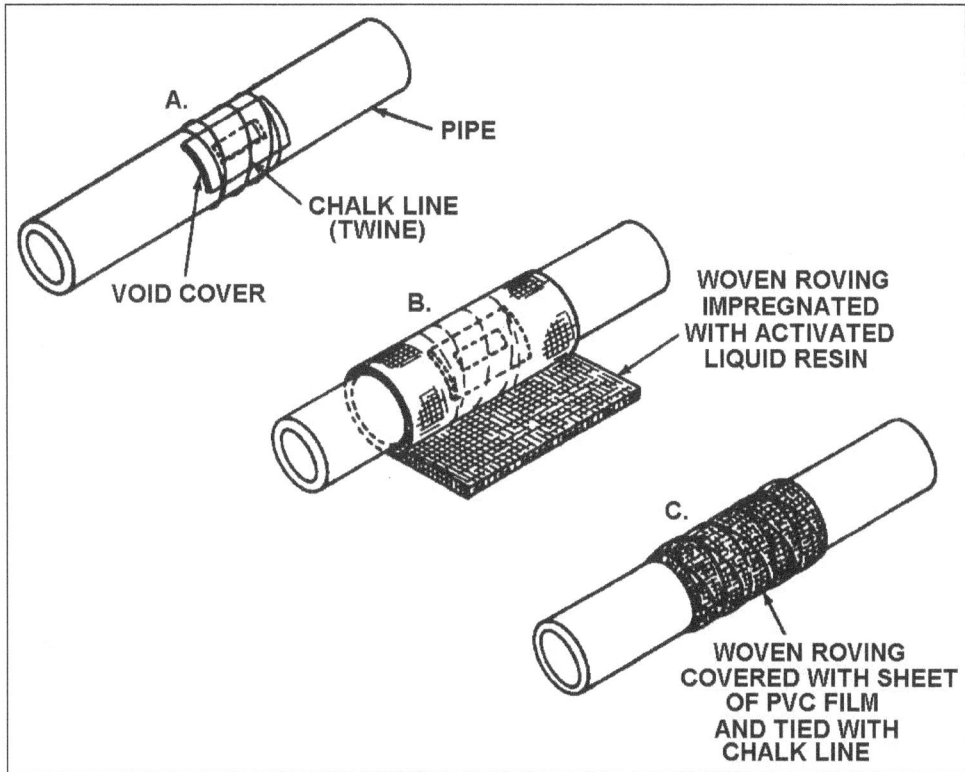

Figure 17-23. Simple Pipe Patch

This page intentionally left blank.

Glossary

SECTION I – ACRONYMS AND ABBREVIATIONS

Acronym	Definition
ABS	American Bureau of Shipping
ACMES	automated COMSEC management and engineering system
AEL	approved equipage/equipment list
AFC	adaptive frequency control
AFFF	aqueous film-forming foam
AKMS	army key management system
ALE	automatic link establishment
AM	amplitude modulation
ANCD	automated net control device
ANSI	American National Standards Institute
AOR-E	Atlantic Ocean Region-East
AOR-W	Atlantic Ocean Region-West
APL	approved parts list
AR	Army regulation
ATTN	attention
AUTODIN	automatic digital network
BC	barge, cargo
BCT	basic combat training
BD	barge, derrick
BGU	basic generation unit
BR	barge, refrigerator
BS	breaking strain; breaking strength
C	celsius; circumference
C2	capability 2
C4ISR	command, control, communications, computers, intelligence, surveillance and reconnaissance
CATS	combined arms training strategy
CB	citizen's band
CBRN	chemical, biological, radiological and nuclear
CBRNE	chemical, biological, radiological, nuclear, high yield explosives
C.E.	compass error
C-E	communications-electronics
CEOI	communications-electronics operation instructions
CELP	code excited linear prediction
CES	coast earth station

CF	causeway ferry
CFD	common fill device
CFR	Code of Federal Regulations
CG	Coast Guard
CINC	Combined Intelligence Center
CMG	course made good
CNR	combat net radio
CO2	carbon dioxide
COG	course over ground
COMDTINST	Commandant, United States Coast Guard Instruction
COML	commercial
comms	communications
COMSAT	comminucations satellite
COMSEC	communications security
CONUS	continental United States
CP	command post
CPR	cardio-pulmonary resuscitation
CSW	crew-served weapon
CW	carrier wave; continuous wave
D	diameter
DA	Department of the Army
DADMS	DMA automated distribution management system
DAJAZA	Department of the Army Judge Advocate
D.C.	District of Columbia
DD	Department of Defense
DDCP	Department of Defense control point
DEV.	deviation
DMA	Defense Mapping Agency
DMAHC	Defense Mapping Agency Hydrographic Center
DMAHTC	Defense Mapping Agency Hydrographic and Topographic Center
DMAODS	Defense Mapping Agency Office of Distribution Services
DNVT	digital non-secure voice terminal
DOD	Department of Defense
DR	dead reckoning
DSC	digital selective calling
DSN	defense switched network
DSVT	digital secure voice terminal
DTD	data transfer device
E	east

EAC	echelon above corps
ECCM	electronic counter-countermeasures
EKMS	electronic key management system
EMP	electromagnetic pulse
EOF	escalation of force
EP	estimated position; electronic protection
EPIRB	emergency position-indicating radio beacon
EPLRS	enhanced position location reporting system
ESA	European Space Agency
EST	engagement skills trainer
ETA	estimated time of arrival
F	fahrenheit; friction
FATH	fathom
FC	floating causeway
FH	frequency hopping
FM	field manual; frequency modulation
Ft	feet
GA	gauge
GDSC	Global Distance Support Center
GI	government issue
GMDSS	global maritime distress and safety system
GPM	gallons per minute
GPS	global positioning system
GT	gross ton
H	hour
HF	high frequency
HQ	headquarters
HQDA	headquarters, Department of the Army
HT	height
Hz	hertz
ICOM	integrated communications security
ID	identifier
IFF	identification friend or foe
IMCO	Inter-Governmental Maritime Consultative Organization
IMO	International Maritime Organization
In	inch
INMARSAT	International Maritime Satellite Organization

INTELSAT	International Telecommunications Satellite Organization
IOR	Indian Ocean Region
ISDN	integrated services digital network
ISO	International Organization for Standardization
ISON	integrated services digital network
JTIDS	joint tactical information distribution system
JLOTS	joint logistics over-the-shore
Kbps	kilobits per second
KC	kilocycles
KDD	key distribution device
kHz	kilohertz
KPE	key processing equipment
KT	kiloton
KTS	knots
LARC-LX	lighter, amphibious resupply cargo, 60-ton
LASH	lighter aboard ship
LBP	length between perpendiculars
LCM	landing craft, mechanized
LCPL	landing craft, personnel, large
LCU	landing craft utility; lightweight computer unit
LCVP	landing craft, vehicle, personnel
LES	land earth station
LF	low frequency
LFX	live fire exercise
LMSR	large,medium speed RO/RO
LNL	letter-number-letter
LOA	length overall
LOD	line of departure
LO/LO	lift on/lift off
LOP	line of position
LOS	line of sight
LOTS	logistics over-the-shore
LS	line size
LST	landing ship tank
LSV	logistics support vessel
LT	large tug
LWL	load waterline

m	minute
MA	mechanical advantage
MAG	magnetic
MARPOL	marine pollution
MF	maritime frequency
MHz	megahertz
MILSTAMP	military standard transportation and movement procedures
MILSTRIP	military standard requisitioning and issue procedures
MILVAN	military-owned demountable container
MOS	military occupational specialty
MPH	miles per hour
MQB	marine qualification board
MQO	marine qualification officer
MSE	mobile subscriber equipment
MSI	maritime safety information
MSR	main supply route
MT	megaton
N	north
NATO	North Atlantic Treaty Organization
NAVICP	naval inventory control point
NCO	noncommissioned officer
NCS	network control station
NE	northeast
NFPA	National Fire Protection Association
NM	nautical mile
NNE	north by northeast
No.	number
NOAA	National Oceanic and Atmospheric Administration
NOS	national ocean survey
NSA	National Security Agency
NW	northwest
OBA	oxygen-breathing apparatus
ODS	Office of Distribution Services
OSUT	one station unit training
OTAR	over-the-air rekey
PAN	personal area network
PCMCIA	Personal Computer Memory Card International Association
PDF	portable document file

PFD	personal floatation device
POL	petroleum, oils and lubricants
POR	Pacific Ocean Region
Psi	pounds per square inch
PSK	phase shift keying
PSTN	public switched telephone network
PT	pusher tug
Pub	publication
PVC	polyvinyl chloride
RBECS	revised battlefield electronics communications system
RC	reserve component
RDG	random data generator
RDS	revised battlefield electronics communications system (RBECS)data transfer device (AN/CYZ-10) (DTD) software
RF	radio frequency
ROE	rules of engagement
RORO	roll on/roll off
R.P.M.	revolutions per minute
RRDF	RO/RO discharge facility
RSS	readiness support system
RSOI	reception, staging, onward movement and integration
R-T	reciever-transmitter
S	south
SALTS	standard automated logistics tool set
SAR	search and rescue
SATCOM	satellite communications
SB	size block
SE	southeast
SES	ship earth station
SF	safety factor; standard form
SHP	strain on hauling part
SINCGARS	single channel ground to air radio system
SL	skill level
SMG	speed made good
SNAP	shipboard non-tactical automated data processing
SOA	speed of advance
SOG	speed over ground
SOI	signal operation instruction
SOP	standard operating procedure
SOS	save our ship

SSB	single side band
ST	small tug
Stbd	starboard
STE	secure terminal equipment
STON	short ton
STU	secure telephone unit
STU-III	secure telephone unit, third generation
SW	southwest
SWL	safe working load
SZ	surface zero
T	true north; time
TAC	tactical
TADDS	training aids, devices, simulators and simulations
T&EO	training and evaluation outline
TB	technical bulletin
TC	Transportation Corps, training circular
TM	technical manual
TNT	trinitrotoluene
TRADOC	Training and Doctrine Command (US Army)
TRI	tri-service
TSK	transmission security key
TW	total weight
UCMJ	Uniform Code of Military Justice
UHF	ultra high frequency
USAML	United States Army marine license
USCG	United States Coast Guard
VDS	vessel defense simulator
VHF	very high frequency
W	west; weight
WinSALTS	Windows standard automated logistics tool set

SECTION II – TERMS

abeam

Directional term meaning at a right angle to the centerline or keel of a vessel.

aft or after at

Near, or toward the stern.

aground

Resting on the bottom (refers only to floating craft).

aids to navigation

Charted marks (such as buoys, beacons, lights, and electronic devices) to assist navigators.

aloft

Above the upper deck, as up in the rigging.

amidships

Usually in the line of the keel, but sometimes halfway between bow and stern; often contracted to "midships."

anchor

Iron device so shaped as to grip the bottom and hold a vessel at her berth by the chain or rope attached.

anchor buoy

A small buoy, usually painted a distinctive color, attached to an anchor by a light line and used to indicate the location of the anchor on the bottom.

anemometer

Instrument for measuring wind velocity.

aneroid

Barometer instrument used to measure atmospheric pressure.

arming

A piece of tallow or soap put in the cavity and over the bottom of a lead line.

assured mobility

Actions that give the force commander the ability to maneuver where and when desired without interruption or delay to achieve the mission. (FM 3-34)

astern

Behind the vessel; in the direction of the stern.

athwart ship

At right angles to the fore-and-aft line of a vessel; across the vessel in a direction at right angles to the keel.

bank

The side of a river channel, especially that part that usually rises above the water level. also a shallow area of shifting sand, gravel, mud, and so forth (for example, a sandbank or mud bank).

bar

A ridge or succession of ridges of sand or other substance, especially a formation extending across the mouth of a river or harbor or off a beach, and which may obstruct navigation.

barrier reef

Name given to reefs separated from the adjacent coast by a channel or lagoon.

batten

Long strip of wood or metal used for various purposes aboard ship.

batten down

To close or make watertight, referring to hatches and cargo.

beacon

A post or buoy placed over a shoal or bank to warn vessels of danger; also a signal mark on land.

beam

The maximum width of a vessel, also called breadth.

becket

Loop of rope with a knot at one end to catch in an eye at the other.

bend

A combination of turns and tucks by which a line is fastened to another one; to tie, as in securing a line to a cleat; to shackle a chain to an anchor; make fast.

berth

A place for securing a vessel, either in the stream or alongside a wharf or other vessel.

bight

Loop or double part of a rope; also, any part of a rope within the ends.

bilge

The lowest interior position of a ship; the water that accumulates in the bilge is called bilge water.

bilge pump

A pump used aboard vessels to remove accumulations of water in the bilge.

binnacle

A compass stand made of nonmagnetic material, serving also to illuminate and protect the compass.

bitter end

Last end of a rope or inboard end of an anchor cable secured to a bitt.

bitts

Pair of heavy posts, set vertically in a deck or on a pier, to which mooring or towing lines are fastened.

block

Grooved pulley or sheave in a frame or shell provided with a hook, eye, or strap, by which it may be attached.

boat hook

A wooden staff with a metal hook and prod at one end used for fending off or holding on when coming alongside a vessel or a wharf. It is also used for picking up small objects from the water.

boom

Spar having many uses, such as the boom for a sail, a boat boom, or a cargo boom.

bow

The forward part of a vessel.

bow anchors

Two heavy anchors carried in the forward part of the vessel and ordinarily used in anchoring.

bowline

A line leading from the bow of a vessel.

breaker

A single breaking wave.

breaker line

The outer limit of the surf.

breast line

A mooring line leading at an angle of about 90 degrees from the fore and aft line of vessel to a wharf or another vessel.

bridge

Raised athwart ships platform from which a vessel is steered and navigated.

broaching to

To be thrown broadside to, in surf, heavy seas, or on the beach.

bulkhead

Partition dividing the interior of a vessel into various compartments.

bulwark

Light plating or wooden extension of the hull above an exposed deck, furnishing protection against weather and loss of material or personnel.

buoys

Floating beacons, moored to the bottom, which by their shape and color convey valuable information as to position (such as channel, anchor, shoal, and rock).

capstan

Vertical revolving drum, spool-shaped, used generally for heaving in towing or mooring lines.

cathead

The outside spool on a winch.

chafe

To wear down by rubbing the surface of a line against a solid object.

chafing gear

A guard of canvas, rope, or similar material placed around spars, lines, or rigging to prevent wear.

chocks

Round or oval holes in a vessel's bulwark, sometimes fitted with rollers, through which hawsers and ropes are passed; also blocks of wood for supporting boats, weights, and so on.

cleat

Wood or metal fitting that has two projecting horns to which a line is secured.

coamings

Sidewall of a hatch projecting above the deck around the perimeter of the hatch to prevent water from going below.

cofferdam

A watertight structure fixed on the side of a vessel for making repairs below the waterline, sometimes constructed on the beach around a part of a vessel.

collision

Bulkhead partition in the forward part of a vessel, of sufficiently heavy material, to stand the great strain if the bow is damaged.

collision mat

Heavy square of canvas, roped and fitted with various lines, which can be hauled under the side to plug a leak or shell hole temporarily.

comber

A wave on the point of breaking. A comber often has a thin line of white water on its crest.

compass

An instrument for determining directions, as by means of a freely rotating magnetized needle that indicates magnetic north.

compass north

An imaginary point toward which the north end of the compass needle actually points; its direction varies with both variation and deviation.

compass rose

A circle graduated in degrees, clockwise from 0 degrees at the reference direction to 360 degrees, and sometimes also marked to show compass points. Compass roses are placed at convenient locations on charts to ease measurement of direction.

cordage

A general form for line of all kinds.

coxswain

The enlisted person in charge of a small craft.

crane derrick

A lifting machine used aboard ship for hoisting heavy weights and swinging boats in and out.

crest

The top of a wave, breaker or swell.

danger angle

An angle (sighted from a vessel), between two charted objects, which if not exceeded will allow the craft to safely pass a known hazard.

danger bearing

A bearing, to a charted object, which is used to ensure that a craft will safely pass a certain known hazard, usually used to determine the point where a change of course can be safely made.

davits

Small cranes on a vessel that are used to hoist and lower boats, especially lifeboats.

deadeye

Wooden block through which holes are pierced to receive a lanyard which is used especially with shrouds and stays.

deadlights

Strong shutters that screw down upon air portholes and keep out water in heavy weather.

dead reckoning

Calculation of ship's position kept by observing a vessel's course and distance by the log.

deviation

Angular difference at the vessel between the direction of magnetic north and compass north. An error of the compass caused by the magnetic influence of the iron and steel within the ship itself.

diesel engine

Type of internal-combustion engine in which air is compressed to a temperature sufficiently high to ignite fuel injected directly into the cylinder.

distribute

To deliver, piece by piece, in turn to members of a group.

dogs

Small, bent metal fittings used to secure watertight doors, hatch covers, manhole covers, and so on to close and fasten as tight as possible.

double-bottom tanks

Watertight tanks formed by placing steel plating a few feet above the skin or outer bottom for the purpose of protecting a vessel if the outer bottom is damaged; used to store oil, water, and so forth.

draft

Single load of cargo; also refers to the depth of water which a vessel requires to float freely; the depth of a vessel from the waterline to the keel; also a sling load of cargo.

dunnage

Loose material placed on the bottom of the hold above the ballast; used to stow cargo.

ebb

Current tidal current flowing out to sea.

engine room

Compartment containing the propulsion machinery of a vessel.

ensign flag

    The emblem of a vessel's nationality.

equator

    That great circle which lies midway between the poles.

estimate

    An opinion or judgment of the nature, character, or quality of something.

eye

    Splice loop spliced in the end of a line.

fairlead

    Fittings or devices used in preserving the direction of line, chain, or wire so that it may be delivered fairly, or on a straight lead, to the sheave, drum, and so on.

fairway

    The ship channel part of the river or harbor where the navigable channel for large vessels lies.

faking

    To lay down rope in long or circular turns (coils) so that each turn of rope overlaps the next one underneath in such a way that the rope is clear for running.

fantail

    Extreme after deck of a vessel; after section of the main deck; upper and round part of the stern.

fast

    Secure; also, secure with a mooring line.

fathom

    A nautical measure equal to 6 feet; used as a measure of depth of water.

fenders

    A device of canvas, wood, or rope used over the side to take the shock of contact between vessel and wharf or other vessel when alongside.

fid

    A tapering pin, usually of wood, used to open the strands of line for splicing.

fire

    Main system permanent fire-control installation for an entire vessel consisting of water pipes, plugs to which hoses are attached, pumps, valves, and controls.

flagstaff

    Small vertical spar at the stern on which the ensign is hoisted while a vessel is at anchor.

flemish down

    To lay out a decorative tight coil of line on deck.

foam crest

    The top of the foaming water that speeds toward the beach after the wave breaks.

fore

    Parts of a vessel at or adjacent to the bow; also parts between the mid-ship section and stern.

fore and aft

    Lengthwise of a ship.

forecastle

    The upper deck forward of the foremast and included in the bow area.

foul

    Entangle or impede.

frames

Skeleton structure, or ribs, of a vessel.

freeboard

Distance from the waterline to the top of the main deck, measured amidships.

gale

Wind with a velocity of 34 to 48 knots.

galley

A vessel's kitchen.

gangway

Passageway or ladder up a ship's side.

gear

General term for a collection of spars, ropes, blocks, and equipment.

gripes

Metal fastenings for securing a boat in its cradle; canvas bands fitted with thimbles in their ends and passed from the davit heads over and under a boat for securing for sea.

ground

Tackle anchor gear.

gunwale

The uppermost continuous strake in a vessel's side; the upper edge of a vessel's hull.

guy

Steadying rope used to support a spar in a horizontal or inclined position.

harbor master

Officer charged with executing regulations respecting the use of a harbor.

hard over

Turning the wheel as far as possible in a given direction.

hatch

Opening in a deck giving access to cargo holds.

hawsepipe

Openings in the forward part of the ship through which the anchor chain runs and where the shank of the anchor is stowed.

hawser

Large line or rope such as mooring or towing line.

headway

A vessel's motion forward or ahead.

heave to

To hold a vessel at sea without way; to check a vessel's way.

heaving line

A light flexible line thrown to another vessel in order to allow a larger line or object to be transferred over.

helm

The machine by which a vessel is steered.

hitch

A tie for fastening by which a line is fastened to another object, either directly or around it, so that it will hold temporarily and can be readily undone.

hold

Space between the lowermost deck and the bottom of a vessel that is used for the stowage of ballast, cargo, and stores.

hull

Framework of a vessel, including all decks, but exclusive of masts, yards, riggings, and all outfit or equipment.

hurricane

A cyclonic storm in the western hemisphere whose winds blow with a velocity of 64 knots or over.

inboard

Toward the centerline of a vessel; also the side next to a wharf or another vessel.

jack staff

Short, vertical flagpole at the bow of a ship from which the union jack is flown when a vessel is at anchor or moored.

jury rig

A term applied to temporary structures, such as masts and rudders, used in an emergency.

jury rudder

Makeshift rudder used to steer a vessel when the rudder is damaged.

kedge

To move a vessel by carrying out a light anchor in a boat, dropping it overboard, and hauling the vessel up to it.

keel

The timber or series of connected plates running from stem to sternpost on the bottom of the centerline of a vessel.

keelson

Any of various fore-and-aft structural members lying above or parallel to the keel in the bottom of a hull.

knot

A unit of speed equal to 1 nautical mile per hour; also any tie or fastening formed with a line.

lanyard

Rope used to make anything fast, especially a short piece reeved through deadeyes, connecting shrouds, stays, and so on, to side of vessel.

latitude

Angular distance north or south from the earth's equator measuring 0degrees at the equator and 90 degrees at either pole.

lazarette

A compartment used for storage purposes in the stern of a vessel.

lead

Line weighted line with markings, which indicate the depth of water, also known as hand lead.

leadsman

One who takes soundings or determines the depth of water by use of a lead line or hand lead.

lee

The side opposite to that from which the wind blows.

leeward

Being in or facing the direction toward which the wind is blowing; (the lee side); being the side opposite the windward.

left-hand propeller

When viewed from astern the propeller that turns counterclockwise while driving the boat ahead.

left rudder

The movement of the rudder to the left of the centerline of the boat.

lifeboat

Small boat of wood, metal, or wood and metal placed aboard a vessel, with standard, prescribed equipment for use in emergencies.

life lines

Lines stretched fore and aft along the decks to give the crew safety against being washed overboard; lines thrown on board a wreck by a lifesaving crew; knotted lines secured to the span of lifeboat davits for hoisting and lowering ropes through to a man overboard.

life jacket

An apparatus of buoyant material, usually kapok, designed to keep a person afloat.

life raft

Raft kept buoyant by cylindrical air chambers, designed to keep survivors of a disaster afloat for rescue.

life ring

Cork ring covered with canvas that is designed to support a person in water.

line-throwing guns

A gun used for lifesaving purposes that throw lines, attached to an eye in the shank of the projectile, from one vessel to another or to the shore; may be mounted or shoulder-type.

list

The inclination of a vessel to one side; as a list to port or a list to starboard.

locker

A chest, box, or compartment to stow things in.

logbook

Book containing the official record of a vessel's activities and other data relevant to its navigation, which furnishes a complete chronological history of the vessel; often called log.

longitude

A position measured as so many degrees east or west of the prime meridian.

lookout

Person stationed above decks for observing and reporting objects seen.

lubber line

A fine black line drawn on an enameled plate inside of the bowl of a magnetic compass, indicating the centerline (along the keel) of the ship.

magnetic compass

Main navigational aid on a vessel.

magnetic course

The angle between magnetic north and the intended track of the vessel over the bottom.

magnetic north

The direction of the magnetic north pole from the ship; the direction in which the compass needle points when not affected by deviations.

main deck

First complete deck running the full length of a vessel.

maneuver

To make a series of changes in direction and position for a specific purpose.

marlinespike

An iron or steel pin that tapers to a sharp point used to splice wire rope.

mast

Long pole or spar rising from the keel through the decks, to sustain yards, booms, sails, and other rigging.

maximum ebb

The greatest speed of an ebb current. the ebb is the tidal current moving away from land or down a tidal stream.

maximum flood

The greatest speed of a flood current. the flood current is the tidal current moving toward land or up a tidal stream.

mayday

General distress call.

meridians

Of longitude great circles of the earth which pass through both poles and are used to establish location in east-west direction.

mess

Group of persons eating together; the meal so taken; to supply with messes or to eat them.

messenger

Light line made fast to a hawser in order to heave the latter in.

messenger line

A small line used to haul in a heavier line.

mix

To combine or blend into one mass; to combine with another.

moored

Lying with both anchors down; tied to a pier or anchor buoy; also to secure a vessel other than by anchoring with a single anchor.

mooring lines

Cables or ropes used to tie up a vessel.

offshore wind

A wind blowing from the land.

outboard

Toward the side of a vessel in relation to the centerline or outside the vessel entirely; also, the side away from a wharf or vessel alongside.

pad eye

Metal eye permanently secured to a deck or bulkhead, to which lines and cables may be secured.

pass

A line to carry a line to or around something, or to reeve through and make fast.

painter

Line in the bow of a small boat for towing or making fast.

pay out

To let out a line or cable secured on board.

pelorus (dumb) compass

Dummy compass used to take sightings, determining the vessel's relative position to another object.

pennant

Small flag of various forms flown on a vessel, in which the long narrow flag is flown at the masthead and the triangular one is flown as a signal; short rope on pendant.

pier

A wharf which projects into a harbor, with water and accommodations for berthing vessels on two or more sides of it.

port side

The left side of a vessel looking forward, indicated by a red running light when underway at night; an opening in a vessel's side; a harbor for cargo operations.

potable water

Drinkable water, meeting standards set by the us public health service.

prow

That part of the bow of a ship above water.

psychrometer

A dry and a wet thermometer placed on one instrument used to determine relative humidity and dew point.

quarter

General area from the middle of a vessel to the extreme stern; also to proceed with the quarter to the wind or sea; to bring the sea or wind first on one quarter and then on the other.

quartermaster

Crewmember having charge of signals, navigation equipment and bridge equipment.

range lights

Two white lights on the mast in the forward part of a vessel; the second range light is in line in back of and higher than the first one.

rat guard

Funnel-like protective device placed over hawsers or mooring lines to keep rats from getting aboard.

reef

A ridge of rock or coral lying at or near the surface of the water which could pose a danger to vessels; also, in sailing vessels, to reduce the area of the sail.

reeve

Pass or thread a rope through a block.

relative humidity

The percentage of water saturation of the air.

relay

To wind a line around a belaying pin or cleat to make it secure or to stop it.

rig

A vessel's upper works; to fit out.

rigging

The ropes of a ship; collective term for all the stays, shrouds, halyards, and lines that support a vessel's mast and booms and operate its movable parts; may be a "standing" or "running" rigging.

right-hand propeller

When viewed from astern, the propeller that turns clockwise while driving the boat ahead.

right rudder

The movement of the rudder to the right of the centerline of the boat.

rudder

Flat structure hung vertically on the sternpost, just aft of the screw, and used to steer a vessel by offering resistance to the water when turned to an angle with the centerline.

rudder amidships

The position of the rudder when it parallels the keel line of the vessel.

rules of the road

The international and national regulations governing sailing of all vessels.

running lights

All lights required to be shown during peacetime by a vessel that is under way.

scope

Length of anchor chain or cable to which a vessel is riding.

screw propeller

Located at a vessel's stern.

screw current

The current caused by action of the propeller in water. it is discharged as a rotary current in the direction opposite to that of the vessel's movement.

scuppers

Small drains in a vessel's bulwark which are located near the deck.

scuttle

To sink a vessel either by boring holes in her bottom or by opening her sea cocks; valves.

sea

Disturbance of the ocean due to the wind (nautical sense).

sea anchor

A device, normally a cone made out of canvas, which is put over (with a line attached) to reduce the vessel's speed over the water and to hold the bow or stern into the wind and sea. it is used when high wind and seas restrict the free navigation of vessels. because of its construction, the sea anchor creates a drag in the water, thus reducing the speed of the vessel.

sea buoy

Last buoy before deep water that marks the channel to a harbor.

sea cock

Valve connecting with the outside sea water in the lower part of vessel which can be used to flood various parts.

set

The direction in which a current flows; also, the direction in which a vessel is forced by the action of current or wind, or the combined force of both.

shackle

U-shaped iron link with a removable pin that is used to make lines or blocks fast.

shaft

Rod transmitting power from a vessel's engine to its propeller.

shaft alley

Watertight passage enclosing the shaft and its supporting bearings; also "shaft tunnel."

sheave

A grooved wheel or pulley inside a block over which a line runs.

sheer

>A sudden change in course; also, the longitudinal upward curve of the deck of a vessel when viewed from the side.

shrouds

>Guy ropes or cables that support a mast by running athwart ship from top of the mast to both sides of the vessel.

skeg

>Wood or metal arm extending abaft the keel with a bearing at its after end. it supports the rudder and protects the propeller.

skin

>The inside or outside of a ship's planking or plating.

snatch block

>Block which can be opened on one side to receive a bight or rope.

soundings

>Depth of water surrounding a vessel which is determined by use of lead line or other equipment.

span

>Piece of wire or rope fastened at each end to a fixture, such as a davit head.

spar

>Pole, such as a mast or boom.

splice

>To join two ropes together by interweaving strands.

spring line

>A mooring line leading at an angle of about 45degrees from the fore and aft line of a vessel to a wharf or another vessel.

stadimeter

>An instrument for measuring the distance to an object when the length or height of the object is known.

stand

>To maintain one's position; to perform duties.

starboard side

>Right side of a vessel looking forward; indicated by a green running light when underway at night.

station bill

>Bill posted in the crew's quarters and other conspicuous places listing the station of the crew at maneuvers and emergency drills; sometimes called muster roll.

stay

>A line or wire running fore and aft, used to support a mast, spar, or funnel; may be "forestay" or "backstay."

steerageway

>Slowest speed at which the rudder will act to change a vessel's course.

stem

>The vertical or nearly vertical forward extension of the keel, to which the forward ends of the strakes are attached.

stern

>The after end (rear) of a vessel.

stern line

>A line leading from the stern of a vessel.

sternpost

Timber or steel bar extending from the keel to main deck at the stern of a vessel.

storm

A marked disturbance in the normal state of the atmosphere; also, a wind with a velocity of 55 to 65 knots.

stow

To put away, to lock up for safekeeping in a proper place.

strake

Continuous line of plates running from bow to stern that contributes to a vessel's skin.

stranded

Of a vessel, run ashore.

strongback

Fore-and-aft spar extending from stem to stern on a lifeboat and serving as a raised spreader for a boat cover; also a strong bar placed across a hatch opening to hold hatch boards or covers.

superstructure

Any structure built above the top full deck.

surf

The breaking swell or waves on a shore or shelving beach; breakers collectively.

surf line

That point offshore, where waves and swells become breakers. the water area from this point to the beach is known as the surf zone.

surf zone

The area between the first break in the swells and the shoreline.

surge

The swell or heave of the sea; to slack off a line.

swell

The unbroken rise and fall of the sea surface persisting after the originating cause of the motion has ceased; a succession of long non-crested waves, as that continuing after a gale or other disturbance some distance away.

swim

To propel oneself in water by natural means.

tackle

A combination of lines and blocks working together and giving a mechanical advantage to assist in lifting or moving.

tail block

Block having a rope about it and an end hanging several feet from it.

tarpaulin

Heavy, treated canvas used as a cover.

tidal current

The flow of water caused by the rise and fall of the tide.

tiller

Bar of iron or wood connected with the rudder head and lead line, usually forward, in which the rudder is moved as desired by the tiller, and quadrant is the form of tiller most frequently used in modern vessels.

topside

Above decks, such as on the weather deck or bridge.

towing

Bitts or towing posts vertical posts on a vessel to which towing or mooring lines are secured.

trim

Difference in draft at the bow and stern of a vessel; manner in which a vessel floats on the water, whether on an even keel or down by the head or stern; shipshape. to adjust a vessel's position in the water by arranging ballast, cargo, and so on. to arrange for sailing; to assume, or cause a vessel to assume, a certain position, or trim, in the water.

trough

The hollow between two wave crests or swells.

turnbuckle

Link threaded on both ends of a short bar that is used to pull objects together.

underway

A vessel is said to be underway when she is not anchored, moored, aground, or beached.

union jack

Flag consisting of the blue star-studded field in the corner of the national ensign, flown at the jack staff by ships at anchor.

variation

Angular difference at the vessel between the direction of true north and magnetic north.

vehicle

Liquid content which acts as a binding and drying agent in paint.

wake

A vessel's track or trail through the water.

watch

Period of time on duty, usually 4 hours in length; the officers and crew who tend the working of a vessel during the same watch.

way

Motion or progress through the water.

weather

Toward the point from which the wind blows; the side toward the wind; the windward.

weather deck

Deck having no overhead protection; uppermost deck.

weigh

To raise the anchor off the bottom.

wharf

Projecting platform of timber, stone, or other material which extends into water deep enough for vessels to be accommodated alongside for loading or unloading.

wheel

The instrument attached to the rudder by which a vessel is steered.

whipping

The lashing of the end of a rope.

winch

A piece of machinery, which operates a shaft, fitted with a drum or drums upon which lines or cables are wound to hoist or haul an object.

windlass

Apparatus in which horizontal or vertical drums or wheels are operated by means of a steam engine or motor for handling heavy anchor chains, hawsers, and so on.

windward

Toward the wind; being in or facing the direction from which the wind is blowing.

wings

Platforms on either side of the bridge.

yard

Spar crossing a mast horizontally.

yardarm

Outer quarter of a horizontal spar attached to the mast athwart ships, equipped with blocks for reeving signal halyards.

References

SOURCES USED
These are the sources quoted or paraphrased in this publication.

ARMY REGULATIONS
AR 25-1. *Army Knowledge Management and Information Technology.* 4 December 2008

AR 25-400-2. *The Army Records Information Management System (ARIMS).* 2 October 2007

AR 56-9. *Watercraft.* 17 March 2010

AR 380-5. *Department of the Army Information Security Program.* 29 September 2000

CODE OF FEDERAL REGULATIONS
Code of Federal Regulations is available at www.gpoaccess.gov/index.html.

CFR Title 33. *Navigation and Navigable Waters*

CFR Title 40. *Protection of Environment*

CFR Title 46. *Shipping*

DEPARTMENT OF THE ARMY FORMS
DA forms are available on the Army Electronic Library (AEL) and the APD web site, www.apd.army.mil.

DA FORM 2028. *Recommended Changes to Publications and Blank Forms*

DA FORM 3068-1. *Marine Service Record*

DA FORM 4640. *Harbor Boat Deck Department Log for Class A&B Vessels* (Available through normal forms supply channels.)

DA FORM 4993. *Harbor Boat Engine Department Log for Class A and C-1 Vessels* (Available through normal forms supply channels.)

DA FORM 5273. *Harbor Boat Deck and Engine Log for Class B Vessels* (Available through normal forms supply channels.)

DEPARTMENT OF THE ARMY PAMPHLETS
DA PAM 350-38. *Standards in Training Commission. 19 November 2012*

DEPARTMENT OF DEFENSE FORMS
DD Forms are available from the Department of Defense Forms Management Program website, http://www.dtic.mil/whs/directives/infomgt/forms/formsprogram.htm.

DD FORM 1348. *DOD Single Line Item Requisition System Document (Manual)* (Available through normal forms supply channels.)

DD FORM 1348M. *DOD Single Line Item Requisition System Document (Mechanical)* (Available through normal forms supply channels.)

DD FORM 1384. *Transportation Control and Movement Document*

DEPARTMENT OF DEFENSE REGULATIONS
DOD Regulation 5200.1-R. *Information Security Program.* January 1997

By Order of the Secretary of the Army:

RAYMOND T. ODIERNO
General, United States Army
Chief of Staff

Official:

[signature]

JOYCE E. MORROW
Administrative Assistant to the
Secretary of the Army
1312302

DISTRIBUTION:

Active Army, Army National Guard, and United States Army Reserve: Not to be distributed.
Electronic media only.

DMA Publications*

DMA Publication 1-PCL. *Portfolio Chart List (Volumes 1&2).*

DMA Publication 1-N-A. *Sailing Directions.*

DMA Publication 9. *American Practical Navigator (Volumes I and II) (also known as "Bowditch").*
    1995

DMA Publication 102 (HO 102). *International Code of Signals.*

DMA Publication 111A and 111B. *List of Lights (Volumes 1 through 7).*

DMA Publication 117A-B. *Radio Navigational Aids.*

DMA Publication. Summary of Corrections (Volumes 1 and 2).

*Available from:

DMA Office of Distribution Services
ATTN: DDCP
Washington, D.C. 20315

Field Manuals

FM 4-01.502. *Army Watercraft Safety.* 1 May 2008

FM 4-25.11. *First Aid.* 23 December 2002

Miscellaneous Publications

**COMDTINST M16672.2C. *Navigation Rules (International – Inland)*

DMAHC-86609. *Chart/Publication Correction Record.*

JP 4-01.6. *JTTP for Joint Logistics Over the Shore (JLOTS).* 5 August 2005

**Available from:

Superintendent of Documents
U.S. Government Printing Office
Washington, D.C. 20402

Soldier Training Publications

STP 55-88K14-SM-TG. *Soldier's Manual, Skill Levels 1/2/3/4 and Trainer's Guide, MOS 88K,*
    Watercraft Operator. 12 December 1991

Standard Form

SF FORM 344. *Multiuse Standard Requisitioning/Issue System Document*

Technical Bulletins

TB 43-0144. *Painting of Watercraft.* 30 November 2005

www.ingramcontent.com/pod-product-compliance
Lightning Source LLC
Chambersburg PA
CBHW061526210326
41521CB00027B/2457

* 9 7 8 1 7 8 2 6 6 5 6 9 4 *